ACTA PHYSICA AUSTRIACA/SUPPLEMENTUM VIII

CONCEPTS IN HADRON PHYSICS

PROCEEDINGS OF THE
X. INTERNATIONALE UNIVERSITÄTSWOCHEN
FÜR KERNPHYSIK 1971 DER KARL-FRANZENS-UNIVERSITÄT
GRAZ, AT SCHLADMING (STEIERMARK, AUSTRIA)
1st MARCH—13th MARCH 1971

SPONSORED BY
BUNDESMINISTERIUM FÜR WISSENSCHAFT UND FORSCHUNG
THE INTERNATIONAL ATOMIC ENERGY AGENCY
STEIERMÄRKISCHE LANDESREGIERUNG
KAMMER DER GEWERBLICHEN WIRTSCHAFT FÜR STEIERMARK AND
THEODOR KÖRNER-STIFTUNGSFONDS
ZUR FÖRDERUNG VON WISSENSCHAFT UND KUNST

EDITED BY
PAUL URBAN
GRAZ

WITH 40 FIGURES

1971

Springer-Verlag Wien GmbH

Organizing Committee:

Chairman: Prof. Dr. Paul Urban
Vorstand des Institutes für Theoretische Physik, Universität Graz

Committee Members: Dr. H. Latal
Dr. F. Widder

Proceedings: Dr. P. Pesec

Secretary: M. Pail

Acta Physica Austriaca / Supplementum I
Weak Interactions and Higher Symmetries
published in 1964

Acta Physica Austriaca / Supplementum II
Quantum Electrodynamics
published in 1965

Acta Physica Austriaca / Supplementum III
Elementary Particle Theories
published in 1966

Acta Physica Austriaca / Supplementum IV
Special Problems in High Energy Physics
published in 1967

Acta Physica Austriaca / Supplementum V
Particles, Currents, Symmetries
published in 1968

Acta Physica Austriaca / Supplementum VI
Particle Physics
published in 1969

Acta Physica Austriaca / Supplementum VII
Developments in High Energy Physics
published in 1970

Originally published by Springer-Verlag / Wien in 1971

Softcover reprint of the hardcover 1st edition 1971

Library of Congress Catalog Card Number 77-170377

ISBN 978-3-7091-8286-4 ISBN 978-3-7091-8284-0 (eBook)
DOI 10.1007/978-3-7091-8284-0

Contents

Ladies and Gentlemen, dear Colleagues!

It is a great pleasure for me to welcome you here in Schladming at the opening ceremony of the 10. Internationale Universitätswochen für Kernphysik der Universität Graz.

I have the special honour to welcome the guests of honour as well as the representatives of press and radio whose presence indicates also the public interest in our school. First of all I welcome Ministerialoberkommissär Dipl. Ing. O. ZELLHOFER, the representative of the Bundesministerium für Wissenschaft und Forschung; the representative of the Karl-Franzens-Universität Graz, Dekan der philosophischen Fakultät Prof. Dr. A. EDER; the representatives of the Steiermärkische Landesregierung Landesrat Prof. Dr. K. JUNGWIRTH, Abgeordneter zum Landtag Direktor H. LAURICH, and Oberregierungsrat Dr. W. HOLZMANN.

Finally I welcome the Mayor of our host city Schladming H. AINHIRN and the Stadtrat of Schladming. To them I would like to express our thanks for their hospitality which they extend every year to us, and their generous cooperation in the organization of the meetings. Especially I want to thank our lecturers and the participants at this school for their coming and wish to all of you a successful and pleasant stay in this beautiful city of Schladming.

PAUL URBAN

PARTICIPANTS

AFEWU K.,University of Marburg,Marburg,Germany
AICHELBURG P.,University of Vienna,Vienna,Austria
ALABISO C.,Istituto di Fisica,Parma,Italy
ALT E.O.,University of Mainz,Mainz,Germany
ANG I.G-L.,University of Heidelberg,Heidelberg,Germany
ARTEAGA-ROMERO N.J.,Lab.de Physique Atomic,Paris,France
ATKINSON D.,University of Bonn,Bonn,Germany

BAGGE E.,Inst.f.Reine und Angewandte Kernphysik,Kiel,Germany
BARKAI D.,ICTP,Trieste,Italy
BARTELS J.,University of Hamburg,Hamburg,Germany
BARTL A.,University of Tübingen,Tübingen,Germany
BARTOLI B.,Lab. Nazionale di Frascati,Frascati,Italy
BAUER W.,University of Bern,Bern,Switzerland
BAUMANN K.,University of Graz,Graz,Austria
BECKER J.D.,Max-Planck Institute,Munich,Germany
BECKER W.,University of Tübingen,Tübingen,Germany
BERKMEN H.I.,Middle East Tech.University,Ankara,Turkey
BERNIS F.,University of Madrid,Madrid,Spain
BERTELMANN H.,University of Vienna,Vienna,Austria
BOGUTA J.,University of Bonn,Bonn,Germany
BÖHM M.,DESY,Hamburg,Germany
BRAMSON B.,Oxford University,Oxford,England
BREGMAN A., Max-Planck Institute,Munich,Germany
BREITENECKER M.,Technische Hochschule,Zürich,Switzerland
BREITENLOHNER P.,Max-Planck Institute,Munich,Germany
BULLA W.,Technische Hochschule,Graz,Austria
BÜMMERSTEDE J., TU.Clausthal,Clausthal Zellerfeld,Germany
BUSCHHORN G.W.,DESY,Hamburg,Germany
BUTERA P.,Istituto di Fisica,Milano,Italy

CASTELL L.,Max-Planck Institute,Munich,Germany
CHRISTIAN R.,University of Vienna,Vienna,Austria

CHRISTILLIN P.,Istituto di Fisica,Pisa,Italy
CICUTA G.M.,Istituto di Fisica,Milano,Italy
CLEYMANS J.,Technische Hochschule,Aachen,Germany
COLEMAN R.A.,ICTP,Trieste,Italy
CONSTANTINESCU F.,University of Mainz,Mainz,Germany

DALLMAN D.P.,Institut für Hochenergiephysik,Vienna,Austria
DANIELS M.,Oxford University,Oxford,England
DEPPERT W.,Inst.f.Reine u.Angewandte Kernphysik,Kiel,Germany
DIBON H.,Institut für Hochenergiephysik,Vienna,Austria
DIN A.,ICTP,Trieste,Italy
DOEBNER H-D.,TU Clausthal,Clausthal-Zellerfeld,Germany
DRAXLER K.,University of Vienna,Vienna,Austria
DRECHSLER W.,Max-Planck Institute,Munich,Germany
DRIESCHNER M.,Max-Planck Institute,Munich,Germany
DRUMMOND I.,DAMTP,Cambridge,England
DUFF M.J.,Imperial College,London,England

EDGEN P.J.,University of Durham,Durham,England
ENFLO B.,University of Stockholm,Stockholm,Sweden
ENRIOTTI-BUTERA M.,Istituto di Fisica,Milano,Italy
ERBER T.,Institute of Technology,Chicago,Illinois,USA

FAKIROV D.,University of Heidelberg,Heidelberg,Germany
FENGLER H.,Technsiche Hochschule,Darmstadt,Germany
FERRARI R.,University of Pisa,Pisa,Italy
FLAMM D.,Institut für Hochenergiephysik,Vienna,Austria
FLUME R.,University of Hamburg,Hamburg,Germany
FRANZEN G.,University of Bonn,Bonn,Germany
FRISHMAN Y.,Weizmann Institute,Rehovot,Israel
FRÖHLICH A.,Institut für Hochenergiephysik,Vienna,Austria

GAUBE G.,University of Graz,Graz,Austria
GORINI V.,University of Marburg,Marburg,Germany
GROSSE H.,University of Vienna,Vienna,Austria
GRÜMM H.,University of Vienna,Vienna,Austria

HAAN O.,University of Heidelberg,Heidelberg,Germany
HABERLER P.,Max-Planck Institute,Munich,Germany
HAMPL B.,University of Graz,Graz,Austria
HAMPL E.,University of Graz,Graz,Austria
HAUBERG T.,Inst.f.Reine u.Angewandte Kernphysik,Kiel,Germany
HEGERFELDT G.C.,University of Göttingen,Göttingen,Germany
HEIMEL H.,Technische Hochschule,Graz,Austria
HEJTMANEK J.,Atominstitut d.Österr.Hochschulen,Vienna,Austria
HENNIG J.,TU Clausthal,Clausthal-Zellerfeld,Germany
HIGATSBERGER M.J.,Österr.Studienges.f.Atomenergie,Vienna,Austria
HORVATH Z.,Eötvös University Budapest,Budapest,Hungary
HUMPERT B.,Cavendish Laboratory,Cambridge,England
HUSKINS J.,Imperial College,London,England

IRO H.,University of Vienna,Vienna,Austria

JÄGER I.,University of Graz,Graz,Austria
JÄGER W.,University of Vienna,Vienna,Austria
JOHANNESSON N.,University of Lund,Lund,Sweden
JUREWICZ A.,Institute for Nuclear Research,Warsaw,Poland

KANTHACK L.,University of Marburg,Marburg,Germany
KINZELBACH W.,Max-Planck Institute,Munich,Germany
KLAUDER J.R.,Bell Tel.Labs.,Murray Hill,New Jersey,USA
KOBLER O.,University of Graz,Graz,Austria
KOEBKE K.,University of Würzburg,Würzburg,Germany
KÖGERLER R.,University of Vienna,Vienna,Austria
KOLLER K.,University of Munich,Munich,Germany
KONETSCHNY W.,Institut für Hochenergiephysik,Vienna,Austria
KOSTERLITZ J.M.,University of Birmingham,Birmingham,England
KRAMMER M.,DESY,Hamburg,Germany
KRIECHBAUM M.,University of Graz,Graz,Austria
KUHN D.,University of Innsbruck,Innsbruck,Austria
KÜHN J.,Max-Planck Institute,Munich,Germany
KÜHNELT H.,University of Karlsruhe,Karlsruhe,Germany
KUMMER W.,Institut für Hochenergiephysik,Vienna,Austria

KUROBOSHI E.,University of Vienna,Vienna,Austria

LAMBACHER H.,University of Graz,Graz,Austria
LANG C.B.,University of Graz,Graz,Austria
LATAL H.,University of Graz,Graz,Austria
LAWRENCE J.K.,University of Vienna,Vienna,Austria
LAZARIDES G.,ICTP,Trieste,Italy
LEADER E.,Westfield College,London,England
LEDINEGG E.,Technische Hochschule,Graz,Austria
LEHMANN H.,University of Hamburg,Hamburg,Germany
LEUTHÄUSER K-D.,Inst.f.Reine u.Angew.Kernphysik,Kiel,Germany
LEVERS R.G.,University of Bonn,Bonn,Germany
LIEBBRANDT G.,Imperial College,London,England
LOGOTHETIDOU A.,University of Hamburg,Hamburg,Germany
LÜCKE W.,TU Clausthal,Clausthal-Zellerfeld,Germany
LUTZENBERGER W.,University of Munich,Munich,Germany

MACKE W.,University of Linz,Linz,Austria
MAJEROTTO W.,Institut für Hochenergiephysik,Vienna,Austria
MAJEWSKI I.,University of Marburg,Marburg,Germany
MANDELKERN M.A.,University of California,Irvine,USA
MARKYTAN M.,Institut für Hochenergiephysik,Vienna,Austria
MAS A.,University of Graz,Graz,Austria
MAYER M.E.,University of California,Irvine,USA
MEHTA N.,University of Groningen,Groningen,Netherlands
MEULDERMANS R.,University of Leuven,Heverlee,Belgium
MILLE H.,University of Marburg,Marburg,Germany
MIRALLES GARCIA F.,University of Munich,Munich,Germany

NAETAR F.,University of Vienna,Vienna,Austria
NILSSON J.,University of Göteborg,Göteborg,Sweden
NINAUS W.,Technische Hochschule,Graz,Austria
NYIRI J.,Central Research Institute,Budapest,Hungary

OBERHUMMER H.,University of Graz,Graz,Austria
OTTER G.,Institut für Hochenergiephysik,Vienna,Austria

PALEV T.,Institute of Physics,Sofia,Bulgaria
PAPOUSEK W.,Technische Hochschule,Graz,Austria
PARRY G.W.,Imperial College,London,England
PASEMAN F.,University of Marburg,Marburg,Germany
PELTZER P.,University of Graz,Graz,Austria
PESEC P.,University of Graz,Graz,Austria
PFEIL W.,University of Bonn,Bonn,Germany
PIETSCH W.,University of Innsbruck,Innsbruck,Austria
PIETSCHMANN H.,University of Vienna,Vienna,Austria
PITTNER L.,University of Graz,Graz,Austria
PLESSAS W.,University of Graz,Graz,Austria
POND P.,Westfield College,London,England
PORTH P.,Institut für Hochenergiephysik,Vienna,Austria

QUARANTA L.,Centre de Physique Theorique,Marseille,France

RAMON MEDRANO M.,University of Madrid,Madrid,Spain
REEH H.,Max-Planck Institute, Munich,Germany
REIN D.,R.W.Technische Hochschule,Aachen,Germany
REK Z.,Institute of Nuclear Research,Warsaw,Poland
RIECKERS A.,University of Tübingen,Tübingen,Germany
RINGHOFER K.,University of Munich,Munich,Germany
RODENBERG R.,R.W.Technische Hochschule,Aachen,Germany
ROHRLICH F.,Syracuse University,Syracuse,USA
RÖMER H.,University of Bonn,Bonn,Germany
ROSENTHAL P.,TU Clausthal,Clausthal-Zellerfeld,Germany
ROTHERY A.,Imperial College,London,England
RUDOLPH E.,Max-Planck Institute,Munich,Germany
RUPERTSBERGER H.,University of Vienna,Vienna,Austria
RUSCHENPLATT G.,University of Hamburg,Hamburg,Germany

SALA GRISO C.,University of Munich,Munich,Germany
SANCHEZ GUILLEN J.,University of Zaragoza,Zaragoza,Spain
SANDHAS W.,University of Mainz,Mainz,Germany
SEILER E.,Max-Planck Institute,Munich,Germany

SEXL R.,University of Vienna,Vienna,Austria
SHAFI Q.,ICTP,Trieste,Italy
SYMONS M.J.,University of Durham,Durham,England
SZASZ G.,University of Mainz,Mainz,Germany

SCHLIEDER S.,Max-Planck Institute,Munich,Germany
SCHMID P.,Institut für Hochenergiephysik,Vienna,Austria
SCHNIZER B.,Technische Hochschule,Graz,Austria
SCHOTT T.,University of Tübingen,Tübingen,Germany
SCHRÖDER U.E.,University of Frankfurt,Frankfurt,Germany
SCHWEDA M.,Institut für Hochenergiephysik,Vienna,Austria
SCHWELA D.,University of Bonn,Bonn,Germany

THIRRING W.,CERN,Geneve,Switzerland
THUN H.J.,DESY,Hamburg,Germany
THURY W.,Technische Hochschule,Graz,Austria
TILGNER H.,University of Marburg,Marburg,Germany
TOMASELLI H.,University of Vienna,Vienna,Austria
TOMBERGER G.,University of Munich,Munich,Germany
TOTON E.T.,University of Vienna,Vienna,Austria
TRÄNKLE E.,Freie Universität Berlin,Berlin,Germany
TUGULEA M.,DUBNA,Moscow,USSR

UNGERER H.,University of Tübingen,Tübingen,Germany
URBAN P.,University of Graz,Graz,Austria
URBANTKE H.,University of Vienna,Vienna,Austria

WACEK I.,Institut für Hochenergiephysik,Vienna,Austria
WALLNER H.,Technische Hochschule,Graz,Austria
WEGENER U.,University of Marburg,Marburg,Germany
WERTH J.,TU Clausthal,Clausthal-Zellerfeld,Germany
WESSEL H.,University of Bonn,Bonn,Germany
WIDDER F.,University of Graz,Graz,Austria
WILLIAMS R.,Imperial College,London,England
WITTE K.,University of Hamburg,Hamburg,Germany
WITTMANN K.,University of Graz,Graz,Austria

ZACHARIASEN F.,Cal.Techn.Pasadena,USA

ZANKEL H.,University of Graz,Graz,Austria

ZINGL H.,University of Graz,Graz,Austria

ZOVKO N.,Institute "Ruder Boskovic",Zagreb,Yugoslavia

ZUR LINDEN E.,Max-Planck Institute,Munich,Germany

10. Internationale Universitätswochen für Kernphysik
Schladming, 1st March—13th March 1971

1. W. Kainz
2. A. Bartl
3. B. Enflo
4. M. Krammer
5. H. Latal
6. P. Porth
7. J. Sanchez-Guillen
8. C. Sala Griso
9. F. Bernis
10. H. Stremnitzer
11. H. Kühnelt
12. M. Kriechbaum
13. W. Lutzenberger
14. R. Ferrari
15. N. Johannesson
16. H. Reeh
17. E. Seiler
18. S. Schlieder
19. K. Draxler
20. G. C. Hegerfeldt
21. W. Deppert
22. W. Kinzelbach
23. H. Römer
24. E. Bagge
25. T. Hauberg
26. M. Daniels
27. K. D. Leuthäuser
28. W. Becker
29. Z. Rek
30. H. Tomaselli
31. H. Ungerer
32. Z. Horvath
33. H. J. Thun
34. B. Bramson
35. M. J. Symons
36. R. Christian
37. H. Iro
38. P. Butera
39. J. Becker
40. D. Fakirov
41. C. Alabiso
42. M. A. Mandelkern
43. O. Haan
44. M. E. Mayer
45. E. Tränkle
46. M. Breitenecker
47. M. Tugulea
48. P. Aichelburg
49. G. W. Parry
50. F. Strauß
51. W. Pfeil
52. H. Bertelmann
53. D. Schwela
54. F. Miralles Garcia
55. A. Jurewicz
56. R. Coleman
57. G. Lazarides
58. J. Cleymans
59. A. Rothery
60. Q. Shafi
61. L. Quaranta
62. P. Breitenlohner
63. E. Leader
64. T. Erber
65. F. Rohrlich
66. H. D. Doebner
67. G. Tomberger
68. W. Konetschny
69. N. Zovko
70. J. K. Lawrence
71. J. Kuhn
72. R. Kögerler
73. W. Jäger
74. E. Kuroboshi
75. Y. Frishman
76. L. Castell
77. P. Pesec
78. P. Urban
79. P. Haberler
80. D. Barkai
81. I. Ang
82. M. Enriotti-Butera
83. H. Lehmann
84. K. Afewu
85. H. Pietschmann
86. J. Nyiri
87. H. Fengler
88. A. Fröhlich
89. W. Majerotto
90. H. I. Berkmen
91. U. E. Schröder
92. E. T. Toton
93. G. W. Buschhorn
94. A. Din
95. H. Lambacher

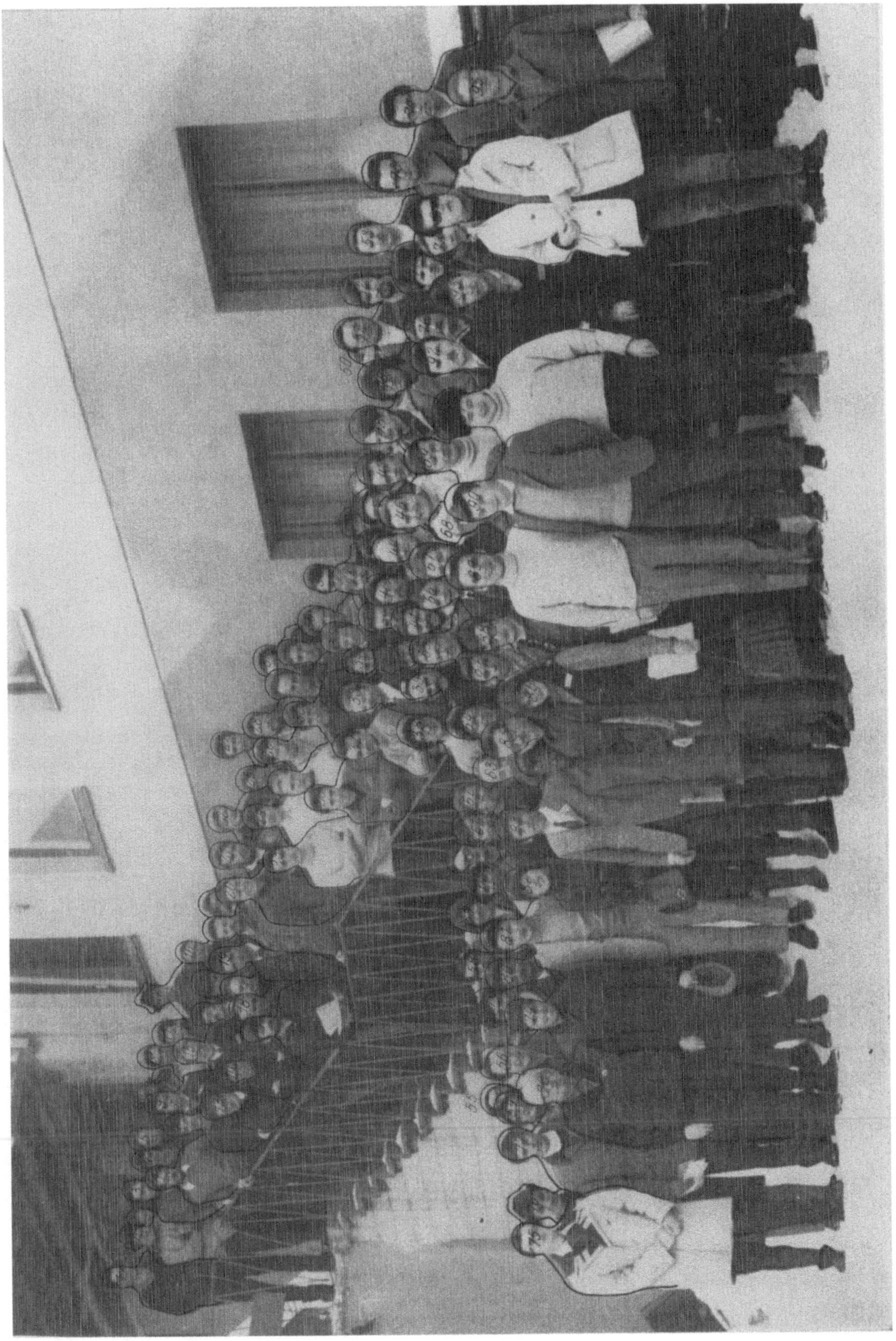

Acta Physica Austriaca, Suppl. VIII, 1—11 (1971)

TEN YEARS OF SCHLADMING
REFLECTION OF THE DEVELOPMENTS IN HIGH ENERGY PHYSICS

BY

PAUL URBAN
Institut für Theoretische Physik der Universität Graz

On the occasion of the tenth anniversary of the "Internationale Universitätswochen für Kernphysik" in Schladming I want to take the opportunity to give a review of the past meetings. Our conferences in Schladming can certainly be viewed as a reflection of the developments in high energy physics in the last ten years. During this time a large amount of new information was generated - some of it helped to clarify open questions and to understand certain phenomena, other parts of it, however, generated new problems. Our Schladming meeting is to my knowledge the first winterschool in elementary particle physics and it might be of interest to note, how this winterschool originated: The idea was born during a skiing weekend of our Institute; as usual where physicists come together, we talked physics during and after skiing. There the thought occurred, that we should extend these discussions by inviting scientists from other institutions and by letting interested young physicists take part in the exchange of ideas.

Therefore in 1962 we organized the first "Universitätskurs für Kernphysik" here in Schladming. This course

lasted for one week, we had 45 participants, out of which 35 were from Austria. I quote these numbers since they are of interest with respect to the further development of this school; in the second meeting, the "II. Internationale Universitätswoche für Kernphysik" already 95 physicists took part, with 4o from Austria - thus we could rightfully call it an international meeting.
Because of the unexpected large interest and the continually expanding scientific subjects covered we eventually had to extend the duration to two weeks. Since then we were happy to welcome an increasingly larger number of participants from all over the world: today there are over 2oo, about 6o of which from Austria. Originally the first week should have been reserved for introductory courses for students, whereas in the second week topical lectures were scheduled. This organization, however, was diluted more and more and our meeting today has more the character of a conference, but we always retain some introductory lectures for students.

Let me now review in some detail the problems covered at our Schladming winterschools. The general subject of the first course, "vectormesons", was chosen because of its topicality, shortly before such particles had been observed experimentally, in accordance with Frazer and Fulco's requirement of their existence to explain the electromagnetic structure of nucleons. At that time, in the year 1962, the properties of these mesons were still somewhat unclear, especially spin and isospin were not confirmed experimentally, today these resonances are one of the basic structures in the increasingly complicated world of elementary particles. We learned to collect them into supermultiplets of higher symmetry groups, they form the basis of the vectormeson dominance model describing the electromagnetic interactions of hadrons and they are confirmed accurately by experiment.

If it was a single subject, which was covered in 1962 then, the second meeting in 1963 was devoted to two topics: firstly to the introduction and application of the Reggepole model in phenomenological high energy physics, and secondly to the foundations and results of axiomatic field theory. In the frame of axiomatic field theory one tries to develop a mathematically exact field theory out of a few basic axioms - fields are here the essential quantities, which have to fulfill certain conditions (as e.g. commutation relations) and out of these fields the physically measurable quantities should be derived or constructed. After the groundbraking papers by Lehmann, Symanzik, Zimmermann and Wightmann, which gave rise to the hope for a quick success of this method, the difficulties however increased. In particular, for a long time it was impossible to construct a theory of truly interacting particles, i.e. one obtained only a trivial S-matrix, the unit matrix. Later on also successes of this kind of field theory were obtained: This we heard a few years later from Professor Klauder, who introduced us to a non-trivial quantum field theory.

The second topic of the "Universitätswoche" 1963 was potential theory and its influence on phenomenological elementary particle physics. Mainly the Reggepole model, which was developed in the frame work of nonrelativistic potential theory where it can be provided, and which began to play an important role in the description of high energy scattering amplitudes, was thoroughly discussed. The method, although not provable in the case of relativistic scattering theory, turned out to be extremely fruitful. It gave us a formalism which was in accordance with some exact principles about the high energy behaviour of scattering amplitudes. Thus the reggeized amplitudes obeyed the Froissart-bound, whereas previously studied one-particle-exchange models had violated this condition. Also some experimental facts could be explained satisfactorily by the Regge theory, but soon serious counterexamples appeared, i.e. experiments which could not be explained by the simple Reggepole model, as e.g. polarization phenomena and a definite structure of some high energy scattering cross sections, which were visible through the increasing accuracy of the experiments. Because of this Regge theory "died" soon after our second meeting. In fact for some time there remained a stillness about this theory, but four years later we could welcome it again among the "living" and introduce it in our winterschool in a somewhat modified and especially refined form. Since then not a year passed, where the theory was not discussed here in some way it became an essential and established part of high energy physics. Of course not all problems in this connection are solved - this year we will learn about some new aspects from Professor Zachariasen.

In the year 1964 again two main topics were covered, which had received general attention shortly before: weak interactions and higher symmetries. Above all I want to mention, that we were informed directly about the then most recent experimental data of the CERN neutrino experiment. In addition the general theory of weak interactions was presented, the first attempts at a renormalization, the peratization, were discussed and also the possibility of a nonlocality in the weak interaction was analyzed. These topics again are of interest this year. So far experiments failed however to confirm the intermediate boson, in any case it should have an enormously large mass. But of remaining value and leading importance, was the then verified CVC hypothesis which allowed a unified frame of the concept of a current (electromagnetic current - weak current). In this respect and also for a classification of strong interactions, the other subject of the meeting, higher symmetries, was extremely important. The point was to establish group theoretical and algebraic methods, which were in general alien to physicists "drilled" to differential and integral equations. I don't have to emphasize the successes of these "new" methods in elementary particle physics, only the thorough discussion of SU(3) at Schladming should be mentioned together with the fact, that the quark model, which was published on February 1, 1964, was introduced to us immediately afterwards at this third meeting in Schladming.

The following meeting in 1965 was devoted to a single topic of current interest, quantum electrodynamics. We tried to give a summary of the status of this theory, the so far one and only renormalizable field theory, which can make enormously exact numerical predictions. Of special interest at that time were the vehement discussions about the "magnitude" of the renormalization constants. This problem is very complex and hard to grasp, and even today it is not known whether these constants are finite or infinite in a consistent field theory of quantum electrodynamics. At this point I would like to commemorate a loyal friend and promoter of our winterschool, whom at that meeting I could welcome for the first time as lecturer: Professor Gunnar Källen. He had been with us right from the beginning, he supported us and gave us confidence in the usefulness of our enterprise, and if not as a lecturer, then as an extremely critical discussion leader and advisor. He participated in our meetings until his tragic death in

October 1968.

From the year 1965 in addition I want to mention the review of the status concerning the agreement between experiment and theory in quantum electrodynamics (especially with regard to the high energy behaviour) by Professor Björken. This form of inventory taking has been repeated every year also at larger international conferences; some details may have changed, some numbers have improved, many experiments became more accurate, but even then this inventory was a documentation of considerable theoretical successes. Also this year we will be concerned with quantum electrodynamical problems, this time, however, with some special topics in connection with external fields and with the electromagnetic structure of hadrons.

In contrast to the fourth meeting with the single subject quantum electrodynamics we tried to give a cross section of the various topics and theories of elementary particle physics during the fifth meeting in 1966. Thus Heisenberg's nonlinear spinortheory was introduced, an attempt at a fundamental field theory of strong interactions. From another point of view Professor Hamilton explained the phenomenological description of the pion-nucleon interaction, a method which is based mainly on the application of partial wave dispersion relations. The increasing importance of higher symmetries was manifested through a number of lectures on this topic - incidentally it should be noted that not only the importance but also the dimension of the basic groups increased: it has been the SU(3) in 1964 and the SU(6) in 1965. Finally in 1967 we became acquainted with U(12). We also heard a lecture about current algebras, a method which starts from algebraic commutation relations for currents. The concept of a current as the fundamental quantity in weak interactions - again quantum electrodynamics played a role in this - became more and more significant since the successes of the CVC theory. The first big success of current algebra, the deriviation of the Adler-Weisberger relation determining the axial vector coupling constants, had occurred in 1965 and was paid due homage at Schladming.

For the sixth meeting we kept the same concept and tried to give a survey over the most important new aspects of elementary particle physics. Out of the application of dispersion theoretical methods to the interaction of particles with higher spin, whose scattering amplitudes show a better converging high energy behaviour, the "superconvergent sumrules" evolved, which yield a number of interesting relations between various coupling constants and masses. Another topic was concerned with the electromagnetic interaction, specifically with the vectormeson dominance model. It rests on the basic assumption that the electromagnetic current of the strongly interacting particles is proportional to the vector meson fields. Its predictions have been and are still compared to the highest degrees of accuracy with experiments at DESY in Hamburg, as a result of which we can regard this model as a useful concept in the description of the electromagnetic structure of hadrons. In addition we heard a lecture on the most recent insights into the analytic structure of the S-matrix, a mathematically extremely complicated matter since we are dealing here with a theory of functions of several complex variables. Also in field theory new developments were notable: by means of generalized functions a method for renormalizing quantized fieldtheories emerged and Professor Rohrlich, whom we may welcome also this year, gave us an introduction in this way to renormalize. Again we are confronted with a new mathematical formalism, at least new in its successful application to physics. Finally I want to mention the beated discussions on the nature of the CP violation, which broke out at that meeting - partially due to the lack of accurate experimental data the question could not be settled then.

The experimental situation regarding CP violation was described in more detail already at the following seventh meeting, Today we know with certainty that CP violation exists, also the then discussed possibility of a superweak interaction has received renewed attention, but we still don't have a fundamental understanding of this phenomenon. In 1968 we heard further lectures on weak interactions, on the theoretical attempts to describe this interaction. Mainly, however, we got a thorough survey about the recently "revived" Reggepole model in strong interactions: how has it changed, the poles had given birth to "daughters" or they "conspired", and in case this failed they made "evasions" - these complicated interrelations between trajectory functions of different poles, or requirements on the residue functions respectively, became necessary in order to explain the intricate variety of experimental data on scattering cross sections at high energies of particles with spin. The introduction of a larger number of "cooperating" Reggepoles and eventually of cuts proved to be successful, the model survived and still is one of the main concepts in high energy phenomenology, according to its significance it was represented also at all the following meetings in Schladming. In 1968 we also gave a survey about the method of effective Lagrangean theories of strong interactions and their relation to the chiral symmetry group SU(2)×SU(2), and in the following eighth meeting in the year 1969 the procedure was extended to SU(3)×SU(3). These methods gave us deep insights into the structure of strong interactions and also yielded models for simple "perturbationtheoretic" computations of practical examples, e.g. nucleon-nucleon scattering in the framework of one-boson-exchange models. The 1969 meeting also was under the sign of the Veneziano model and the success of the finite energy sum rules, about which topics we heard lectures which aroused subsequently a good deal of discussion. There a synthesis is achieved between the description of scattering amplitudes by means of resonances at low energies and through Reggepoles at high energies - this synthesis then gave rise to the concept of duality which we discussed in detail at our meeting of last year. Again the Reggepole model was also of special significance, in the framework of the dual resonance model the trajectories should be parallel straight lines: a fact which is confirmed at least to a certain degree by the experiments and which was investigated also theoretically in a few lectures at that Schladming conference. The lecture by Professor Klauder, mentioned before, dealt with a completely different subject; he succeeded in acquainting us with abstract quantum field theory in a clear way and also gave us hope to learn to understand and construct realistic model field theories with interaction in the near future. These methods touch the basis of our mathematical understanding of operators in physical Hilbert spaces.

Out of last year's Schladming school on developments in high energy physics I already mentioned the discussion about the problem of duality. In addition much Regge phenomenology was in the program, especially the necessity and possibility of the generation of Regge cuts was treated. If Reggepoles were already hard to grasp, now the situation became even more complex and difficult by postulating these cuts. Last year we also heard about non-linear techniques and exact rules, which can be applied in general S-matrix theory or can be derived from it. These results form serious criteria for any ambitious model theory of strong interactions, which should fulfill both unitarity and crossing symmetry. This problem, i.e. the construction of such exact models is still interesting today and not yet finished. Finally the comprehensive survey about the results of current algebra should be mentioned, a field which remains to be of great significance.

For this year's tenth meeting not much is left to tell about, since I frequently referred to the coming attractions in the course of my review. Again we will study general concepts of the physics of hadrons, the further development of the Reggepole model, the Venezianomodel and our understanding of the electromagnetic structure of hadrons. But also quantum electrodynamics and weak interactions will have their say, special emphasis again will be on abstract field theory.

So far some facts about ten years of Schladming winter-school and ten years of high energy physics. Although the main point of our meetings was on the theoretical side, we always made efforts to elucidate the connection with experiment and thus repeatedly invited wellknown experimenters for talks, as e.g. on the neutrino experiment, on CP violation and again this year on the electromagnetic structure of hadrons. Only in this way meaningful theoretical research can be achieved. On the one hand this meeting is meant for specialists, which may discuss their problems and get acquainted with the results and open questions of their colleagues. On the other hand through review talks and "pedagogical" lectures it also appeals to students and younger scientists, which above all have the opportunity to present their most recent work in the form of seminars.

In addition fields at the borderline of elementary particle physics were discussed, as e.g. some mathematical lectures or talks on general relativity, since again and again it occurred that these "borderline fields" were suddenly at the center of physical interest. We always tried to bring out the lectures as quick as possible in printed form, during the meeting as lecture notes and afterwards as proceedings in a supplement to Acta Physica Austriaca. The proceedings of the first meeting were just home mimeographed lecture notes and were out of print in such a short time, that we had to publish the second meeting in a number of Acta Physica Austriaca and since 1964 the collection in a special supplement proved to be of advantage. For some years about 2oo participants already attend these "Universitätswochen für Kernphysik", this also being proof of the significance of this meeting, which has earned itself an established place in the calendar of such schools. In conclusion I would like to express our sincerest hope that in the future we still may be informed about the most recent developments in physics and may discuss them in these beautiful surroundings.

Acta Physica Austriaca, Suppl. VIII, 12–20 (1971)

HIGH ENERGY PHYSICS AND BIG SCIENCE

BY

WALTER THIRRING
CERN, Geneva

It is an honour and a pleasure for me to address this audience to celebrate the loth anniversary of the Winter School in Schladming. Let me begin by thanking everybody who has helped to make this Winter School such a successful enterprise, in particular the Austrian Authorities for having carried the financial burden and Professor Urban and his collaborators for their enthusiasm and devotion which have contributed so much to the development of high energy physics in Austria and in Europe. Since the main theme of the School has always been elementary particle physics, I will speak about its relation to science as a whole. In order to do so, let me go back to the beginning of my scientific association with my friend and teacher, Professor Urban, which is more than a quarter of a century old. It was just after the war when he introduced me to meson theory and in fact at that time the science of elementary particle physics was just beginning. I still remember that after having calculated for a year or so various cross sections by perturbation theory, which was fashionable at that time, I asked him about the status of our experimental knowledge concerning these cross sections.

He told me that the situation regarding comparison with experiment was rather lousy. This was indeed an understatement because the π-meson had not been discovered; its discovery and artificial production followed only shortly afterwards. Ever since, there has been a tremendous development in our knowledge of elementary particles; a world of which before we could see only a very small fraction has been revealed to our eyes. However, since I do not want to look at the situation only from the point of view of an expert* in elementary particle physics, I should hasten to say that also in other branches of natural science, new horizons have been opened up in the past quarter of a century. For instance, in the 50s we learnt a great deal about living organisms. The plan of how they are built is written in a large molecule, the DNA. It is written in a language which has words of three letters and there are four different in its alphabet. The characteristics of the creature are laid down in about 10^9 words. Clearly, this great discovery is full of potentialities for the future. In the 60s, astronomy made the headlines and one can characterize the most remarkable findings perhaps by saying that nuclear matter which so far seemed to exist only in the form of tiny droplets making up the atomic nucleus also seems to exist in the form of balls of about 10 km radii which represent collapsed stars. These macroscopic accumulations of strongly interacting particles can give rise to phenomena which even the wildest science fiction writer has not yet dreamt about. These are just a few examples and I do not by any means want to indicate that I consider other sciences as dead and not full of excitement. In view of all

---

* According to the classical definition of Hilbert, an expert is somebody whose spiritual horizon has shrunk to a point and this point is then what is called the point of view.

these impressive discoveries one may ask whether our exploration of the subnuclear world is more or less important than anything other sciences have to offer. Since the question of importance and relevance is largely a matter of personal taste, the question cannot be answered simply by yes or no, and we have to leave it to history to make a final judgment. However, there is another aspect of elementary particle physics which distinguishes it from the other sciences mentioned, and which seems to me essential. It was the first science which, by necessity, had to grow into big science and it has taught us the following:

a) If one has all the tools of modern technology at one's disposal, one can solve many problems which before appeared completely insoluble and which were many orders of magnitude away from what seemed feasible. This fact which was so spectacularly illustrated by the first landing on the moon something we meet in elementary particle physics over and over again. If somebody had predicted twenty years ago that it would be feasible to measure the magnetic moment of the μ-meson with an accuracy of one part in 10^7 one would have considered this as pure utopia.

b) Although big science is a costly enterprise, it is relatively cheaper than small science. This is a simple consequence of the fact that large machines can work more efficiently. For instance, when accelerators increase in size by a factor 10 and in cost by about a factor 6, it must also be remembered that while their energy is increased ten-fold, the intensity of many beams is increased by two orders of magnitude, so that the price per secondary particle decreases by a factor about 100. To give you a similar example, when I was in Bern in the early 50s, there was an emulsion group with four scanners looking for τ-mesons. It took them three month until they finally found the world's 8the τ-meson. Clearly, if science had been pursued on this level, it would have been much more expensive to accumulate our present knowledge. At CERN with a budget which is perhaps 500 times as big we get 1000 τ-mesons per second at the same time as a dozen other experiments are running.

c) Thirdly, high energy physics has shown us what political problems one has to resolve in Europe if one wants to compete with the USA or the USSR. Here, high energy physics was a project of just the right size for Europe in its present political configuration to undertake. High energy physics was large enough to be too expensive for the individual countries, so scientists had to collaborate or to quit the field, but it was not as expensive as space research so that the economic interests of the various countries became incompatible and hampered the functioning of the European collaboration. On the other hand, the equipment needed was not as small as, for instance, in molecular biology, so that people would prefer to do it on an national scale rather than get involved in all sorts of political problems to do it internationally. Nevertheless, the history of the 300 GeV project has shown clearly the weakness of Europe in big science and that it is only by unlikely coincidence that the authorities and scientists of all countries concerned are both willing and able to put down at the same time the money for a certain project.

Since the political and the financial problems of big science are closely connected, I shall make a few remarks concerning how expensive science is. In either direction extreme statements can be made: like, on the one hand, that millions are spent on pure science and wasted because nothing useful comes out whilst one third of the world population is still famine stricken. On the other hand, one may argue that most cultures spent large fractions of their national product on various architectonic monuments, whereas today we have the keys in our hand to open up the secrets of nature and yet we can hardly persuade our governments to spend a few per mille on this enterprise. Many more arguments of this kind could be given since it is always easy to find unpopular expenditure for which more money or real need for which less money is spent than for science. This sort of partial argument certainly does not help to decide on an admittedly rather difficult question. The value of science is partly of an esthetical, educational or philosophical nature and partly of direct economical value. To estimate the former values in terms of money to be spent on it, is a rather subjective matter and also, the expenditure for science cannot easily be compared with the expenditure for, say, arts, since the needs are completely different. The economical value can be more easily assessed by looking at the expenditure for research in industry. Since industry wants to maximize its profits, it is reasonable to expect that its expenditure for research corresponds to what it is worth to the industry. To quote one example which is probably typical for industry: Philips, about 7% of the turnover is spent on research and development and 1-2% on fundamental research. These figures will clearly represent an upper limit for what a country as a whole will spend on science because an industry which lives on the product of science profits more directly from it. In fact, the amount industrialized countries spend on pure research is in the region of 2-3 per mille of the national product. This is one order of magnitude less than industry and one can give an admittedly somewhat primitive argument that this figure is not too much. In fact, it means that all the money which has ever been spent on pure research is much less than the annual increase of the national product of the industrialized countries which is, after all, a fruit of scientific discoveries. Therefore, from an investment point of view pure research is justified. The amount of money spent on applied research and development is naturally more and differs from country to country. It depends on whether a country has large military developments and is stronger in aircraft or, say, chemical industry. Fundamental research is more connected with the University system. Therefore, in a civilized country there should be a more direct relation between national product and pure research than applied research and development. That Europe spends too little on the latter is another matter which I am not obliged to discuss here. In any case, I shall not argue in the following with the figures for pure research, but assume them to be a reasonable compromise between the desirable and the possible. Let us take these numbers and see how expensive big science really is, by looking at their consequences for Austria[x].

In CERN we have now a rather clear picture of its budget for the 70s and the Austrian annual contribution to CERN I and II together will amount to 60 Mil. Aust. Sch. If we assume that another 50% of that is spent in Institutes and Universities in Austria, we arrive at a total expense for this discipline of 90 Mil. Aust. Sch. per annum. In 1970 the Austrian gross national product[xx] was 350.10^9 Aust.Sch.

---

x Since the CERN contribution is proportional to the national income, these comparisons are not too different for the various CERN Member States.

xx Sometimes different figures are quoted which refer to the net national product. The percentages should then be changed accordingly.

If we take a 4% increase of the real national income per year and exempt $2^{0}/_{00}$ for pure research, we conclude that high energy physics will consume in this decade 12-10% of the money available for pure research. Although the total amount seems large, the percentage may seem appropriate considering the large machinery which is necessary for this kind of science. Nevertheless, let me say a few words to justify this figure. Even admitting the value of research, one might argue why not concentrate on these areas which are of direct practical value and why this investment into pure knowledge? To this argument one has to object that science is a structure which does not grow according to laws originating from economical needs. Newton's theory of gravitation was never a big money maker and yet without it, the theory of electricity could not have evolved, which is so vital for our present way of life. Quantum theory turned out to be extremely important for advanced technology, whereas the theory of relativity has not borne any economical fruit as yet. But physics is a logical unity where one part cannot be dismissed without destroying the whole edifice. Only by having the whole picture in one's mind can one appreciate the significance of the natural phenomena. Let me illustrate this point by a story about Einstein. When I met Einstein for the first time in Princeton, I came as a visitor from the University of Bern. He told me that he was always delighted to see people from Bern since the city had played such a large role in his life. In particular, Einstein said that he had learnt a great deal by watching the bears in the "ditch of the bears" in Bern. He noticed that generally they walked on four feet and then they could see only what was right in front of their nose. Only if they made an effort could they stand up on their hind legs, but once they got a broad view of the situation, they were much more efficient in finding the food. He said that in this respect physicists reminded him very much of the bears because only at rare instances are they in a position to get a sufficiently elevated view of the situation to see where the important things are.

Let me close by a few remarks concerning the future. It seems clear to me that with all the technological possibilities we have at our disposal, great discoveries are due to come. Not only do we believe that the large accelerators under construction will bring us closer to a solution of the riddle of elementary particles, but also in other sciences I think are we on the verge of substantial progress once we learn to use the possibilities which are opening up. Having seen in elementary particle physics improvements of many orders of magnitude, I just cannot believe that the missing factor of 5 or so in the magnifying powers of electro-microscopics for identifying atoms is insurmountable. Once this is achieved, one may, for instance, be able to read directly the message written down in the DNA. Since present computer technology is just about ready to store 10^{10} digits, we will be in a position to handle and to analyse this information. Certainly, this will require a great effort and molecular biology will sooner or later also grow into big science. In astronomy, with an observatory on the space platform one will certainly put out of business astronomers on the ground, who can see only through two narrow windows in the frequency scale of electromagnetic radiation. But again, orbital laboratories belong to big science. These examples can easily be added to: for instance, in nuclear physics, the production of sizeable amounts of super-heavy elements will also be a major effort. The real problem of science in Europe seems to me to be how shall we be able to cope with these challenging fields which require large scale collaboration. The individual countries are too small to deal effectively with these branches of science and it will be difficult to co-ordinate the intentions and desires of various countries and various scientific disciplines. Without something like a European Research Council which plans and supervises science as well as directs the spending of the research money on a European level, I can hardly see that we will get anywhere.

These considerations have led me somewhat astray from the main theme of this Winter School, namely elementary physics. Here, we are in the fortunate position that European unity exists for us, and thanks to this we are in a competitive position on a world-wide scale. Let me close by thanking again the Authorities who by their support and their understanding made all this possible.

Acta Physica Austriaca, Suppl. VIII, 21–49 (1971)

A SIMPLE APPROACH TO THE REGGEISATION OF PHOTOPRODUCTION*

BY

E. LEADER
Westfield College
London, England

INTRODUCTION

The theory of the Reggeisation of photonic processes is in a very peculiar state, despite the appearance of many papers on the subject. The problem of Gauge Invariance, the question of whether the π pole is kinematical or dynamical and the singularity of the Regge residue combine to produce a situation in which the usual "rules for Reggeisation" lead to unsatisfactory results.

There is almost universal agreement [1] as to what the result must be, i.e. almost everybody agrees on the final form of the Regge Pole contributions to the physical scattering amplitudes, but there is a wide difference of viewpoint [2] as to what must be done in order to produce this "physically acceptable" result.

---

*Lecture given at X. Internationale Universitätswochen für Kernphysik, Schladming, March 1 - March 13, 1971.

The main point of these lectures will be to show that there is a simple, straight-forward way to achieve the desired result. We shall concentrate on photoproduction and demonstrate that the imposition of:

i) Gauge Invariance together with
ii) the absence of fixed poles in the complex J-plane

is sufficient to provide an essentially unique scheme for Reggeisation. Moreover, it will be possible to restrict assumptions of Regge asymptotic bahaviour to amplitude which are strictly analagous to those used in spinless scattering.

To start with we shall briefly summarise the sort of difficulty one gets into when one tries to Reggeise photoproduction. We shall then look at a very simple example -- photoproduction from spinless "nucleons" -- and study in detail the role played by Gauge Invariance. This will establish, what is really self-evident, that the π pole is a dynamical effect. The spinless nucleon example is then Reggeised and it is shown how the absence of fixed J-plane poles specifies the Regge residue uniquely. Finally we turn to the real world of spin 1/2 nucleons and show that the entire procedure goes through with very little modification.

DIFFICULTIES WITH PHOTOPRODUCTION

Consider

$$\gamma + p \rightarrow \pi^+ + n$$

It is well known the Born terms shown below are not individually Gauge invariant.

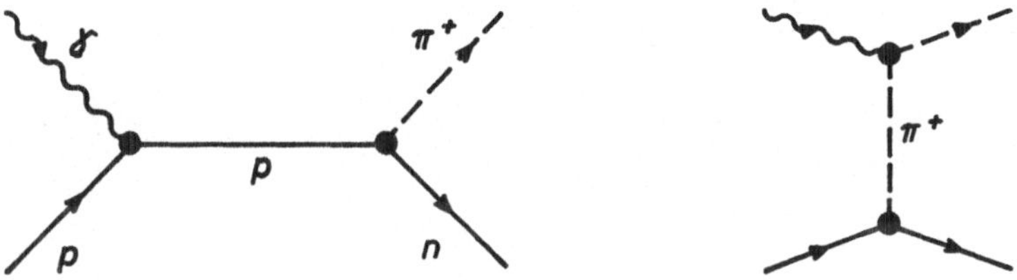

Fig. 1.

Only the sum of both Born terms satisfies Gauge invariance. Yet when we Reggeise we are only interested (at least for small t) in the t-channel Regge Poles. So we would include π_R, the π Regge Pole, but not the nucleon Regge Pole. Thus the π Regge Pole must be Gauge invariant by itself. However, as $t\rightarrow\mu^2$, we must have $\pi_R\rightarrow\pi$ i.e.

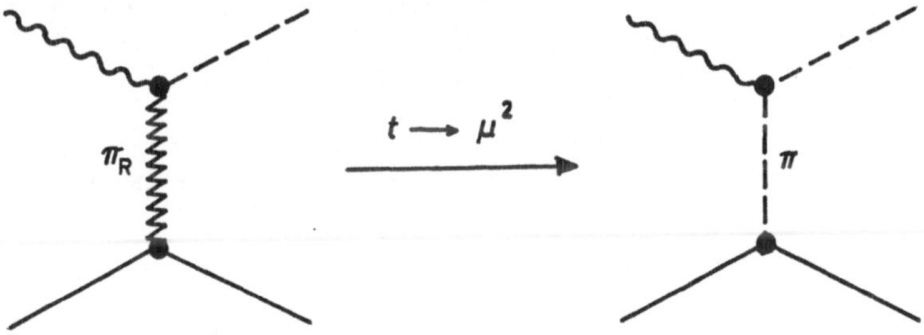

Fig. 2.

which contradicts the above assertion that the π pole is by itself not Gauge invariant.

This is the first problematic feature of the Reggeisation procedure. We shall see later that this apparent contradiction is easily resolved, and that the above argument is really somewhat misleading.

The second problem arises as a result of the fact that the photon can only have helicity ± 1. Naively one would expect the pion Regge Pole to contribute to the t-channel helicity amplitude $f^{(t)}_{\frac{1}{2}\frac{1}{2};10}$ in the form

$$f^{(t)}_{\frac{1}{2}\frac{1}{2};10} \propto \frac{t\alpha(t)}{\sin\pi\alpha(t)} \tag{1}$$

where $\alpha(t)$ is the π trajectory function, with $\alpha(\mu^2)=0$, and the factor $\alpha(t)$ in the numerator is a "sense-nonsense" factor representing the inability of a spin zero object to cause a spin flip of one unit (from the photon to the outgoing π^+). But then the amplitude in (1) has no singularity at all at $t=\mu^2$. However, helicity amplitudes for a process

$$A + B \to C + D$$

are known to have kinematic singularities at thresholds

$$t = (m_A + m_C)^2$$

and at pseudo-thresholds

$$t = (m_A - m_C)^2 \ .$$

Since in photoproduction $m_A=m_\gamma=0$ and $m_C=\mu$, we have a pseudo-threshold singularity at $t=\mu^2$, and in fact (1) must be modified to

$$f^{(t)}_{\frac{1}{2}\frac{1}{2};10} \propto \frac{1}{t-\mu^2} \cdot \frac{t\alpha(t)}{\sin\pi\alpha(t)} \ .$$

But this makes the pole at $t=\mu^2$ look as if it is kinematic in origin rather than dynamic. It suggests that if we could exchange a heavy π, call it π^1 with mass μ^1, we would still find a pole at $t=\mu^2$! This is clearly an absurd result and we shall see later that this argument is once again quite misleading. In fact the π pole is certainly a dynamic effect.

SPINLESS NUCLEONS

We consider now in some detail the process

$$\gamma + p \rightarrow \pi^+ + n$$

in which we ignore the spin of the nucleons. The kinematics are as shown:

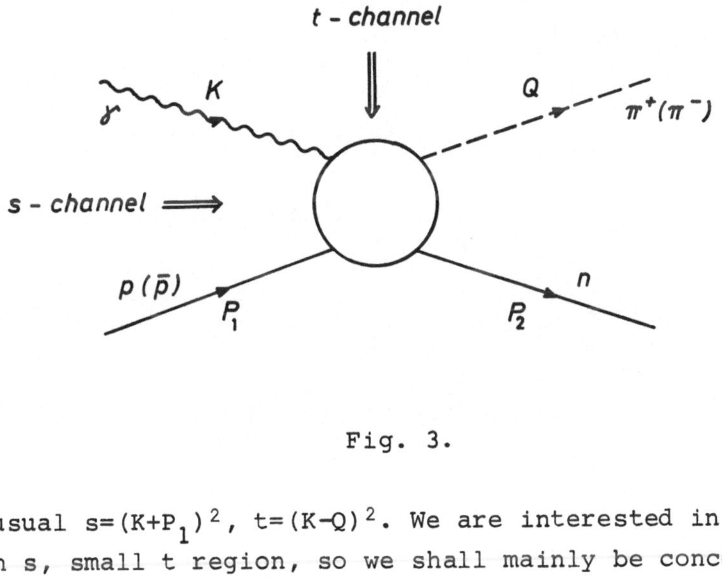

Fig. 3.

As usual $s=(K+P_1)^2$, $t=(K-Q)^2$. We are interested in the high s, small t region, so we shall mainly be concerned with the t-channel process

$$\gamma + \pi^- \rightarrow \bar{p} + n \quad .$$

In the C.M. of the t-channel the energy and momenta are labelled as follows:

$$\begin{aligned}
\gamma &: K = (k,\underline{k}), && K^2 = 0, \quad k = |\underline{k}| \\
\pi^- &: q = -Q = (q_o,-\underline{k}), && q^2 = \mu^2 \\
\bar{p} &: P = -P_1 = (E, \underline{p}), && P^2 = m^2 \\
n &: P' = P_2 = (E,-\underline{p}), && P'^2 = m^2
\end{aligned} \tag{2}$$

We then have:

$$\begin{aligned}
k &= \frac{t-\mu^2}{2\sqrt{t}}, \quad q_o = \frac{t+\mu^2}{2\sqrt{t}} \\
p &\equiv |\underline{p}| = \sqrt{t/4-m^2} \\
E &= \frac{1}{2}\sqrt{t}
\end{aligned} \tag{3}$$

The scattering angle between the photon and the anti-proton, is given, in the t-channel C.M. by

$$\cos\theta_t \equiv z = \frac{1}{2kp}\left[s-m^2+\frac{1}{2}(t-\mu^2)\right]. \tag{4}$$

The amplitude for the process is given by

$$T = \varepsilon_\mu T_\mu \tag{5}$$

where ε_μ is the photon polarisation vector. T_μ can be expanded into invariant amplitudes as follows:

$$T_\mu = AP_\mu - BQ_\mu \quad . \tag{6}$$

The possibility of a term of the form CK_μ is irrelevant, since $\varepsilon.K=0$ implies that it will not contribute to the physical amplitude [3].

Gauge Invariance requires that under the substitution

$$\varepsilon_\mu \rightarrow \varepsilon_\mu + \lambda K_\mu$$

for arbitrary λ, T should not change. i.e. we require

$$K_\mu T_\mu = 0 \quad . \tag{7}$$

From (6) then we require

$$0 = AK.P - BK.Q$$

or

$$(s-m^2)A = (t-\mu^2)B \quad . \tag{8}$$

Note that the inclusion of a term CK_μ in (6) would not affect (8). Thus Gauge invariance relates the invariant amplitudes A and B. Since A and B can be shown to be analytic except for dynamical singularities, (8) shows us that generally A must vanish at $t=\mu^2$ and B at $s=m^2$. The only possibility for A not to vanish at $t=\mu^2$ is if B has a dynamical singularity at that same point. The vanishing of A at $t=\mu^2$ is a consequence of the kinematics. The non-vanishing of A at $t=\mu^2$ is a consequence of the dynamics.

If we evaluate the Born terms shown in Fig. 1 by usual Feynman rules we get: (aside from irrelevant numerical factors)

B^π = contribution to B from Born terms

$$= \frac{e_\pi g}{t-\mu^2} \tag{9}$$

and

A^p = contribution to A from Born terms

$$= \frac{e_p g}{s-m^2} \quad . \tag{10}$$

Eq. (8) is then satisfied provided

$$e_{\pi^-} = e_{\bar{p}} = e$$

i.e. provided that charge is conserved. This is of course a well-known consequence of Gauge invariance.

We can thus write

$$A = \frac{eg}{s-m^2} + A'$$

$$B = \frac{eg}{t-\mu^2} + B' \qquad (11)$$

where $A'(s-m^2) = B'(t-\mu^2)$ and where A', B' are regular at $s=m^2$ and $t=\mu^2$ respectively. Thus A', B' must have a zero at $t=\mu^2$ and $s=m^2$ respectively. Finally then we can write

$$A = \frac{eg}{s-m^2} + (t-\mu^2)\bar{A}$$

$$B = \frac{eg}{t-\mu^2} + (s-m^2)\bar{A} \qquad (12)$$

where $\bar{A}$ is analytic except for dynamical singularities, and its dynamical singularities do not include the points $s=m^2$ or $t=\mu^2$.

The crucial point is the following. Insofar as A is concerned the consequence of the existence of a dynamical pole from π exchange is exactly expressed by the fact that A has a piece $eg/(s-m^2)$ which does not vanish at $t=\mu^2$. Moreover this statement about A contains the full information about the π pole, since if A has a piece $eg/(s-m^2)$ we can deduce from (8) that B has a piece $eg/(t-\mu)^2$.

The reason why this point is so important is that there is only one independent amplitude in the problem. Thus when we come to look at the helicity amplitudes we shall see that only A is involved. Since A does not have a pole at $t=\mu^2$ we must be careful how we interpret poles at $t=\mu^2$ in the helicity amplitudes. We shall see that there is no problem of misinterpretation provided we realise that the dynamical consequence of π exchange is completely contained in the statements

$$A \xrightarrow{t\to\mu^2} \frac{eg}{s-m^2} \quad . \tag{13}$$

Let us now consider the helicity amplitudes for the t-channel process. Remembering that our "nucleons" have zero spin we see that there are two possible amplitudes $f_{\pm 1}^{(t)}$ corresponding to the two possible photon helicities, and these amplitudes differ from each other by only a phase, as a result of parity conservation. So we can simply work with $f_1^{(t)}$.

Choosing ε_μ to correspond to a photon of helicity +1, and using (5) and (6) one finds (up to irrelevant numerical factors)

$$f_1^{(t)} = p \sin \theta_t \, A \quad . \tag{14}$$

As remarked earlier $f_1^{(t)}$ does not depend on B.

Now

$$p \sin \theta_t = \frac{\sqrt{\phi}}{t-\mu^2} \tag{15}$$

where $\phi=-t(s-m^2)^2+(t-\mu^2)(\mu^2 m^2-st)$, so that $p \sin \theta_t$ has a pole, of kinematic origin, at $t=\mu^2$ (provided $s\neq m^2$).

This fact has led to the incorrect statement that the pole in $f_1^{(t)}$ at $t=\mu^2$ is kinematic. This is of course wrong, as we have shown above, since in general A has a zero, of kinematic origin, at $t=\mu^2$. Only if there is genuine dynamic π exchange can A have a non-vanishing piece at $t=\mu^2$, and only then is $f_1^{(t)}$ singular at $t=\mu^2$. Thus the pole in $f_1^{(t)}$ at $t=\mu^2$ is certainly dynamic. Indeed, if we consider the process

$$\gamma p \to \pi^0 p$$

in which there is no π exchange Feynman diagram, then we find A=O at $t=\mu^2$ and $f_1^{(t)}$ is finite there.

We may also comment here on the question of Gauge invariance. It is clear that any model we make for $f_1^{(t)}$ will be acceptable and in complete accord with Gauge invariance, provided only that it corresponds to an A whose structure is compatible with (12). B would then automatically have its correct form since it would be calculated from (12) or (8). We could only run into trouble if we tried to construct an independent model for B. But in the present reaction B simply does not appear in the expressions for the helicity amplitudes. Hence making say a Regge model for the helicity amplitude can never lead to a contradiction with Gauge invariance. (We shall see later that this remains true even when we introduce the nucleon spin into the problem.

In this connection Fig. 2 is quite misleading. It gives the impression that the "Regge Pole Feynman diagram" must reduce at $t=\mu^2$ to the Feynman diagram for π exchange. If this were indeed so then we would really have a paradox of Gauge invariance, as suggested in the discussion on page 23. But there is no such thing as a "Regge Pole Feynman diagram". What we actually mean in Fig. 2 is that the contribution of π_R to the helicity amplitude must reduce at $t=\mu^2$ to the contribution of an elementary π exchange. We have seen above that there is no contradiction with Gauge invariance in this requirement.

REGGEISATION FOR SPINLESS NUCLEONS

We turn now to the actual problem of Reggeisation and examine the question of singular residues. The partial-wave expansion of $f_1^{(t)}$ is given by

$$f_1^{(t)} = \sum_{J=1} (2J+1) f^J_{(t)} d^J_{10}(z)$$

$$= \sin\theta_t \sum_{J=1} \frac{(2J+1)}{\sqrt{J(J+1)}} f^J_{(t)} P'_J(z) \; . \tag{16}$$

From (14) and (16) we can write

$$A = \sum_{J=1} (2J+1) a^J(t) P'_J(z) \tag{17}$$

where

$$a^J = \frac{1}{p} \cdot \frac{1}{\sqrt{J(J+1)}} \cdot f^J \; . \tag{18}$$

If we proceed to Reggeise in the usual, rather blind fashion, we would assume that a^J possesses a moving pole in the complex J-plane at $J=\alpha(t)$. i.e.

$$a^J(t) = \frac{\beta(t)}{j-\alpha(t)} + R(J,t)$$

where $R(J,t)$ is analytic in J. Application of the Sommerfeld-Watson transform would then lead to

$$A \overset{s\to\infty}{\propto} \frac{(2\alpha+1)\beta(t)\alpha(t)}{\sin\pi\alpha(t)} z^{\alpha(t)-1} \tag{19}$$

where the factor $\alpha(t)$ in the numerator comes from $P'_\alpha(z)$.
Then as $t\to\mu^2$, $\alpha(t)\to 0$ and

$$A(s\to\infty, t\to\mu^2) \propto \frac{\beta(t)\alpha(t)}{\sin\pi\alpha(t)} \cdot \frac{1}{z} \tag{20}$$

and since from (4)

$$\frac{1}{z} \propto t - \mu^2$$

we get that

$$A \to 0 \quad \text{as} \quad t \to \mu^2 \ . \tag{21}$$

Thus $A \to 0$ and consequently $f_1^{(t)}$ is not singular at $t=\mu^2$, whereas it ought to be. The remedy is to assume that the residue function $\beta(t)$ is itself singular at $t=\mu^2$. This is contrary to the assumption always made in other reactions, that residues are non-singular functions, and it has been the cause of much argument.

We shall now show that a study of the J-plane structure of a^J leads to a natural, unique form for the residue function.

Inverting (17) gives, for $J>1$,

$$a^J = \frac{1}{2J(J+1)} \int_{-1}^{+1} \sin^2 \theta \, P_J'(\theta) A \, d(\cos \theta) \ . \tag{22}$$

We transform this into a Froissart-Gribov type expression by writing a fixed t dispersion relation for A:

$$A(t,s) = \frac{1}{\pi} \int \frac{\rho(t,s')}{s'-s} ds' \qquad + \text{(term in u)} \tag{23}$$

(In the following we shall ignore the dispersion integral in u since its only effect is to produce a signature factor, which we can just insert at the end of the calculation. What follows holds for the even signature amplitudes.)

Putting

$$\frac{1}{s'-s} = \frac{1}{2kp} \frac{1}{z'-z} \tag{24}$$

where

$$z' = \frac{1}{2kp}\left[s'-m^2 + \frac{1}{2}(t-\mu^2)\right] \tag{25}$$

we can carry out the integration in (22) and get

$$a^J = \frac{1}{2kp}\cdot\frac{1}{\pi J}\int [z'Q_J(z')-Q_{J+1}(z')]\rho(t,s')ds' . \tag{26}$$

Note that by z' we shall always mean z'(s') as given by (25). Eq. (26) is valid initially for integer J>1. However, we now continue into the complex J-plane using (26) to define the continuation.

Let us now put in the information about A implied in (12). It is clear that we can write

$$\rho(t,s') \doteq -\pi eg\ \delta(s'-m^2)+(t-\mu^2)\bar{\rho}(t,s')\theta[s'-(m+\mu)^2] . \tag{27}$$

Then (26) becomes

$$a^J = \frac{1}{2kpJ\pi}\{-\pi eg[z'(m^2)Q_J(z'(m^2))-Q_{J+1}(z'(m^2))]$$

$$+ (t-\mu^2)\int [z'Q_J(z')-Q_{J+1}(z')]\bar{\rho}(t,s')ds'\} \tag{28}$$

where

$$z'(m^2) \equiv z'(s'=m^2)$$

$$= \frac{t-\mu^2}{4kp} = \frac{\sqrt{t}}{2p} . \tag{29}$$

Now let us compare (28) to the analogous expression for a reaction involving only spinless particles and without any massless particles. We would have

$$a^J = \frac{1}{2kp}\int Q_J(z')\rho(t,s')ds' \tag{30}$$

from which we usually deduce two things:

(i) Threshold behaviour of a^J: namely that as k or $p \to 0$, $z' \to \infty$ for all s', so that

$$Q_J(z') \overset{z' \to \infty}{\propto} (z')^{-J-1}$$
$$\propto (kp)^{J+1}$$

giving the threshold behaviour

$$a^J \propto (kp)^J \quad \text{as} \quad k \quad \text{or} \quad p \to 0. \tag{31}$$

(ii) Regge Poles: We assume that

$$\rho(t,s') \overset{s' \to \infty}{\sim} (s')^{\alpha(t)}$$

so that as J is moved to the left in the complex J-plane we reach a point

$$J = \alpha(t)$$

at which the integral in (30) diverges, and this is supposed to cause a pole in a^J at $J=\alpha(t)$. The essential point which we wish to stress here is that the Regge Pole is assumed to rise from a divergence of the ds' integral at large s'. This means that we can cut off any finite segment of the integral without affecting the existence and position of the Regge Pole.

Let us now return to eq. (28) and examine to what extent the photoproduction a^J differes from the a^J of a spinless, massive reaction.

Firstly let us look at the threshold behaviour when $k \to 0$, i.e. when $t \to \mu^2$. We see immediately that it is completely different from the spinless, massive case. Because of the masslessness of the photon, the Born term involves $z'(m^2)$ which, by (29), does not $\to \infty$ as $t \to \mu^2$. Thus the Born term contributes a piece which is J independent. i.e.

$$a^J_{BORN} \propto (kp)^{-1} \quad \text{as} \quad k \to 0 \ . \tag{32}$$

The integral in (28) contributes a piece, which, because of the explicit $(t-\mu^2)$ factor outside it, gives

$$a^J_{REST} \propto k(kp)^{J-1} \quad \text{as} \quad k \to 0 \ . \tag{33}$$

Thus the threshold behaviour is quite different and we must bear this in mind when we come to discuss the behaviour of the Regge residue.

Secondly it is clear that if a^J has a pole at $J=\alpha(t)$ then it must come from the divergence of the integral in (28), since by separating out the Born term in a^J we cannot have altered the convergence properties at $s'\to\infty$.

Thus in strict analogy with the spinless case will shall put

$$\int [z'Q_J(z')-Q_{J+1}(z')]\bar{\rho}(t,s')ds' = $$

$$= \left(\frac{2kp}{s_o}\right)^J \left\{\frac{\bar{\beta}(t)}{J-\alpha(t)} + R(t,J)\right\} \tag{34}$$

where s_o is a scale factor, $R(t,J)$ is a function regular in J, and we have factored out the threshold behaviour, so that $\bar{\beta}(t)$ should be a regular function of t. Once we have made the Regge ansatz in the form of (34) we shall see that everything else follows more or less automatically. We would claim that (34) is the only sensible way to introduce the Regge Pole and we emphasise its strong analogy with the spinless case.

Lastly, we see that there is an overall factor of $1/J$ in (28) which would imply that a^J has a fixed pole at $J=0$. There is no absolutely convincing argument as to why a^J should not have a fixed pole. Indeed some authors claim that there should be a fixed pole at $J=0$. However there is really no proof one way or the other at present [4]. We shall therefore <u>assume</u> that there are no fixed poles in a^J and we shall later give arguments to show that we are probably correct in this assertion.

Putting (34) into (28) gives

$$a^J = \frac{1}{2kp\pi J}\{-\pi eg[z'(m^2)Q_J(z'(m^2))-Q_{J+1}(z'(m^2))] + \\ + (t-\mu^2)\left(\frac{2kp}{s_o}\right)^J\left[\frac{\bar{\beta}(t)}{J-\alpha(t)} + R(t,J)\right]\} . \tag{35}$$

We write

$$\frac{1}{J}\cdot\frac{1}{J-\alpha(t)} = \frac{1}{\alpha(t)}\left[\frac{1}{J-\alpha(t)} - \frac{1}{J}\right] \tag{36}$$

and gather terms in (35) which have 1/J as coefficient. Thus

$$a^J = \frac{1}{2\pi kp}\{-\frac{1}{J}[\pi eg[z'(m^2)Q_J(z'(m^2))-Q_{J+1}(z'(m^2))] + \\ + (t-\mu^2)\left(\frac{2kp}{s_o}\right)^J\left(\frac{\bar{\beta}(t)}{\alpha(t)} - R(t,J)\right)] \\ + \frac{(t-\mu^2)}{\alpha(t)}\left(\frac{2kp}{s_o}\right)^J\frac{\bar{\beta}(t)}{J-\alpha(t)}\} . \tag{37}$$

We eliminate the fixed pole at J=0 by demanding that its residue vanish. Thus, using

$$\underset{J\to 0}{L}\,[z\,Q_J(z)-Q_{J+1}(z)] = 1$$

we require that

$$\frac{t-\mu^2}{\alpha(t)}\,\bar{\beta}(t) = -\pi eg + (t-\mu^2)R(t,0) . \tag{38}$$

Now $\bar{\beta}(t)$ and $R(t,0)$ from their definition in (34) are non-singular. Then (38) is consistent if, and only if

$$\alpha(\mu^2) = 0$$

Thus, we have proved that if a^J has only a moving pole in the J-plane then its trajectory must pass through J=0 at $t=\mu^2$, i.e. we have proved that the trajectory can be identified with a Reggeised π.

It is this remarkable fact that leads us to believe that it is correct to eliminate the fixed pole at J=0. What we are in effect doing, is intimately connected with the usual picture of a Regge Pole in field theory. There, in analogy with potential scattering, one assumes that the crossed Born term plus ladder graphs as shown in Fig. 4 give rise to the Regge assymptotic behaviour.

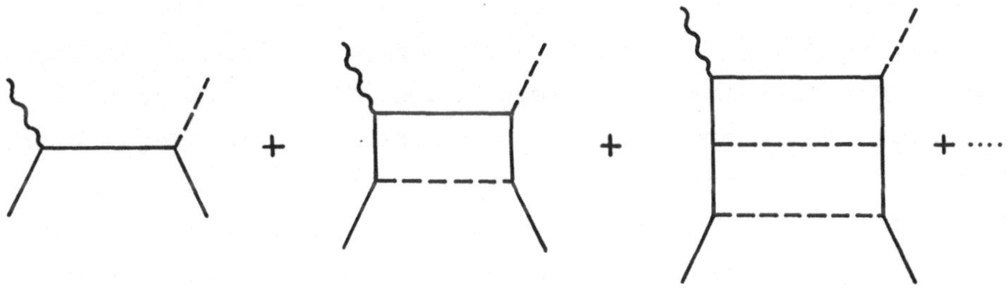

Fig. 4.

The crossed Born term always gives rise to a fixed pole in the complex J-plane. For spinless reactions, and in potential scattering it is a pole at J=-1. But in reactions with spinning particles it can occur at J=0 or even at positive integer values of J. If the amplitude Reggeises then the fixed pole from the Born term becomes a moving pole, as a result of adding the 4th and higher order diagrams. The mechanism is illustrated by writing

$$\frac{1}{J-\alpha} = \frac{1}{J} + \frac{\alpha}{J^2} + \frac{\alpha^2}{J^3} + \ldots \qquad (39)$$

in which each term comes from the successive ladder diagrams, with the 1/J term coming from the Born diagram.

If we write (39) in the form

$$\frac{1}{J-\alpha} = \frac{1}{J}\{1+\alpha[\frac{1}{J} + \frac{\alpha}{J^2} + \ldots]\} \qquad (40)$$

then from (28) we can identify

$$(t-\mu^2)\int [z' \, Q_J(z')-Q_{J+1}(z')]\bar{\rho}(t,s')ds'$$

$$\sim \alpha[\frac{1}{J} + \frac{\alpha}{J^2} + \ldots] \; . \qquad (41)$$

Our assumption that there is no fixed pole at J=0 can now be seen to be equivalent to asserting that the crossed Born term is the driving term in a ladder diagram sequence representing the Regge Pole. This seems to be a very reasonable and satisfactory picture of the Regge Pole. Moreover the possibility of having α=o at $t=\mu^2$ is seen to be directly due to the vanishing of the L.H.S. of (41) at $t=\mu^2$ which in turn is ultimately due to Gauge invariance (see eq. (12).

Thus our approach to the Reggeisation appears to be quite coherent and intuitively satisfactory.

Returning now to (37) and (38) we see that the Regge Pole contribution to a^J is given by

$$a^J_{Regge} = \frac{1}{2kp} \cdot (\frac{2kp}{s_o})^J \frac{\gamma(t)}{J-\alpha(t)} \qquad (42)$$

where

$$\gamma(t) = - eg + \frac{1}{\pi}(t-\mu^2)R(t,0) \qquad (43)$$

is a regular function.

SPINLESS NUCLEONS: ASYMPTOTIC BEHAVIOUR

In order to compute the asymptotic behaviour we must apply the Sommerfeld-Watson transformation to (17). To this end we first replace P_J' by the Mandelstam function

$$P_J'(z) = -\frac{1}{\pi}\tan \pi J \, . \, Q'_{-J-1}(z) \, . \tag{44}$$

Next, since a^J is non-singular at J=0 and since

$$P_o' = P_o' = 0$$

we can extend the sum over J to include J=0. Thus we have

$$A = \sum_{J=0} (2J+1)a^J(t)P_J'(z) \, . \tag{45}$$

Carrying out the S-W transform in the usual way and picking up the contribution of the Regge Pole as given in (42), gives

$$A_{Regge} = \frac{\pi(2\alpha+1)}{2kp} \cdot \frac{\gamma(t)}{\sin \pi\alpha(t)} \cdot \left(\frac{2kp}{s_o}\right)^{\alpha} P_\alpha'(-z) \, . \tag{46}$$

The formula (46) gives a valid representation of A when $|z|\to\infty$. The physical asymptotic region for photoproduction, $s\to\infty$ at fixed t, corresponds from (4), to $|z|\to\infty$ except when t=o, in which case z=o independent of s. This particular problem, that $s\to\infty$ at t=o does not correspond to $|z|\to\infty$ is an old problem in Regge Theory [5] and has nothing specific to do with photoproduction. It is solved by the introduction of daughter trajectories in the usual way, so we shall henceforth ignore it.

On the other hand we note that at fixed $s\neq m^2$, $t\to\mu^2$ also makes $|z|\to\infty$. So we should also expect (46) to give an accurate representation of A for any $s\neq m^2$ as $t\to\mu^2$.

Putting in the asymptotic form of $P_\alpha'(z)$ for $|z|\to\infty$ we get

$$A \sim g(\alpha)\,\frac{1}{2kp}\,\frac{\gamma(t)}{\cos\,\pi\alpha(t)}\,\left(\frac{2kp}{s_o}\right)^{\alpha} z^{\alpha-1} \qquad (47)$$

where

$$g(\alpha) = \frac{(2\alpha+1)2^{\alpha}\sqrt{\pi}\;\Gamma(1-\alpha)}{\Gamma(1/2-\alpha)}\,e^{i\pi(\alpha-1)} .$$

Let us check what happens if we let $t\to\mu^2$. We get

$$A = -\;\gamma(\mu^2)\left[\frac{1}{2kpz}\right]_{t=\mu^2}$$

$$= -\;\frac{\gamma(\mu^2)}{s-m^2} \text{ from (4)}$$

$$= \frac{eg}{s-m^2} \text{ from (43)}$$

which according to (12) is the exact result. Thus, as expected our asymptotic representation valid for $|z|\to\infty$ is exact as $t\to\mu^2$ for all s, big or small. From (14) we thus get the exact expression for $f_1^{(t)}$ at $t=\mu^2$.

To get the asymptotic form when $s\to\infty$ at fixed small t we put

$$z \sim \frac{s}{2kp}$$

in (47) and get

$$A \overset{s\to\infty}{\sim} g(\alpha)\,\frac{\gamma(t)}{\cos\,\pi\alpha(t)}\,\frac{1}{s}\cdot\left(\frac{s}{s_o}\right)^{\alpha(t)} \qquad (48)$$

with $\gamma(t)$ given by (43).

This completes our treatment of photoproduction on spinless "nucleons". We have succeeded in obtaining a sensible and consistent scheme of Reggeisation which is free from ad hoc assumptions about singular residues.

REAL PHOTOPRODUCTION

We consider now the reaction

$$\gamma + p \to \pi^+ + n$$

in which we take into account the spin of the nucleons. The method of Reggeisation developed for the spinless nucleon case applies equally well and we shall therefore not present the argument in so much detail.

The scattering amplitude is now written as

$$T = \varepsilon_\mu \, \bar{u}(-\underline{p}) J_\mu \, v(\underline{p})$$

with

$$J_\mu = B_1 \, i\gamma_5 \, \gamma_\mu \, \gamma \cdot K + B_2 \, i\gamma_5 (P'_\mu - P_\mu)$$
$$- B_3 \, 2i\gamma_5 \, q_\mu + B_5 \, \gamma_5 \, \gamma_\mu$$
$$+ B_6 \, \gamma_5 \, \gamma \cdot K \, \tfrac{1}{2}(P'_\mu - P_\mu) - B_8 \, \gamma_5 \, \gamma \cdot K \, q_\mu \quad . \tag{49}$$

The peculiar labelling of the invariant functions B_1, B_2, B_3, B_5, B_6, B_8 is done so as to agree with the functions used by Ball and Jacob [6].

The six invariant functions satisfy two Gauge invariance conditions:

$$(s-u)B_2 = 2(t-\mu^2)B_3 \tag{50}$$

and

$$B_5 + \tfrac{1}{4}(s-u)B_6 = \tfrac{1}{2}(t-\mu^2)B_8 \quad . \tag{51}$$

It is useful to rewrite (51) as

$$B_5 + kpz \, B_6 = \tfrac{1}{2}(t-\mu^2)B_8 \quad . \tag{52}$$

The Born contributions to the B_i are:

$$B_1 = - \frac{eg}{s-m^2}$$

$$B_2 = \frac{eg}{s-m^2}$$

$$B_3 = \frac{1}{2}\frac{eg}{s-m^2} + \frac{eg}{t-\mu^2} \;. \qquad (53)$$

The B_i are supposed to have only dynamic singularities, so from (50) B_3 must vanish along the line s=u.

If we put

$$B_3 = (s-u)B_3'$$

then from (53) the Born terms of B_3' are

$$B_3'{}_{BORN} = \frac{eg}{2}\frac{1}{s-m^2} \cdot \frac{1}{t-\mu^2}$$

and from (50) we can now write

$$B_2 = \frac{eg}{s-m^2} + (t-\mu^2)\bar{B}_2 \qquad (54)$$

where $\bar{B}_2$ has only dynamical singularities, and these do not include $s=m^2$ or $t=\mu^2$. Eq. (54) is the exact analogue of (12) i.e. B_2 plays the role of A.

Thus we can say that B_2 has, in general, a kinematic zero at $t=\mu^2$. Only if there is dynamic π exchange, i.e. B_3 has a pole at $t=\mu^2$, can B_2 be finite at $t=\mu^2$. Once again the full content of dynamic π exchange is contained in the requirement

$$B_2 \xrightarrow{t\to\mu^2} \frac{eg}{s-m^2} \;. \qquad (55)$$

The second Gauge equation (52) will simply be used as it is, to eliminate the combination B_5+kpz B_6 which occurs in the helicity amplitudes.

The helicity amplitudes in the t-channel C.M. are labelled $f_{\bar{p}p;\gamma\pi}$ where the subscripts refer to the helicities.

There are only four independent amplitudes which are taken as

$$f_{1/2\,1/2;1},\ f_{-1/2-1/2;1},\ f_{1/2-1/2;1},\ f_{-1/2\,1/2;1}\ .$$

The usual parity-conserving helicity amplitudes [7] $f^{\pm}$ are formed, and in the notation of ref.[6] are called

$$F_1 = f^{+}_{1/2\,1/2;\,1} \qquad F_2 = f^{-}_{1/2\,1/2;1}$$

$$F_3 = f^{+}_{1/2-1/2;1} \qquad F_4 = f^{-}_{1/2-1/2;1}\ .$$

The F_i have the following partial-wave expansions:

$$\begin{aligned}
F_1 &= \sum_{J=1} (2J+1)F_1^{\,J}\ P_J'(z)\\
F_2 &= \sum_{J=1} (2J+1)F_2^{\,J}\ P_J'(z)\\
F_3 &= \sum_{J=1} \frac{2J+1}{J(J+1)}\{F_3^{\,J}(P_J'+z\ P_J'')-F_4^{\,J}\ P_J''\}\\
F_4 &= \sum_{J=1} \frac{2J+1}{J(J+1)}\{F_4^{\,J}(P_J'+z\ P_J'')-F_3^{\,J}\ P_J''\}
\end{aligned} \qquad (56)$$

where the $F_i^{\,J}$ are related to the actual parity conserving partial wave helicity amplitudes by

$$\begin{aligned}
F_1^{\,J} &= -\frac{1}{\sqrt{J(J+1)}}\,F^{(+)J}_{1/2\,1/2;1}\\
F_2^{\,J} &= -\frac{1}{\sqrt{J(J+1)}}\,F^{(-)J}_{1/2\,1/2\,;1}\\
F_{3,4}^{\,J} &= F^{(\pm)J}_{1/2-1/2\,;1}\ .
\end{aligned} \qquad (57)$$

The quantum numbers corresponding to these amplitudes are:

$F_1^{\,J}$: Triplet $P = (-1)^J$

$F_2^{\,J}$: Singlet $P = (-1)^{J+1}$

$F_3^{\,J}$: Triplet $P = (-1)^J$

F_4^J: Triplet $P = (-1)^{J+1}$.

We wish now to study the structure of the Froissart-Gribov formulae for the F_i^J. To this end we first invert eqs. (56), giving the F_i^J as Legendre projections over the F_i. We then express the F_i in terms of the B_i as follows (up to an unimportant numerical factor)

$$
\begin{aligned}
F_1 &= -\frac{Ek}{m} B_1 \\
F_2 &= \frac{p}{m}\{k(B_1 - m\, B_6) + 2E\, B_2\} \\
F_3 &= k(B_1 + \frac{p^2}{m} B_6) \\
F_4 &= -\frac{p}{m}(B_5 + kpz\, B_6) \\
&= -\frac{p(t-\mu^2)}{2m} B_8 \text{ from (52)}
\end{aligned}
\tag{58}
$$

and finally writing fixed t dispersion relations for the B_i in the form

$$
\begin{aligned}
B_{6,8} &= \frac{1}{\pi} \int \frac{b_{6,8}(t,s')ds'}{s'-s} + (\text{term in } u) \\
B_2 &= \frac{eg}{s-m^2} + \frac{t-\mu^2}{\pi} \int \frac{\bar{b}_2(t,s')ds'}{s'-s} + (\text{term in } u) \\
B_1 &= -\frac{eg}{s-m^2} + \frac{1}{\pi} \int \frac{\bar{b}_1(t,s')ds'}{s'-s} + (\text{term in } u)
\end{aligned}
\tag{59}
$$

we can carry out the angular integrations and end up with the analogue of (28). To simplify the writing we use the notation

$$
\begin{aligned}
\bar{Q}_J(z) &= z\, Q_J(z) - Q_{J+1} \\
\bar{\bar{Q}}_J(z) &= (J+1) z\, Q_J(z) - Q_{J+1}
\end{aligned}
\tag{60}
$$

and we ignore the u terms, concentrating on the even signature amplitudes.

We have then

$$F_1{}^J = -\frac{Ek}{mJ}\{\pi eg\, \bar{Q}_J(z'(m^2))$$

$$+ \int \bar{Q}_J(z')\bar{b}_1(t,s')ds'\}$$

$$F_2{}^J = \frac{p}{mJ}\{k[\pi eg\, \bar{Q}_J(z'(m^2)) + \int Q_J(z')(\bar{b}_1 - mb_6)ds']$$

$$- E[\pi eg\, \bar{Q}_J(z'(m^2)) - (t-\mu^2)\int \bar{Q}_J(z')\bar{b}_2\, ds']\}$$

$$F_3{}^J = \frac{k}{J}\{\pi eg\, \bar{\bar{Q}}_J(z'(m^2)) + \int \bar{\bar{Q}}_J(z')(\bar{b}_1 + \frac{p^2}{m}b_6)ds'\}$$

$$- \frac{p(t-\mu^2)}{2}\int Q_J(z')b_8\, ds'$$

$$F_4 J = k\{\pi eg\, Q_J(z'(m^2)) + \int Q_J(z')(\bar{b}_1 + \frac{p^2}{m}b_6)ds'\}$$

$$- \frac{p(t-\mu^2)}{2\, m^2\, J}\int \bar{\bar{Q}}_J(z')b_8\, ds' \quad . \tag{61}$$

From (58) or from its inverse, it is easy to see that only B_2 can contain an object with the quantum numbers of the π. Eliminating the fixed pole at J=0 in $F_1{}^J$ and $F_3{}^J$ gives two conditions:

$$\int \bar{b}_1(t,s')ds' = -\pi eg \tag{62}$$

and

$$\int b_6(t,s)ds' = 0$$

when we have used the fact that

$$\underset{J\to 0}{L}\, \bar{\bar{Q}}_J(z) = \underset{J\to 0}{L}\, \bar{Q}_J(z) = 1 \; .$$

Strictly speaking, (62) cannot be true for all t since the asymptotic behaviour of B_1 and B_6, if dominated by Regge Poles (e.g. A_2) would imply that the integrals in

(62) diverge as soon as one of the trajectories, α_{A_2} become greater than 0. However, this does not invalidate the method. One must simply replace the integrals in (62) by Regge Pole type expressions analogous to (34) and then the method proceeds without difficulty. Since, for the purpose of these lectures, we are mainly concerned with the Reggeisation of the π, we shall ignore this problem and pretend that (62) holds for all t [8].

We now make the π Regge Pole ansatz in the B_2^J in a part of F_2^J in a form analogous to (34), namely

$$\int \bar{Q}_J(z')\, \bar{b}_2(t,s')ds' = \left(\frac{2kp}{s_o}\right)^J \left[\frac{\bar{\beta}(t)}{J-\alpha(t)} + R(t,J)\right] . \qquad (63)$$

Then using (62) and eliminating the fixed pole at J=0 in F_2^J we get ultimately the analogue of (38), i.e.

$$\frac{t-\mu^2}{\alpha(t)}\, \bar{\beta}(t) = -\pi eg + (t-\mu^2)R(t,o) .$$

Everything follows exactly as in the case of spinless nucleons, and we end up with the contribution to F_2 from the Reggeised π, valid for $s\to\infty$

$$F_2^{\pi R} \sim g(\alpha) \cdot \frac{2pE}{m} \cdot \frac{\gamma(t)}{\cos \pi\alpha(t)} \cdot \frac{1}{s}\left(\frac{s}{s_o}\right)^{\alpha(t)} \qquad (64)$$

where, as before eq. (43)

$$\gamma(t) = -eg + \frac{t-\mu^2}{\pi} R(t,o)$$

and $g(\alpha)$ is given after eq. (47) on page 4o.

Once again if we consider $t\to\mu^2$ and arbitrary $s\neq m^2$ we get back the exact result for F_2 at $t=\mu^2$.

Thus the entire Reggeisation procedure developed for the spinless nucleon case carries over unchanged to the case of real nucleons. There are no ambiguities and the structure of the π_R residue function emerges automatically.

CONCLUSION

In these lectures we have concentrated on the pedagogic aspects of the problems of Reggeising photoproduction. We have presented a method for carrying out the Reggeisation which is closely analogous to the methods used for spinless reactions and in which the form of the residue function emerges naturally as a consequence of the assumption that only moving singularities exist in the complex J-plane.

We have not discussed further applications at all. At present we are investigating from a field theoretic point of view whether our "picture" of the Reggeised π is supported by Feynman diagrams analysis. We are also using Feynman diagrams to try to understand the structure of the other Regge Poles in the problem. In particular there seems to be fairly convincing theoretical evidence for the existence of a J-plane singularity with indefinite parity. Whether this is a pair of completely degenerate Regge Poles of opposite parity or whether it is simply a cut is not yet clear. It is encouraging to note that from a phenomenological point of view such a singularity is most welcome [9].

REFERENCES

1. The original formula was given by
   (a) G. Zweig, Nuovo Cimento, 32, 689 (1964).
   Others who "agree" are
   (b) S. Frautschi and L. Jones, Phys. Rev. 163, 1820 (1967);
   (c) J. P. Adler and M. Capdeville, Nucl. Phys. B3, 637 (1967);
   (d) J. S. Ball, W. R. Frazer and M. Jacob, P.R.L. 20, 518 (1968);
   (e) K. Dietz and W. Korth, Phys. Letts. 26B, 394 (1968);
   (f) F. S. Henyey, Phys. Rev. 170, 1619 (1968);
   (g) T. Ebata and K. E. Lassila, Phys. Rev. 183, 1425 (1969);
   (h) H. F. Jones, M. D. Scadron and F. D. Gault, Nuovo Cimento 66A, 424 (1970);
   (i) P. D. B. Collins and F. D. Gault, Durham University Preprint, March 1970.
   The main "dissenter" is
   (j) N. Dombey, Phys. Letts. 30B, 646 (1969).
2. Compare e.g. Refs. 1(a), 1(g), 1(h) and 1(i).
3. We are working in the Lorentz gauge. It should be noted, when we come later to talk about Feynman diagrams, that a one-to-one correspondence between a diagram and a mathematical expression exists if and only if the Gauge is specified. The diagrams have no independent meaning until the Gauge is chosen. This fact is often overlooked and can lead to confusion.

4. The strongest argument against a fixed pole is a phenomenological one, (which is thus really relevant to photoproduction from nucleons with spin, and not to the artificial case being considered here) namely that if a fixed pole at J=0 exists we would get $d\sigma/dt \propto s^{-2}$, whereas experimentally $s^2\, d\sigma/dt$ seems to be decreasing with increasing s.
5. For a review of the problem of Daughters and Conspiracies see: E. Leader, Boulder Summer School Lectures (1969).
6. J. S. Ball and M. Jacob, Nuovo Cimento 54A, 620 (1968).
7. M. Gell-Mann, M. L. Goldberger, F. E. Low, E. Marx and F. Zachariasen, Phys. Rev. 133B, 145 (1964).
8. It is of interest to note that eqs. (55) when added together, are exactly equivalent to the superconverence relation, of M. B. Halpern, Phys. Rev. 160, 1441 (1967).
9. See Ref. 1(e). Further work along these lines is being carried out in collaboration with J. Hanzal and P. G. Williams.

Acta Physica Austriaca, Suppl. VIII, 50–90 (1971)

THEORY AND PRACTICE OF COMPLEX REGGE POLES*

BY

F. ZACHARIASEN**
CERN, Geneva, Switzerland

1. INTRODUCTION

In the form in which it was originally put forward, Regge phenomenology contained a wealth of predictions. If Regge poles alone dominated scattering amplitudes at high energies, then the simple form taken by a single Regge pole term, together with factorization of Regge residues, permitted one to parametrize scattering amplitudes with only very few unknowns, and also dictated very specific relations between scattering amplitudes describing different processes.

As it has turned out, however, this admirable vision of high-energy phenomenology has failed. It is simply not true that in general Regge poles alone dominate; other kinds of j-plane singularities, specifically branch points, are also present and are important.

---

*Lecture given at X. Internationale Universitätswochen für Kernphysik, Schladming, March 1 - March 13, 1971.

**J.S. Guggenheim Memorial Foundation Fellow, on leave from Calif. Inst. of Technology, Pasadena, Calif., USA.

If cuts, as well as poles, are necessary to parametrize high-energy amplitudes, what is left of the predictive power of Regge theory? In the absence of any knowledge about the cuts, one would at first expect almost nothing. The fact that an unknown j-plane cut has an arbitrary function of two variables (t and j) as its discontinuity means that one can fit essentially any scattering amplitude (which is, after all, also only a function of two variables) with a cut. Furthermore, because factorization fails in the presence of cuts, no predictions remain which relate different scattering processes to each other. Essentially the only prediction which obviously remains is that scattering amplitudes, at a given value of t, behave like powers of s at large s (apart from logarithmic factors).

Therefore, if Regge theory is not to be discarded as basically vacuous, it is necessary to discover some theoretical limitations on the cuts. What then do we know about j-plane cuts?

That cuts should, theoretically, be expected to exist was first shown by Mandelstam [1,2]. He found, by studying a particular class of Feynman graphs, that if two Regge poles with trajectories $\alpha_1(t)$ and $\alpha_2(t)$ existed, then one also had a cut with a branch point at $\alpha_c(t) = \alpha_1(t/4)+\alpha_2(t/4)-1$. In particular, if one of the two poles (say α_2) were the Pomeranchon (assuming that it indeed exists) with $\alpha_P(o) = 1$, we find $\alpha_c(o) = \alpha_1(o)$, and the cut and pole α_1 have the same quantum numbers; that is, they occur in the same amplitude. In addition, since the slope of the branch point is less than the slope of the pole $\alpha_1(t)$, it lies above the pole for $t<o$ but below for $t>o$.

This branch point, formed by the interference of two poles, may well exist. But is it the highest branch point? For example, if a pole $\alpha_1(t)$ and the Pomeranchon $\alpha_P(t)$ generate a cut with branch point at $\alpha_c(t) = \alpha_1(t/4)+\alpha_P(t/4)-1$, then the cut itself together with another Pomeranchon should, by the same argument, be expected to generate another cut with branch point at $\alpha_c^{(2)}(t)=\alpha_c(t/4)+\alpha_P(t/4)-1$; this second cut has even smaller slope than the first one, yet still satisfies $\alpha_c^{(2)}(o)=\alpha_1(o)$, so it lies above the first cut for $t<o$.

Such arguments may be repeated, making plausible the existence of a sequence of branch points with smaller and smaller slopes, all crossing the original trajectory $\alpha_1(t)$ at $t=o$.

To summarize, then, we can say only the following about the position of branch points in the j-plane: given a pole $\alpha_1(t)$, there exists a branch point $\alpha_c(t)$ with the same quantum numbers such that $\alpha_c(o)=\alpha_1(o)$, and which lies above $\alpha_1(t)$ for $t<o$. Even this feeble conclusion, we should note, depends on the existence of a Pomeranchon with $\alpha_P(o)=1$; if $\alpha_P(o)$ is slightly less than one (as some models suggest) then the pole and cut cross at some finite negative value of t. The various cases are illustrated in Fig. 1.

Can there be other cuts as well as these? Of course there can. Specific models yielding different cuts exist. For example, one such model gives a flat cut associated with each pole $\alpha_1(t)$ crossing the pole at $t=o$: $\alpha_c(t) = \alpha_1(o)$ [3]. The only thing we can do is, optimistically, to believe that the only cuts which exist are those for which we can find specific theoretical arguments.

The most important characteristic of a cut, which we would like to know if we are to be able to make any experimental predictions, is the nature of the branch point. The first Mandelstam cut, generated by two Regge poles, turns out to be logarithmic [1,2]; that is, the t-channel partial wave amplitude has the form $T(t,j) \sim \log(j-\alpha_c(t))$. Successive cuts, however, have more complicated types of singularities [2]. Other models [3] lead to cuts with square root singularities: $T(t,j) \sim \sqrt{j-\alpha_c(t)}$.

Obviously, we can say nothing which is model independent.

The same is true about the third characteristic of cuts which we would like to know; that is the actual value of the discontinuity across the cut. Any statement we can make is model dependent: the discontinuity of the first Mandelstam cut can be calculated in terms of the poles (and their residues) which generate it. The same is the case for the cut in other specific models; however the answers in the various models are all different.

To summarize, perhaps the only thing which we can safely say on (nearly) model independent grounds is that there exists a cut associated with each Regge pole, occuring in the same amplitude, which is flatter than the pole and which crosses it at or near t=o. All other details, such as the type of branch point or the value of the discontinuity, are model dependent.

If we wish not to commit ourselves to specific [and doubtless wrong [4]] models, is there then anything left to say? Can we make any phenomenological predictions at all? The answer is yes, as we shall see below. What we shall do is to accept from the foregoing discussion only the (probably) reliable statement about the position of the branch point. Then, using only this input, we shall on the basis of general (and incontrovertible) principles such as unitarity argue that it may still be possible to make certain non-trivial phenomenological predictions. These predictions will, of course, be less detailed than those which could be made on the basis of specific models; but they will, in contrast, have a chance of being right.

2. THE FORM OF THE t-CHANNEL PARTIAL WAVE AMPLITUDE

The crucial principle which we must maintain is t-channel unitarity. (This, incidentally, is entirely ignored in the specific cut models now commonly in use).

We assume that the t-channel partial wave amplitude [5] $T(t,j)$ has a branch point at $j=\alpha_c(t)$. Because $T(t,j)$ must satisfy unitarity, it is reasonable to suppose that the corresponding D function $D(t,j)$ also contains a branch point at $j=\alpha_c(t)$. In the neighbourhood of the branch point, we may write the contribution of the branch point in D explicitly; the remainder of D is then smooth in j and, near $j=\alpha_c$, we may expand it in powers of j. Then we may write, near $j=\alpha_c$,

$$D(t,j) = j-\alpha_o(t) - D_c(t,j) \qquad (2.1)$$

where $D_c(t,j)$ contains the cut. (This may be thought of as a sort of effective range expansion in j. Since the normalization of D is arbitrary, we choose the coefficient of j equal to one.)

Let us, by means of some examples and other arguments, understand why this form is to be expected, and make clear what is meant by the function D_c [6].

First, suppose we start with a model for the partial wave amplitude containing a pole at $\alpha_{in}(t)=\alpha_0+\alpha_1 t$ (α_0 and α_1 are real), and a cut, added to each other. (As, for example, in the absorption model). We write as the input partial wave amplitude

$$T_{in}(t,j) = \frac{\beta_{in}(t)}{j-\alpha_{in}(t)} + T_c(t,j) \tag{2.2}$$

where $T_c(t,j)$ is the (explicit) cut term in the model. This amplitude is manifestly not unitary. Suppose we unitarize it, by, say the K-matrix method. Then we write the unitarized amplitude as

$$T(t,j) = \frac{T_{in}(t,j)}{1-i\rho(t)T_{in}(t,j)} \tag{2.3}$$

where $\rho(t)$ is a phase space factor. (Note that ρ is real for t above threshold, and imaginary for t below threshold). Then we find

$$T(t,j) = \frac{\beta_{in}+(j-\alpha_{in})T_c}{j-\alpha_{in}-i\rho(\beta_{in}+(j-\alpha_{in})T_c)} \tag{2.4}$$

so that

$$D(t,j) = j-\alpha_o(t) - D_c(t,j) \tag{2.5}$$

where

$$\alpha_o(t) = \alpha_{in}(t) + i\rho(t)\beta_{in}(t)$$

and where

$$D_c(t,j) = i\rho(t)(j-\alpha_o(t))T_c(t,j) \tag{2.6}$$

If the cut is absent (T_c=o), then the resulting trajectory in the unitarized amplitude is simply $\alpha_0=\alpha_{in}+i\rho\beta_{in}$; i.e. Im $\alpha_0(t)=\rho(t)\beta_{in}(t)$ for t above threshold, a well known result.

Thus one may think of $\alpha_0(t)$ as a sort of unperturbed trajectory, which is altered (through unitarity) by the cut into a new trajectory $\alpha(t)$. The new trajectory, of course, is the solution to

$$D(t, \alpha(t)) = o . \tag{2.7}$$

Secondly, suppose we unitarize the amplitude of

Eq. (2.2) by a slightly more sophisticated method, say the N over D method. We choose N to be [7]

$$N(t,j) = \beta_{in}(t) + (j-\alpha_{in}(t))T_c(t,j) \tag{2.8}$$

Then it follows that D is given by

$$D(t,j) = j-\alpha_{in} - \frac{1}{\pi}\int_{t_o}^{\infty} \frac{\rho(t')dt'}{t'-t}(\beta_{in}(t')-(j-\alpha_{in}(t')) \times \times T_c(t',j)) \tag{2.9}$$

so that we now identify

$$\alpha_0(t) = \alpha_{in}(t) + \frac{1}{\pi}\int_{t_o}^{\infty} \frac{\rho(t')dt'}{t'-t}\beta_{in}(t') \tag{2.10}$$

and

$$D_c(t,j) = \frac{1}{\pi}\int_{t_o}^{\infty} \frac{\rho(t')dt'}{t'-t}(j-\alpha_{in}(t'))T_c(t',j). \tag{2.11}$$

The same interpretation as before is natural.

As a third illustration, and confirmation, of the form (2.1) let us revert to potential theory. Recall the situation when two Regge poles $\alpha_0(t)$ and $\alpha_1(t)$, cross [8]. The D function is a quadratic form in j near the collision point, and may be written as

$$D(t,j) = \det \begin{vmatrix} j-\alpha_0(t) & \varepsilon_1(t) \\ \varepsilon_1(t) & j-\alpha_1(t) \end{vmatrix} \tag{2.12}$$

Evidently solutions of $D(t,\alpha)=o$ are either two real poles or a complex conjugate pair. Thus when two poles collide, they become a complex conjugate pair of poles [8]. One may interpret ε_1 as the "coupling" of the two poles, and one may say that an unperturbed trajectory α_1 is perturbed by the presence of another trajectory α_1 into a new trajectory α which is the solution of $D(t,\alpha)=0$.

Next suppose the unperturbed pole α_0 collides with a family of other poles α_i, $i=1...\infty$ [9].

In analogy with Eq. (2.12), we write

$$D(t,j) = \det \begin{vmatrix} j-\alpha_0 & \varepsilon_1 & \varepsilon_2 & \cdots \\ \varepsilon_1 & j-\alpha_1 & 0 & \cdots \\ \varepsilon_2 & 0 & j-\alpha_2 & \cdots \\ \vdots & \vdots & \vdots & \end{vmatrix}$$

$$= (j-\alpha_0 - \sum_{i=1}^{\infty} \frac{\varepsilon_i^2}{j-\alpha_i}) \exp \sum_{i=1}^{\infty} \log(j-\alpha_i). \tag{2.13}$$

(We have here, for simplicity, neglecting "coupling" between the "other" trajectories α_i). Now let the family of poles α_i all be parallel: we write $\alpha_i(t)=\alpha_1(t)-\delta(i-1)$. Then we let δ, the spacing, approach zero and call $\alpha_1(t)=\alpha_c(t)$. We have thus constructed the situation of an unperturbed pole α_0 perturbed by the presence of a continuum of parallel other poles; that is, by a cut. The D function [dropping the irrelevant exponential in Eq. (2.13)] is [9]

$$D(t,j) = j-\alpha_0(t) - \int_{-\infty}^{\alpha_c} \frac{\varepsilon^2(t,j')}{j-j'}\, dj'. \tag{2.14}$$

Again, we recover the basic form (2.1).

As a final confirmation of Eq. (2.1), we may merely note that in two explicit dynamical models containing both poles and cuts this form appears. In all versions of the multiperipheral, or multi Regge, model, one finds [10] (at least approximately)

$$D(t,j) = j-\alpha_0(t)-\beta_0(t)\log(j-\alpha_c(t)) \tag{2.15}$$

while in the Carlitz-Kislinger model [3] one finds

$$D(t,j) = j-\alpha_0(t) - \beta_0(t)\sqrt{j-\alpha_c(t)} \;. \qquad (2.16)$$

We take it then that the form (2.1) for the D-function near $j=\alpha_c$ may be believed on a model independent basis. As we have seen, the way to interpret Eq. (2.1) is to think of an unperturbed Regge pole $\alpha_0(t)$ colliding with the cut contained in $D_c(t,j)$, and thus becoming altered into a new pole $\alpha(t)$, satisfying $D(t,\alpha(t))=o$. This interpretation is then entirely analogous to that of Eq. (2.12), where one thinks of the original pole α_0 colliding with the pole α_1, and becoming altered to a new pole α. The only feature of the cut which we need to know in order to make this interpretation is the position of the branch point: this allows us to say that $\alpha_0(t)$ and $\alpha_c(t)$ "collide" at, or near, t=o.

What do we learn from this about the behaviour of the output poles, the solutions to $D(t,\alpha)=o$? Let us take the two specific models, Eqs. (2.15) and (2.16) as illustrations.

In the logarithmic cut example, there is one pole on the physical sheet, satisfying the equation

$$\alpha = \alpha_0 + \beta_0 \log(\alpha-\alpha_c)$$

We presume that α_0 and α_c cross at t=o. Since the logarithm blows up as $\alpha\to\alpha_c$, the pole can never pass through the cut, but must instead approach it asymptotically. The situation is illustrated in Fig. 2.

There are also poles on all unphysical sheets, satisfying

$$\alpha_n = \alpha_0 + \beta_0(\log(\alpha_n-\alpha_c) + in\;\pi)$$

for $n=\pm1, \pm2 \ldots$. Evidently these occur in complex conjugate pairs, $\alpha_n=\alpha^*_{-n}$, and their imaginary part never vanishes. These complex poles can pass the cut, as sketched in Fig. 2.

A further comment of interest is the following [10]. The residue of the physical sheet pole is proportional to $(\partial D/\partial j)^{-1}|_{j=\alpha}$; that is, to $(\alpha-\alpha_c/\alpha-\alpha_c-\beta_0)$. Thus as $\alpha\to\alpha_c$, the residue becomes very small.

Hence the physical sheet pole becomes weak, and therefore unimportant, as t becomes more and more negative. This same situation does not obtain for the unphysical sheet poles. There, since Im $\alpha\neq 0$ $\alpha-\alpha_c$ does not become small, so these poles are not, in general, weakly coupled. Thus, crudely speaking, the partial wave amplitude at negative t may be described as containing the cut and a (on the nearest unphysical sheets) complex conjugate pair of poles, moving through the cut, plus a weakly coupled (on the physical sheet) pole just above the cut. Poles on further unphysical sheets ($n\neq\pm1,0$) are, presumably, too far away to be of much interest.

The square root example is also easily analyzed [9]. There are now two sheets, and two poles α_+ and α_-, given by

$$\alpha_\pm = \alpha_0 + \beta_0^2/2 \pm \sqrt{\beta_0^4/4+\beta_0^2(\alpha_0-\alpha_c)}\ .$$

Again we assume α_0 and α_c cross at t=o. Poles with $\mathrm{Re}\sqrt{\alpha}>0$ are on the physical sheet, and poles with $\mathrm{Re}\sqrt{\alpha}<0$ are on the unphysical sheet.

At $t\to+\infty$, $\alpha_\pm$ approach α_0, the unperturbed pole, and α_+ lies on the physical sheet, while α_- is on the unphysical sheet. As t decreases, α_+ and α_- decrease too. At a negative (or zero if β_0 vanishes at t=o) value of t, the two poles collide at a negative (or zero) value of $j-\alpha_c$. They are by this time both on the same sheet, α_+ having passed through the cut onto the unphysical sheet, or α_- having passed through the cut onto the physical sheet, depending on the sign of β_0. Below the point of collision, the two poles are a complex conjugate pair. This situation is illustrated in Fig. 3.

Thus the partial wave amplitude contains, at negative t, the cut and a complex conjugate pair of poles (on either the unphysical or physical sheet).

The effective situation is therefore very similar to the logarithmic case, and indeed one can convince oneself that the same is true for any type of cut: the partial wave amplitude will consist of a cut, plus complex conjugate pairs of poles on various sheets, plus (possibly) real poles.

3. THE FORM OF THE SCATTERING AMPLITUDE AT HIGH ENERGY [11]

We learned in Section 2 what we may expect, independently of any model, for the t-channel partial wave amplitude; namely it contains a cut, complex pairs of poles, and (perhaps) real poles. Let us for simplicity suppose it to contain a cut and a single complex pair. (If there are in reality several complex pairs, the one we shall select is that with smallest imaginary part. If there are real poles, they will appear in the usual way). Thus we assume that $T(t,j)$ has a branch point at $j=\alpha_c(t)$, and poles at $j=\alpha_\pm(t)=\alpha_R(t)\pm i\alpha_I(t)$.

Let us exhibit the poles explicitly, by writing

$$T(t,j) = \frac{f(t,j)}{(j-\alpha_+(t))(j-\alpha_-(t))} \ . \tag{3.1}$$

The function f contains the cut, and the poles are on an unphysical sheet if $f(t,\alpha_\pm(t))=0$.

The high-energy scattering amplitude is constructed in the usual way by means of a Mellin transform:

$$T(s,t) = \frac{1}{2\pi i}\int_{c-i\infty}^{c+i\infty} \frac{s^j f(t,j)}{(j-\alpha_+)(j-\alpha_-)}\, dj \tag{3.2}$$

where C is larger than any singularity in T(t,j). We can now deform the contour in Eq. (3.2) to obtain

$$T(s,t) = \bar{\beta}_+ s^{\alpha_+} + \bar{\beta}_- s^{\alpha_-} + \frac{1}{\pi}\int_{-\infty}^{\alpha_C} \frac{s^j \, \mathrm{Im}\, f(t,j)}{(j-\alpha_+)(j-\alpha_-)}\, dj \qquad (3.3)$$

where 2i Im f(t,j) is the discontinuity across the cut in f(t,j), and where

$$\bar{\beta}_\pm = \pm \frac{f(t,\alpha_\pm)}{\alpha_+-\alpha_-} . \qquad (3.4)$$

Hence $\bar{\beta}_\pm$=o if the poles are on an unphysical sheet; and in general $\bar{\beta}_+=(\bar{\beta}_-)^*$.

It is worth pointing out explicitly certain qualitative features of Eq. (3.3), and most particularly of the integral therein. This integral, evidently, exhibits the cut contribution to T(s,t). The weight function for the cut -- that is, the discontinuity across the cut -- is Im $f/[(j-\alpha_R)^2+\alpha_I^2]$. If α_I is small, therefore, the weight function has a large Breit-Wigner like peak at $j=\alpha_R$. This tends to emphasize the $j=\alpha_R$ region of the integral, and suggests that the energy dependence ought to be approximately s^{α_R}, unless s is very large. If s is very large, however, then clearly the dominant behaviour will be s^{α_C}. How large is large, clearly, depends on the size of α_I, as well as on the form of Im f.

The size of α_I is, of course, proportional to the strength of the cut, as we have seen in Section 2; that is, α_I must be proportional to Im f. Therefore, as the cut becomes very weak, and vanishes, the integral in Eq. (3.3) approaches

$$\left(\frac{\mathrm{Im}\, f(t,\alpha_R)}{\alpha_I}\right)_{\alpha_I=0} s^{\alpha_R} ,$$

and we recover the original single real pole with $\alpha_R=\alpha_0$, which we would have had if there were no cut at all.

For very weak cuts -- i.e. for very small α_I, we may conveniently rewrite Eq. (3.3) in the form

$$T(s,t) = \bar{\beta}_+ s^{\alpha_+} + \bar{\beta}_- s^{\alpha_-} + \frac{s^{\alpha_R}}{\pi} \int_{-\infty}^{\alpha_c} \frac{\text{Im } f(t,j)}{(j-\alpha_R)^2+\alpha_I^2} dj +$$

$$+ \frac{1}{\pi} \int_{-\infty}^{\alpha_c} \frac{\text{Im } f(t,j)(s^j - s^{\alpha_R})}{(j-\alpha_R)^2+\alpha_I{}^2} dj .$$

This is in a form relevant to the absorption model: it attempts to approximate the remaining integral by the explicit absorption model cut, in which the cut weight function is a smooth function of j. Obviously, this is a dubious procedure except in the limit $\alpha_I \to 0$; for any finite cut strength, i.e. for any finite α_I, the integrand is certainly not smooth. It appears, then, that the absorption model is likely to make sense only in the limit in which there is no cut at all.

Let us return now to the general form (3.3). Our discussion has made it evident what approximation we should make: we will assume that the function Im f(t,j) is relatively smooth over a range in j of the order of α_I around α_R. Then we may write

$$T(s,t) \approx \bar{\beta}_+ s^{\alpha_+} + \bar{\beta}_- s^{\alpha_-} + \frac{\text{Im } f(t,\alpha_R)}{\pi} \int_{-\infty}^{\alpha_c} \frac{s^j}{(j-\alpha_R)^2+\alpha_I^2} dj . \tag{3.4}$$

The similarity to the Breit-Wigner approximation is evident. There, one assumes that once one takes out a complex pole in the energy, the remaining weight function of the cut in the energy plane varies smoothly -- thus one ends up with a single complex pole in energy to describe the contributions of a cut in the energy.

Here we do exactly the same thing with j instead of energy, and a complex pair of poles instead of a single one.

The integral in Eq. (3.4) can be evaluated explicitly, and we find

$$T(s,t) = \bar{\beta}_+ s^{\alpha_+} + \bar{\beta}_- s^{\alpha_-} + \frac{\operatorname{Im} f(t,\alpha_R)}{2\pi i\, \alpha_I} [s^{\alpha_+} \operatorname{Ei}(-\alpha_+ \log s) - $$

$$- s^{\alpha_-} \operatorname{Ei}(-\alpha_- \log s)]. \qquad (3.5)$$

The accuracy of this approximation depends, of course, on the function Im f, and on the size of α_I as well as on s. Representative examples are shown in Fig. 4. The Ei function has a simple representation:

$$\operatorname{Ei}(-z) = \gamma + \log z + \sum_{n=1}^{\infty} \frac{(-z)^n}{n\, n!} \qquad (3.6)$$

where γ is Euler's constant.
Using this representation, we may, if $|\alpha_\pm \log s| \lesssim 1$, finally express T as a pair of complex poles alone, in complete analogy to the Breit Wigner expression for an amplitude containing a resonance. We obtain

$$T(s,t) = \beta_+ s^{\alpha_+} + \beta_- s^{\alpha_-} \qquad (3.7)$$

where

$$\beta_\pm = \bar{\beta}_\pm + \frac{(\phi \pm i(\gamma + \log R)) \operatorname{Im} f(t,\alpha_R)}{2\pi\, \alpha_I} \qquad (3.8)$$

and we write $\alpha_+ - \alpha_c = R \exp i\phi$. Thus $\beta_+ = \beta_-^*$.

Again the error involved in replacing Eq. (3.5) by Eq. (3.7) depends, for a given s, on the size of α_I and of $(\alpha_R - \alpha_c)$ log s. An example is shown in Fig. 5. As before, for sufficiently large s, the approximation breaks down; but the smaller α_I, the farther it goes.

Because this stage of the approximations involves $(\alpha_R - \alpha_c)$ log s as well as α_I, it should be noted that in some situations it may be more accurate to stop with Eq. (3.5), which is still an explicit form for the amplitude not involving detailed knowledge of the cut, and not to continue on to Eq. (3.7). Equation (3.5) requires only that α_I is small; Eq. (3.7) requires $|(\alpha_R-\alpha_c) \log s|$ to be small as well.

The essential assumption, leading to Eq. (3.4), was that Im f(t,j) had no unusual behaviour near $j=\alpha_R$. This assumption can be somewhat relaxed, without invalidating Eq. (3.8), as follows.

We expand Im f(t,j) in powers of j, around $j=\alpha_R$:

$$\text{Im } f(t,j) = \sum_{n=0}^{\infty} a_n (j-\alpha_R)^n .$$

Inserting this expansion into the integral in Eq. (3.3) then yields

$$\sum_{n=0}^{\infty} a_n \; s^{\alpha_R} \left(s\frac{\partial}{\partial s}\right)^n \left(\frac{1}{\pi} \int_{-\infty}^{\alpha_c-\alpha_R} \frac{s^{j'}}{j'^2+\alpha_I^2}\, dj'\right). \tag{3.9}$$

From this, in so far as the approximation of the Ei function we used before is valid, we readily again obtain Eq. (3.7), only now Eq. (3.8) is replaced by

$$\beta_\pm = \bar{\beta}_\pm + \frac{(\phi\mp i(\gamma+\log R))\,\text{Im } f(t,\alpha_\pm)}{2\pi\ \alpha_I} . \tag{3.10}$$

We have, then, two somewhat complementary formulae. If Im f varies, but $|(\alpha_R-\alpha_c)\log s|$ is small, we may use Eq. (3.7). However, if Im f is smooth, but $|(\alpha_R-\alpha_c)\log s|$ is not so small, we can use Eq. (3.5). If both are valid, then of course we use Eq. (3.7).

In any event, subject to the stated assumptions, let us assume that we can obtain the form (3.7). The amplitude is thus described (to a high degree of accuracy over a sizeable range of s if α_I is fairly small) as a pair of complex conjugate Regge poles and nothing else. In this moderate energy range, the entire effect of the cut, no matter what the detailed behaviour of its discontinuity, looks just like the (Breit-Wigner like) complex pair of

Regge poles. This, we emphasize, is true whether or not the poles lie on the physical sheet. The accuracy of the representation (or rather the range of s over which it is an accurate representation) depends on the size of α_I, and of $|(\alpha_R-\alpha_c)\log s|$; and it also depends on the smoothness of Im f in the vicinity of α_R. Finally, it is also important to note that the representation of the cut as a complex pair is far more accurate than to represent it simply by a single real pole -- that is, by the limiting case in which the cut is absent altogether.

The conclusion, then, is that the presence of a cut does not necessarily leave Regge theory with zero predictive power, even if very little is known about the cut. All that the cut accomplishes, in the moderate energy region at least, is to replace each real Regge pole which one would have dealt with in the absence of cuts, by a complex pair. The number of parameters is therefore only doubled: whereas before, in a pure pole model, we had a real trajectory $\alpha(t)$ and a real residue $\beta(t)$ we now have a complex trajectory $\alpha(t)=\alpha_R(t)+i\,\alpha_I(t)$ and a complex residue $\beta(t)=|\beta(t)|e^{i\phi\beta(t)}$.

We may exhibit the correspondence to the single real pole explicitly. The amplitude of Eq. (3.7), for a complex pair of poles, may be written in the following simple form: (we now re-introduce the signature).

For even signature

$$T(s,t) = \gamma_+ s^{\alpha_+} e^{-i\pi\alpha_+/2} + \gamma_- s^{\alpha_-} e^{-i\pi\alpha_-/2} = |\gamma| s^{\alpha_R} e^{-i\pi\alpha_R/2} \times F(s,t) \tag{3.11}$$

where the "correction factor" F is

$$F(s,t) = \cos(\alpha_I \log s+\phi_\gamma) \cosh \frac{\pi\alpha_I}{2} + i \sin(\alpha_I \log s+\phi_\gamma) \sinh \frac{\alpha\pi_I}{2} . \tag{3.12}$$

For odd signature

$$T(s,t) = i|\gamma|s^{\alpha_R} e^{-i\pi\alpha_R/2} \times F(s,t) \tag{3.13}$$

with the same "correction factor" F.

It should always be kept in mind that this correction factor only represents, approximately, the existence of the cut. For that reason the residues in Eqs. (3.11) and (3.13) do not factor. Only the residue of a true pole factors; the complex "poles" we deal with here are not really poles of the partial wave amplitude (through they may include contributions from physical sheet poles if there are any) but are only approximations to a cut.

For the same reason it is possible for the complex poles alone to produce polarization; this simply reflects the fact that a cut can produce polarization. Again, complex poles can produce "cross-over zeros" without the residue vanishing; (this can happen because F vanishes, not $|\beta|$. This is merely saying that a cut can give a cross-over zero. And so forth.

We have now completed our theoretical development. The result we have obtained is that if the cut is fairly weak, so that α_I is small, then even though nothing is known about the cut there are nevertheless phenomenological predictions. Indeed, an entire Regge pole plus Regge cut combination, up to some cut-off energy (the weaker the cut the higher the cut-off) is describable to a high degree of accuracy by only a complex conjugate pair of poles, and nothing else.

The following section will be devoted to some detailed illustrations of the use of this result in specific phenomenological situations, to conclude this section, however, let us make a few general observations, as follows.

3.1 Total Cross-Sections

As we have seen, the contribution of a given complex Regge pair of even signature to the imaginary part of a forward elastic amplitude will be

$$+|\gamma(o)|\, s^{\alpha_R(o)} \, \mathrm{Im}\left(e^{-i\pi\alpha_R(o)/2} F(s,o)\right) \tag{3.14}$$

(For odd signature, an extra factor of i is inserted in the bracket). Using the form given for F in Eq. (3.12), this may be rewritten as a contribution to the total cross-section of

$$|\gamma(o)|\, s^{\alpha_R(o)-1} \{\cos[\alpha_I(o)\log s + \phi_\gamma(o)] \times \times \sin\frac{\pi\alpha_R}{2} \cosh\frac{\pi\alpha_I}{2} - \sin[\alpha_I(o)\log s + \phi_\gamma(o)] \times \times \cos\frac{\pi\alpha_R}{2} \sinh\frac{\pi\alpha_I}{2}\} . \tag{3.15}$$

At first sight, therefore, there would appear to be an oscillation with increasing log s, of period $2\pi/\alpha_I(o)$, assuming that $\alpha_I(o)$=o [12]. Several cautionary remarks are, however, necessary [13]. First, the approximation of replacing the cut by a complex pair failed when s became sufficiently large; in fact by referring back to the original integral for the cut, Eq. (3.5), it is easily seen that it necessarily fails before the oscillations in Eq. (4.2) begin. The cut term actually goes monotonically down, and does not oscillate at all. Oscillations can exist only if there are physical sheet complex poles (i.e. if $\bar{\beta}_\pm \neq o$).

Secondly, while oscillatory terms, if they exist, do not cause any trouble for lower lying Regge poles (for which $\alpha_R(o)<1$) it is obviously not possible for the Pomeranchon itself to exhibit such behaviour; if it did, we would eventually find negative total cross-sections. For the Pomeranchon, therefore (if it is also describable, at least approximately, as a complex pair) either $\alpha_I(o)=o$ or the complex pole is on an unphysical sheet.

3.2 Forward Diffraction Peaks

The question of how the complex pole idea applies to the Pomeranchon may as well be faced immediately, since it's been brought up in the context of total cross-sections. Let us list the possibilities.

i) The Pomeranchon is simply a real pole -- perhaps like the physical sheet pole in the logarithmic cut model [14]. In this event the total cross-section is the normal Pomeron contribution plus, perhaps, damped oscillatory contributions from lower lying complex pairs.

ii) The Pomeranchon is a complex pair, either with $\alpha_I(o)=o$ or on an unphysical sheet to avoid negative cross-sections. In this event, we can see from Eq. (3.11) that an elastic differential cross-section looks like

$$\frac{d\sigma}{dt} = \left(\frac{d\tau}{dt}\right)_{\alpha_I=o} |F(s,t)|^2$$

and we note

$$|F(s,t)|^2 = 1 + \sinh^2 \pi \alpha_I/2 - \sin^2(\alpha_I \log s+\phi_\gamma)$$

The correction factor thus introduces minima in $d\sigma/dt$ which move toward $t=o$ with log s [15].

iii) The Pomeranchon is something entirely different. The entire question of what constitutes diffraction scattering has, of course, always been present in Regge theory. We do not escape the uncertainty here; as a result, detailed phenomenological analysis of elastic processes where Pomerons exist, are subject to some ambiguity.

4. APPLICATIONS

4.1 π-N Charge Exchange Scattering [16]

We shall begin with this reaction for the usual reasons: it is rather well measured, and, more important, it is theoretically very clean because only a single important trajectory, the ρ, can be exchanged.

There are two scalar amplitudes in this process, the usual $A'^{(-)}$ and $B^{(-)}$. In conformity with the foregoing discussions, these are parametrized by a single pair of complex poles, representing the ρ trajectory and its associated cut. We write

$$A'^{(-)} = \gamma_{+A} \left(\frac{1-e^{-i\pi\alpha_+}}{\sin \pi\alpha_+}\right) \left(\frac{s}{s_o}\right)^{\alpha_+} + \gamma_{-A} \left(\frac{1-e^{-i\pi\alpha_-}}{\sin \pi\alpha_-}\right) \left(\frac{s}{s_o}\right)^{\alpha_-} \quad (4.1)$$

and

$$B^{(-)} = \alpha_+\gamma_{+B} \left(\frac{1-e^{-i\pi\alpha_+}}{\sin \pi\alpha_+}\right) \left(\frac{s}{s_o}\right)^{\alpha_+-1} + \alpha_-\gamma_{-B} \left(\frac{1-e^{-i\pi\alpha_-}}{\sin \pi\alpha_-}\right) \left(\frac{s}{s_o}\right)^{\alpha_--1} . \quad (4.2)$$

The factors α_+ and α_- in the residues of the B amplitude are conventional and reflect the P'_ℓ appearing in the cross channel partial wave expansion for B.

For α_R the usual straight line trajectory is chosen: $\alpha_R(t) = a+bt$; we expect that as in the simple real pole situation we will find $a \sim 0.5$ and $b \sim 1.0$.

Two models are chosen for α_I, corresponding (roughly) to the square root and log cases. We try both $\alpha_I(t)=g\sqrt{-t}$ and $\alpha_I(t)=g$ (a constant).

Finally, for the residues we choose forms similar to the normal ones; specifically $\gamma_{+A}=h_o\, e^{h_1 t}(\alpha_+ +1)e^{i\phi_{+A}}$ and $\gamma_{+B}=d_o\, e^{d_1 t}(\alpha_+ +1)e^{i\phi_{+B}}$. (Of course $\gamma_{-A}=\gamma^*_{+A}$ and $\gamma_{-B}=\gamma^*_{+B}$). The phases are parametrized by either $\phi_{+A} = \pi\sqrt{-t}\cdot(\gamma_0 \mp \gamma_1 t)$ or by $\phi_{+A}=\pi(\gamma_0+\gamma_1 t)$ corresponding to the two choices for α_I, with similar expressions for ϕ_{+B}.

There are two manifestations of the complex poles to look for, neither of which can be understood with a real ρ pole alone. These are polarization and the cross-over phenomenon. Only these allow us to fix α_I and the phases; the remaining data, on $d\sigma/dt$ and $\sigma_T(\pi^- p)-\sigma_T(\pi^+ p)$, provide only negative constraints, in that the excellent fits to these which can be obtained with a real pole alone must not be seriously disturbed.

Results for all this are shown in Figs. 6 and 7, and the corresponding parameters are listed in reference 16, for the "square root" and "constant" models of α_I mentioned earlier. The best fit trajectories in the two cases are

$$\alpha^\rho_\pm(t) = 0.53 + 1.02 \pm 0.20\, i\sqrt{-t} \qquad (4.3)$$

and

$$\alpha^\rho_\pm(t) = 0.50 + 0.95 \pm 0.088\, i. \qquad (4.4)$$

The fits are excellent, though of course the miserable quality of the polarization data precludes us from a serious test of the forms of α_I. It may be noted, also, that a cross-over is indeed obtained; however, its position is not sensitive to the parameters and may be placed anywhere in the range $t=-0.15$ (GeV)^2 to $t=-0.5$ (GeV)^2 without affecting the other fits.

4.2 π-N Backward Scattering [17]

We limit ourselves here to two trajectories, the N and the Δ, and confine our attention, for the time being, to the "square root" model. Fits are made to backward differential cross-sections for the three processes $\pi^- p \to p\pi^-$, $\pi^+ p \to p\pi^+$ and $\pi^- p \to n\pi^0$.

There are, now, many parameters (22 to be exact). Best fits are shown in Figs. 8, 9 and 10, and the corresponding trajectories are, for the nucleon,

$$\alpha_{\pm}^{N}(u) = -0.42 - 0.03\,W + 0.54\,W^2 \pm 1.04\,W\sqrt{0.27\,W^2 - 0.03\,W + 0.04} \qquad (4.5)$$

and for the Δ

$$\alpha_{\pm}^{\Delta}(u) = 0.07 + 0.08\,W + 0.44\,W^2 \pm 0.94\,W\sqrt{0.22\,W^2 + 0.08\,W - 0.08} \qquad (4.6)$$

Points to be emphasized are that one can obtain excellent fits to the dip in backward charge exchange scattering, and that the trajectories given in Eqs. (4.5) and (4.6) have no MacDowell partners.

4.3 π-N Elastic Scattering

As yet results for these processes do not exist. When they do, they will be plagued by the uncertainty, already referred to at the end of Section 3, arising from the Pomeron. The most straightforward thing to do is simply to permit the Pomeron to be a complex pair as well, and see what happens; this is what is being done in the fits to forward π-N elastic scattering. (As mentioned before, there are of course some constraints on the Pomeron; for example, if it is on the physical sheet, it must have $\alpha_I=o$ at $t=o$. These constraints do not seriously affect the data fitting, however).

4.4 Other Pseudoscalar Meson-Baryon Scattering Processes: Line Reversal

Next we discuss processes such as K-N scattering, and hyperon production in K-N or π-N collisions.

One thing to be done here is, of course, simply to make detailed fits in complete analogy to those we have already discussed for π-N scattering. Before going on to do this, however, some more general comments are in order, having to do with the relation of exotic processes and their line reversed associates [18]. As examples, think of the exotic charge exchange process $K^+n\to K^o p$ and its line reversed partner $\bar{K}^o n\to K^- p$, or the exotic hypercharge exchange process $K^- p\to\pi^-\Sigma^+$ together with its line reversed associate $\pi^+ p\to K^+\Sigma^+$.

It is, apparently, experimentally the fact that the ratio R(s,t) of an exotic differential cross-section to the differential cross-section of its line reversed partner is always greater than one [19]. It is also quite difficult to understand this fact within pure Regge pole models, and even within Regge models containing absorptive cuts. The point is that the reality of exotic amplitudes requires, through duality, exchange degeneracy of the Regge trajectories contributing to the high energy form of the amplitude (ρ-A_2 for charge exchange, and K^*-K^{**} for hypercharge exchange); indeed, this exchange degeneracy seems well confirmed experimentally, at least in some cases [20]. But, with exchange degeneracy, a pure pole model yields $R(s,t)\equiv 1$, and including absorptive cuts gives $R(s,t)<1$.

This apparent difficulty is easily resolved when one replaces the (incorrect) absorptive cut model by the more realistic complex pole picture. With exact exchange degeneracy, we may write for the exotic amplitude the representation

$$T_E(s,t) = \frac{1}{\pi}\int_{-\infty}^{\alpha_c} \frac{\mathrm{Im}\ f(t,j)\, s^j}{(j-\alpha_+)(j-\alpha_-)}\, dj$$

as in Eq. (3.3). (We assume the complex pole to be on the unphysical sheet, for now). For the line-reversed non-exotic amplitude, we obtain, in contrast

$$T_N(s,t) = \frac{1}{\pi}\int_{-\infty}^{\alpha_c} \frac{\mathrm{Im}\ f(t,j)\, s^i\, e^{-i\pi j}}{(j-\alpha_+)(j-\alpha_-)}\, dj$$

with the same weight function Im f. Now, obviously, if Im f is smoothly varying around α_R over a region of j of order α_I, we get $|T_N|<|T_E|$, and hence find $R(s,t)>1$.

Detailed fits to the (relatively poor) data on R(s,t) for both Σ and Λ production (the pairs $K^-p\to\pi^-\Sigma^+$ and $\pi^+p\to K^+\Sigma^+$, and $K^-n\to\pi^-\Lambda$ and $\pi^-p\to K^0\Lambda$) can be made, and these provide adequate fits to the (also poor) polarization data available as well [18]. The resulting exchange degenerate K^*-K^{**} trajectory is either (square root case)

$$\alpha_\pm^{K^*}(t) = 0.4 + 1.0\ t \pm 0.6\ i\ \sqrt{-t} \qquad (4.7)$$

or (constant case)

$$\alpha_\pm^{K^*}(t) = 0.4 + 1.0\ t \pm 0.2\ i\ (1-2t). \qquad (4.8)$$

It is interesting to note that the α_I for the K^*-K^{**} is noticeably larger than that for the ρ-A_2 found earlier; this is quite consistent with the fact that R(s,t) for charge exchange is smaller than for hypercharge exchange, and tends to decrease more rapidly toward one as the energy increases.

As to detailed fits to, for example, K-N data, these have not yet been done. Obviously it will be interesting to see how these work, in order to check that the same α_I for the ρ as we obtained before works here too.

4.5 Baryon-Baryon Scattering

Here, as yet, no fits exist; however, a fit to the n-p CEX scattering is under way, and hopefully, will be available soon [21].

4.6 Vector-Meson Production

The only fit existing so far under this heading is a fit to the reaction $\pi^+ p \to \pi^0 \Delta^{++}$ [22]. This reaction is a natural one to choose, since one believes that it, like π-N CEX is dominated by ρ exchange alone; it should therefore provide another check on the ρ-parameters found earlier.

Unfortunately, the functional form chosen here for $\alpha_I(t)$ is not quite identical to either the "square root" or "constant" models; however, it is not too different from the "constant" case. The best fit turns out to give

$$\alpha^{\rho}_{\pm}(t) = 0.57 + 0.96\ t \pm \frac{0.274\ i}{1.14-t}\ . \tag{4.9}$$

For very small t, we have here an α_I=0.24 while in the "constant" case before we got α_I=0.09. It is not clear whether this is or is not a serious discrepancy; perhaps it is to be attributed to the different choice of functional forms, and to different choices of the other parameters. Fits obtained to $d\sigma/dt$ and to the spin density matrix elements are shown in Figs. 11 and 12.

4.7 Photoproduction

One fit, so far, exists here [23], to near forward charged pion photoproduction. The dominant trajectories are assumed to be the pion and the A_2. Fits are made to the differential cross-section and to the asymmetry parameters for pion production by linearly polarized photons on polarized nucleons. Best fits are shown in Figs. 13, 14 and 15; the corresponding trajectories are

$$\alpha_{\pm}^{\pi}(t) = -0.02 + 1.0\,t \pm 1.03\,i \qquad (4.10)$$

and

$$\alpha_{\pm}^{A_2}(t) = 0.25 + 1.0\,t. \qquad (4.11)$$

The results are quite insensitive to the A_2; the pion plays by far the dominant role. For this reason an α_I for the A_2 has been ignored.

The value of α_I for the pion seems abnormally large, in comparison with those found earlier for other trajectories. Indeed, one is entitled to wonder whether, with such a large α_I the entire approximation of replacing the pole-cut combination with simply a complex pair of poles is not invalidated. For this reason it is worth nothing explicitly that the residue functions in this fit are explicitly assumed to be real; the phases ϕ_β are set equal to zero. It may well be that if this (somewhat artificial) constraint is relaxed, a smaller α_I, more in keeping with our earlier results, will be obtained.

REFERENCES AND FOOTNOTES

1. S. Mandelstam, Nuovo Cimento 3o, 1127 (1963).
2. R. Eden et al., "The Analytic S-matrix" (Cambridge Univ. Press, 1966).

3. R. Carlitz and M. Kislinger, Phys. Rev. Letters 24, 186 (1970). It should be emphasized that the cut occurring in this model may well be the same as that suggested by Mandelstam (ref. 1). For example, if $\alpha_P(t)=1$, then the first Mandelstam cut is also flat and crosses the pole at t=o. Or, the identification may be more subtle. It might be that the C-K cut coincides with the ultimate Mandelstam cut which is, perhaps, flat. Or it might be that the C-K cut is not really flat, but only appears so in the simple model in which they first obtained it. In any case it is highly unattractive to believe that the C-K and Mandelstam cuts really represent different phenomena.
4. In fact, we shall show that the specific values for the cut discontinuities assumed in these models are unlikely ever to be correct. Thus all cut models in use so far are, in fact, wrong. The predictions of such models may be specific; they are also wrong.
5. The partial wave amplitude is of a given signature. It should be emphasized that we presume there to be a cut, of the same quantum numbers and the same signature, associated with each Regge pole of a given signature. Generally, however, the signature is merely a notational complication; we shall therefore normally ignore it, and mention it only where its inclusion is not entirely obvious or where it makes some important difference.
6. Evidently, D_c is not uniquely defined by Eq. (2.1). The following examples illustrate what is meant by it.
7. In order to keep N real on the right-hand cut, α_{in} must be a polynomial in t, which we take to be linear. In order to assure that α_{out} is a zero of D, and not a pole of N, N and D are both multiplied by $(j-\alpha_{in})$; this, of course, does not alter their analyticity properties.

8. Hung Cheng, Phys. Rev. 130, 1283 (1963).
9. P. Kaus and F. Zachariasen, Phys. Rev. D1, 2962 (1970).
10. W. R. Frazer and C. M. Mehta, Phys. Rev. D1, 696 (1970);
J. S. Ball and G. Marchesini, Phys. Rev. 188, 2508 (1969);
G. F. Chew and D. R. Snider, UCRL 20033, July 1970.
11. J. S. Ball, G. Marchesini and F. Zachariasen, Phys. Letters 31B, 583 (1970).
12. G. Chew and D. Snider, Phys. Letters 31B, 71 (1970). It should be noted that the model which these authors base their discussion on has a logarithmic cut; the complex poles are therefore always on unphysical sheets. Hence oscillating terms, the total cross-sections cannot occur.
13. See Ref. 11.
14. See G. F. Chew and D. R. Snider (Ref. 10). They interpret the physical sheet pole in a logarithmic cut model as the Pomeron, and the nearest unphysical sheet complex pair as the P'.
15. The suggestion that the Pomeron is a complex pair was made first by Freund and Oehme, Phys. Rev. Letters 10, 459 (1963). They suggest, in addition, that $\alpha_P=1$, a constant, for the Pomeron, in an attempt to have (nearly) non shrinking diffraction peaks without violating t-channel unitarity.
16. B. R. Desai, P. E. Kaus, R. T. Park and F. Zachariasen, Phys. Rev. Letters 25, 1389 (1970).
17. R. T. Park and P. E. Kaus, private communication.
18. D. P. Roy, J. Kwiecinski, B. R. Desai and F. Zachariasen, to be published.

19. K. Lai and J. Louie, Nuclear Phys. B19, 205 (1970).
20. A. Firestone et al., Phys. Rev. Letters 25, 958 (1970).
21. D. P. Roy, private communication.
22. P. Butera, M. Enriotti and G. Marchesini, University of Milan preprint, December (1970).
23. J. S. Ball, H. J. W. Müller and K. B. Pal, preprint UCRL 20057, August (1970).

FIGURE CAPTIONS

Fig. 1a: A trajectory, and the sequence of branch points generated by its repeated interference with the Pomeron, if $\alpha_P(0)=1$.

1b: A trajectory, and the sequence of branch points generated by its repeated interference with the Pomeron, if $\alpha_P(0)<1$.

Fig. 2 : The unperturbed or input trajectory, the branch point, and the output physical and first sheet poles, in a logarithmic cut model.

Fig. 3 : The unperturbed or input trajectory, the branch point, and the two output trajectories, in a square root cut model.

Fig. 4a: Comparison of Eqs. (3.3) and (3.5) in a square root cut model. For $\alpha_c-\alpha_R=-0.3$, $\alpha_I=0.2$ and Im $f=\sqrt{-\ell}$. (From Ref. 11).

4b: Comparison of Eqs. (3.3) and (3.5) in a logarithmic cut model, so that Im $f=(\ell-\alpha_+)(\ell-\alpha_-)/(\ell-\alpha_0-g\log(-\ell))^2+\pi^2g^2$ with $g=0.05$ and $\alpha_0=-0.35$. (From Ref. 11).

Fig. 5 : Comparison of Eqs. (3.5) and (3.7), for $\alpha_c-\alpha_R=-0.3$, $\alpha_I=0.2$.

Fig. 6 : Fits to π-N CEX data for the "square root" model. (From Ref. 16).

Fig. 7 : Fits to π-N CEX data for the "constant" model. (From Ref. 16).

Fig. 8 : Fits to $d\sigma/dt$ for $\pi^+ p \to p\pi^+$. (From Ref. 17).

Fig. 9 : Fits to $d\sigma/dt$ for $\pi^- p \to p\pi^-$. (From Ref. 17).

Fig.10 : Fits to $d\sigma/dt$ for $\pi^- p \to n\pi^0$. (From Ref. 17).

Fig.11 : Fits to $d\sigma/dt$ for $\pi^+ p \to p^0 \Delta^{++}$. (From Ref. 22).

Fig.12 : Fits to density matrix element for $\pi^+ p \to p^0 \Delta^{++}$. (From Ref. 22).

Fig.13 : Fits to $d\sigma/dt$ for $\gamma p \to \pi^+ n$. (From Ref. 23).

Fig.14 : Fit (solid line) to polarized photon asymmetry parameter. The dotted line shows the perturbation theory prediction. (From Ref. 23).

Fig.15 : Fit to left-right asymmetry parameter A. (From Ref. 23).

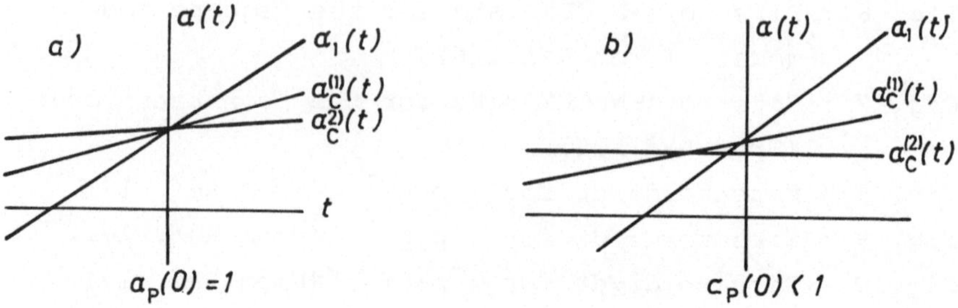

Fig. 1.

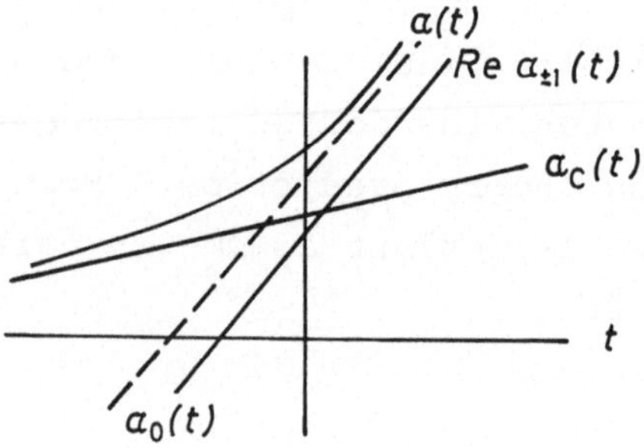

Fig. 2.

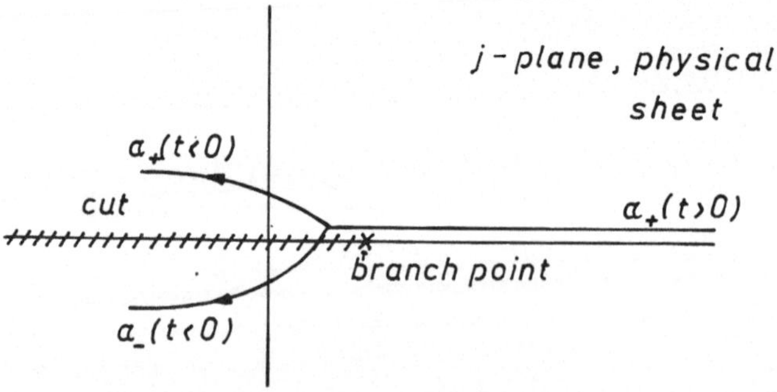

Fig. 3.

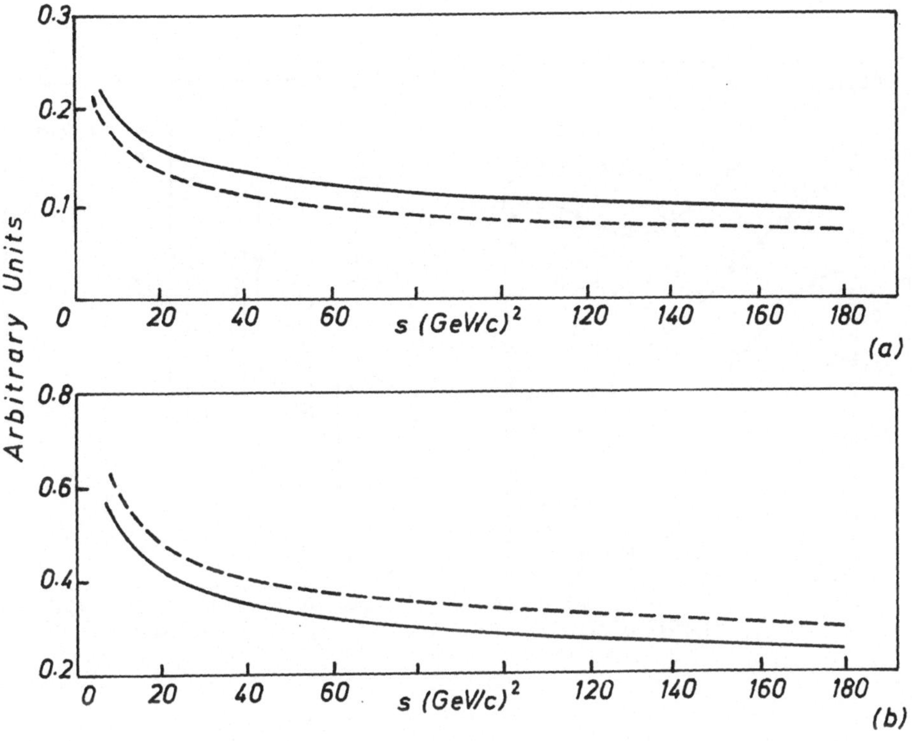
Arbitrary Units
0.3
0.2
0.1
0
20
40
60
s (GeV/c)^2
120
140
160
180
(a)
0.8
0.6
0.4
0.2
0
20
40
60
s (GeV/c)^2
120
140
160
180
(b)

Fig. 4.

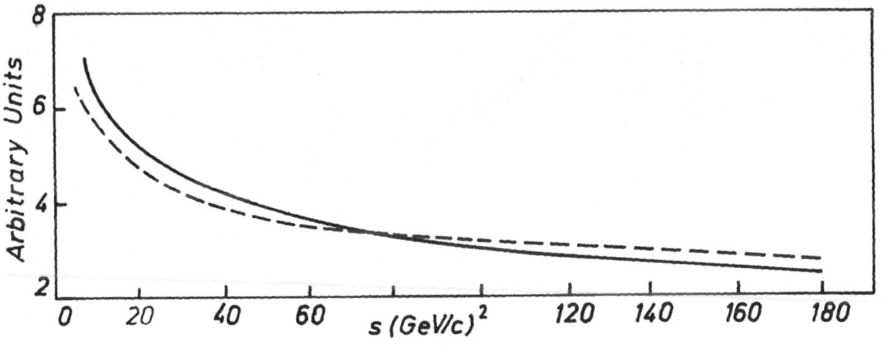
Arbitrary Units
8
6
4
2
0
20
40
60
s (GeV/c)^2
120
140
160
180

Fig. 5.

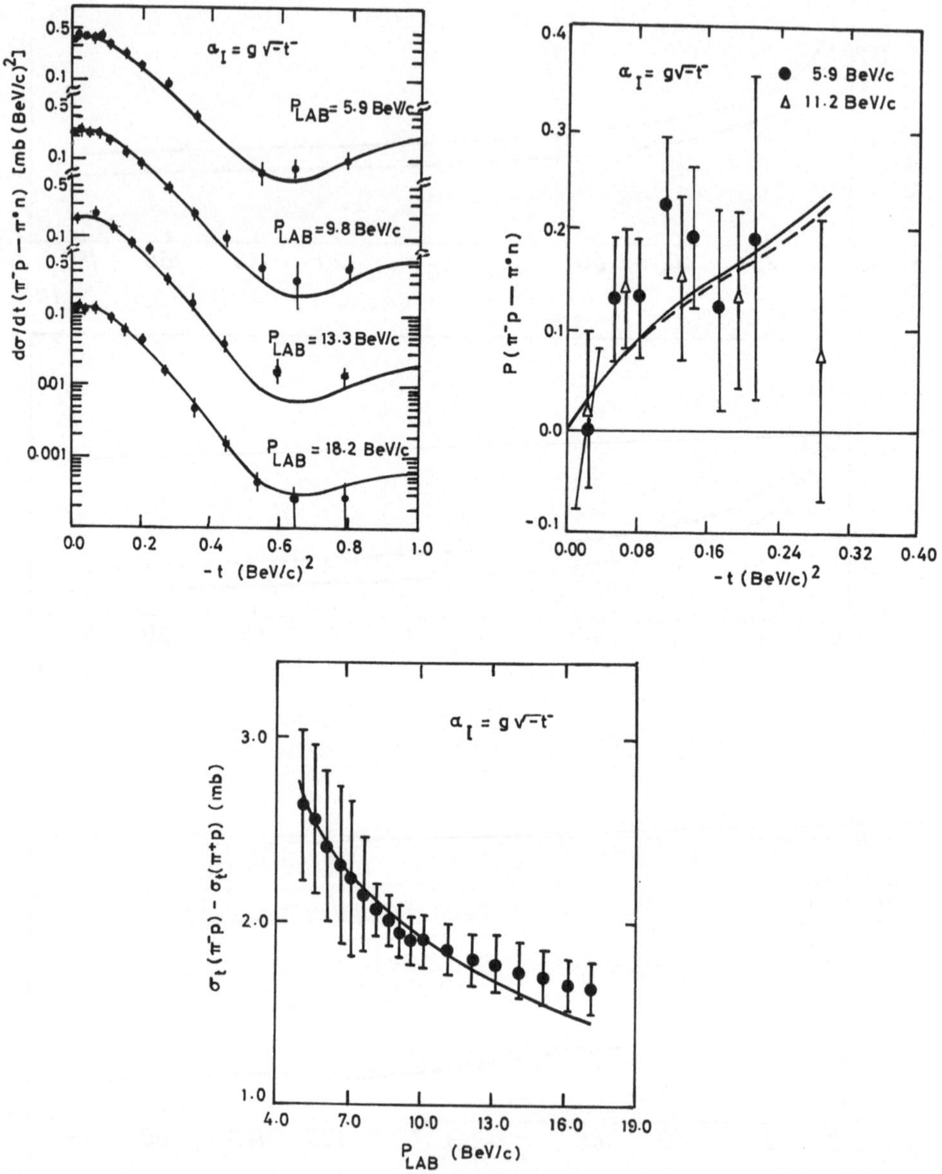
$\alpha_I = g\sqrt{-t}$
P_{LAB} = 5.9 BeV/c
P_{LAB} = 9.8 BeV/c
P_{LAB} = 13.3 BeV/c
P_{LAB} = 18.2 BeV/c
$d\sigma/dt\,(\pi^- p \to \pi^0 n)$ [mb (BeV/c)^2]
$-t$ (BeV/c)^2
5.9 BeV/c
11.2 BeV/c
$P(\pi^- p \to \pi^0 n)$
$\sigma_t(\pi^- p) - \sigma_t(\pi^+ p)$ (mb)
P_{LAB} (BeV/c)

Fig. 6.

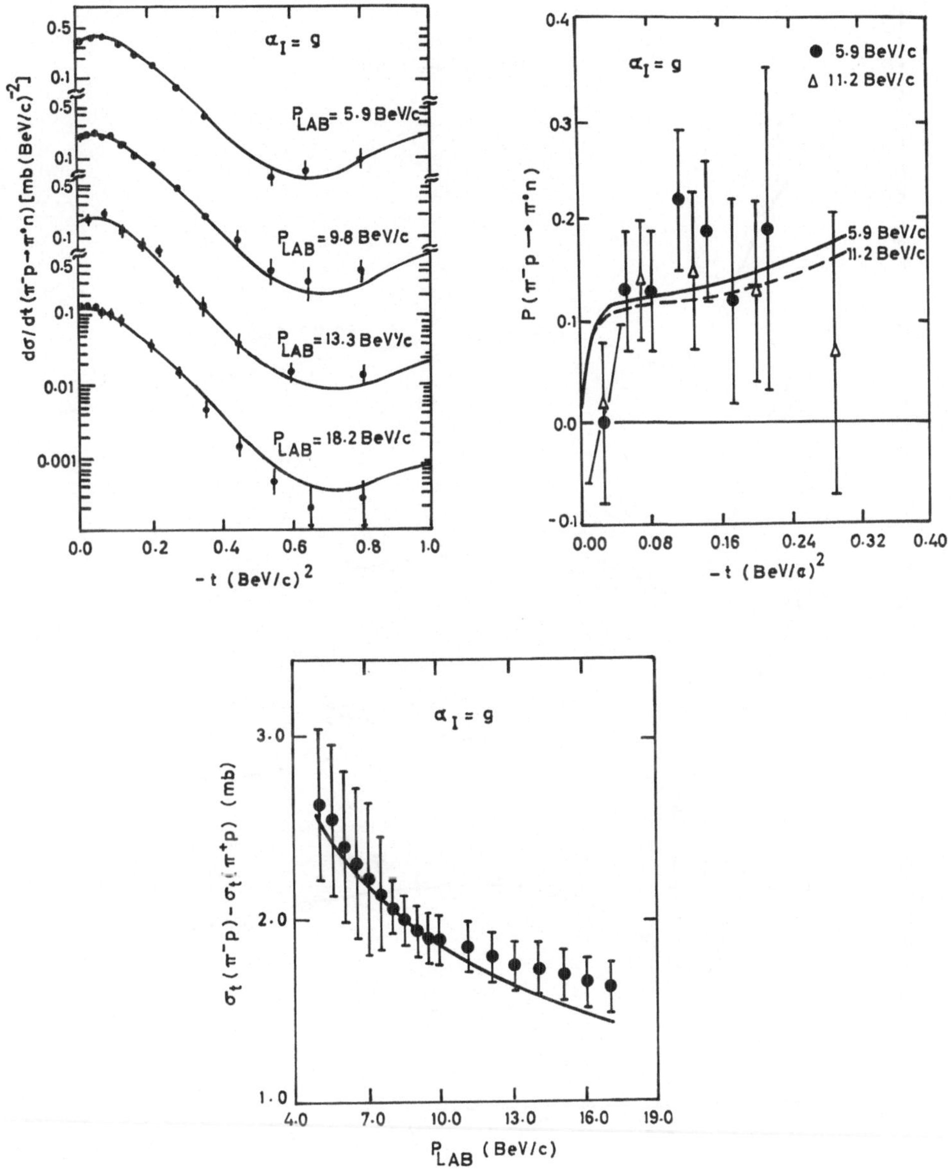
$\alpha_I = g$
P_{LAB} = 5.9 BeV/c
P_{LAB} = 9.8 BeV/c
P_{LAB} = 13.3 BeV/c
P_{LAB} = 18.2 BeV/c
dσ/dt (π⁻p → π°n) [mb (BeV/c)⁻²]
0.5
0.1
0.01
0.001
0.0
0.2
0.4
0.6
0.8
1.0
− t (BeV/c)²
5.9 BeV/c
11.2 BeV/c
P (π⁻p → π°n)
0.4
0.3
0.2
0.1
0.0
-0.1
0.00
0.08
0.16
0.24
0.32
0.40
σt (π⁻p) − σt (π⁺p) (mb)
3.0
2.0
1.0
4.0
7.0
10.0
13.0
16.0
19.0
P_{LAB} (BeV/c)

Fig. 7.

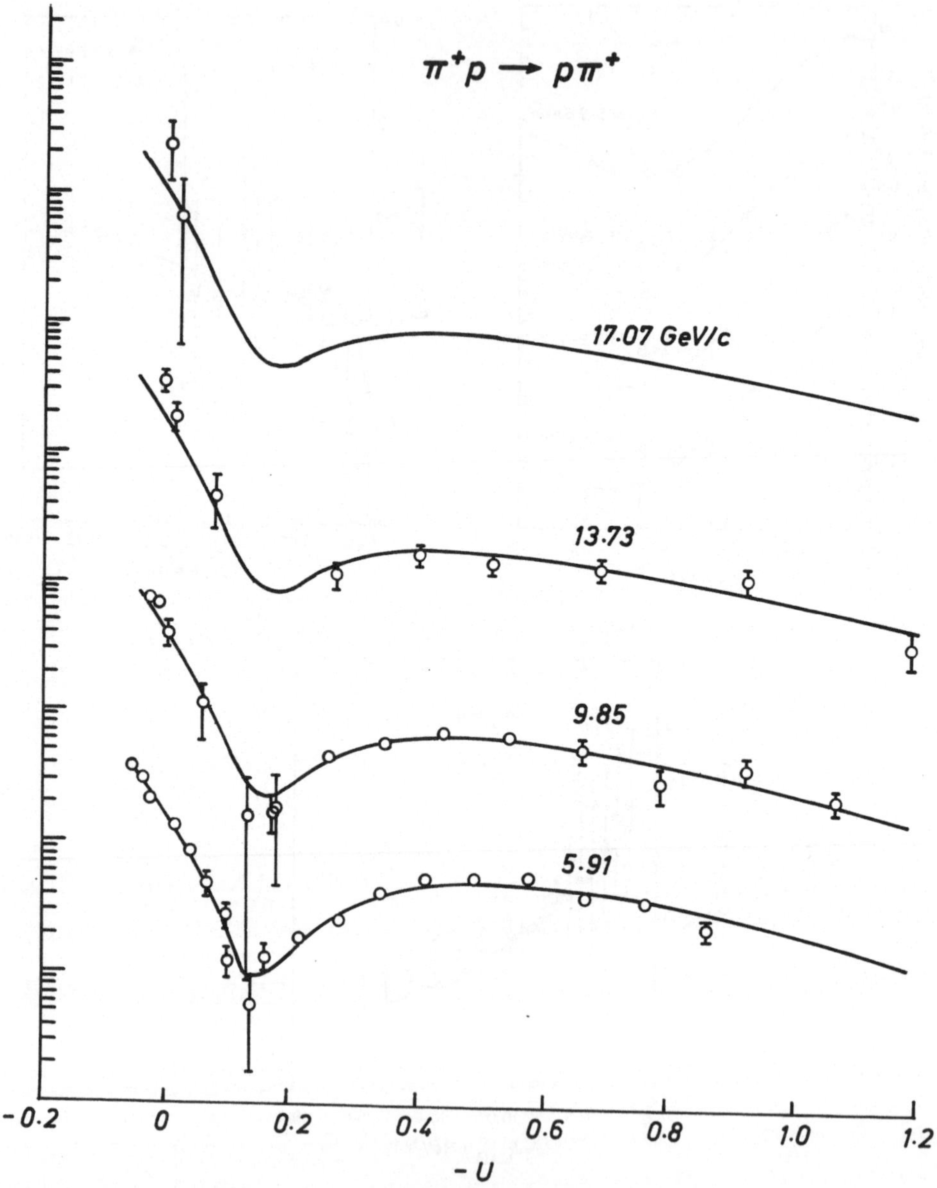

Fig. 8.

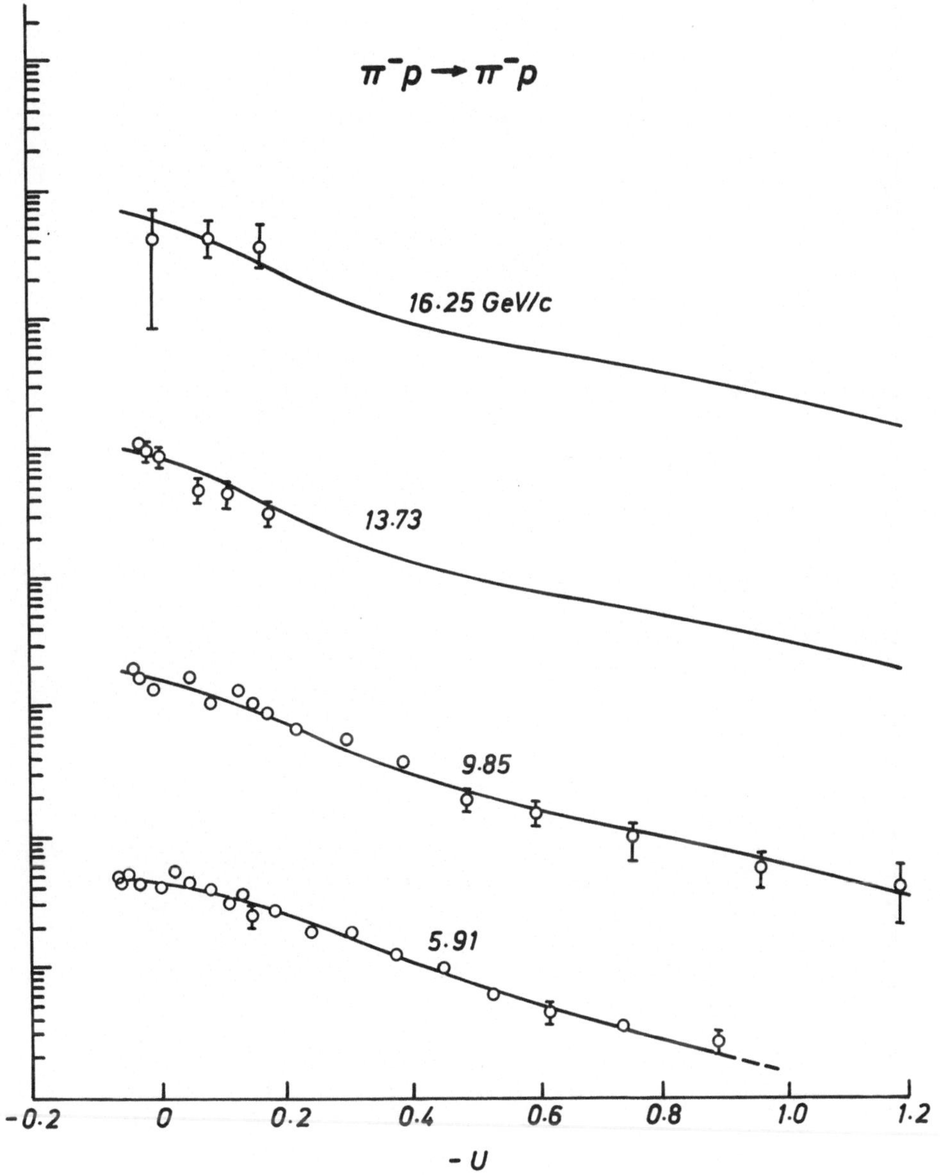

Fig. 9.

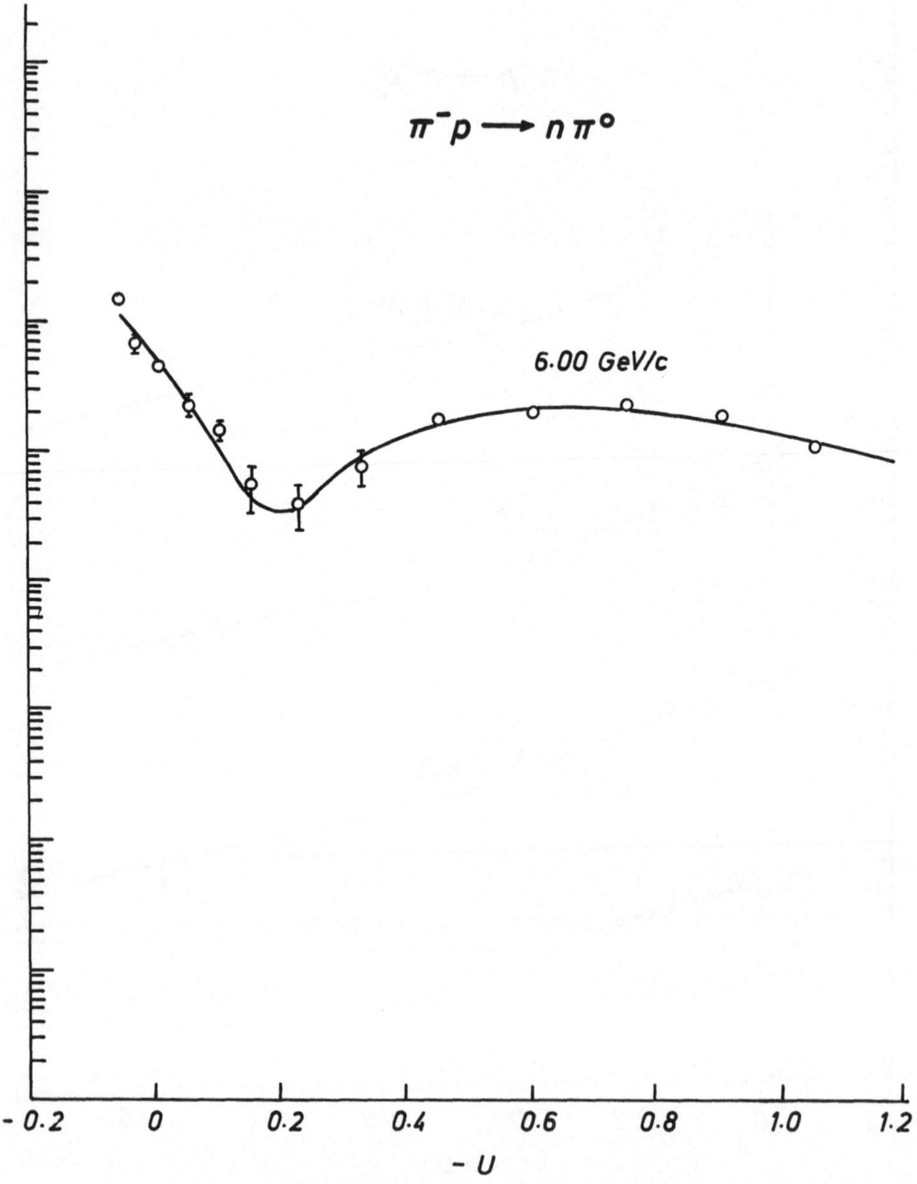

Fig. 10.

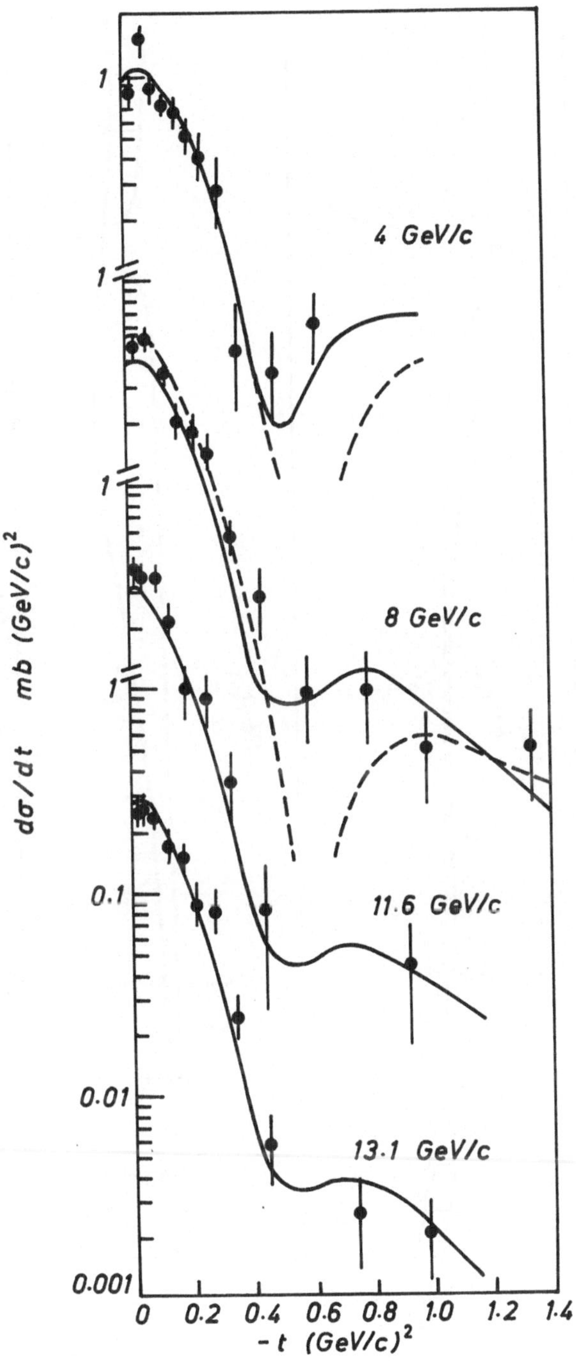
4 GeV/c
8 GeV/c
11.6 GeV/c
13.1 GeV/c
dσ/dt mb (GeV/c)²
-t (GeV/c)²
1
1
1
1
0.1
0.01
0.001
0
0.2
0.4
0.6
0.8
1.0
1.2
1.4

Fig. 11.

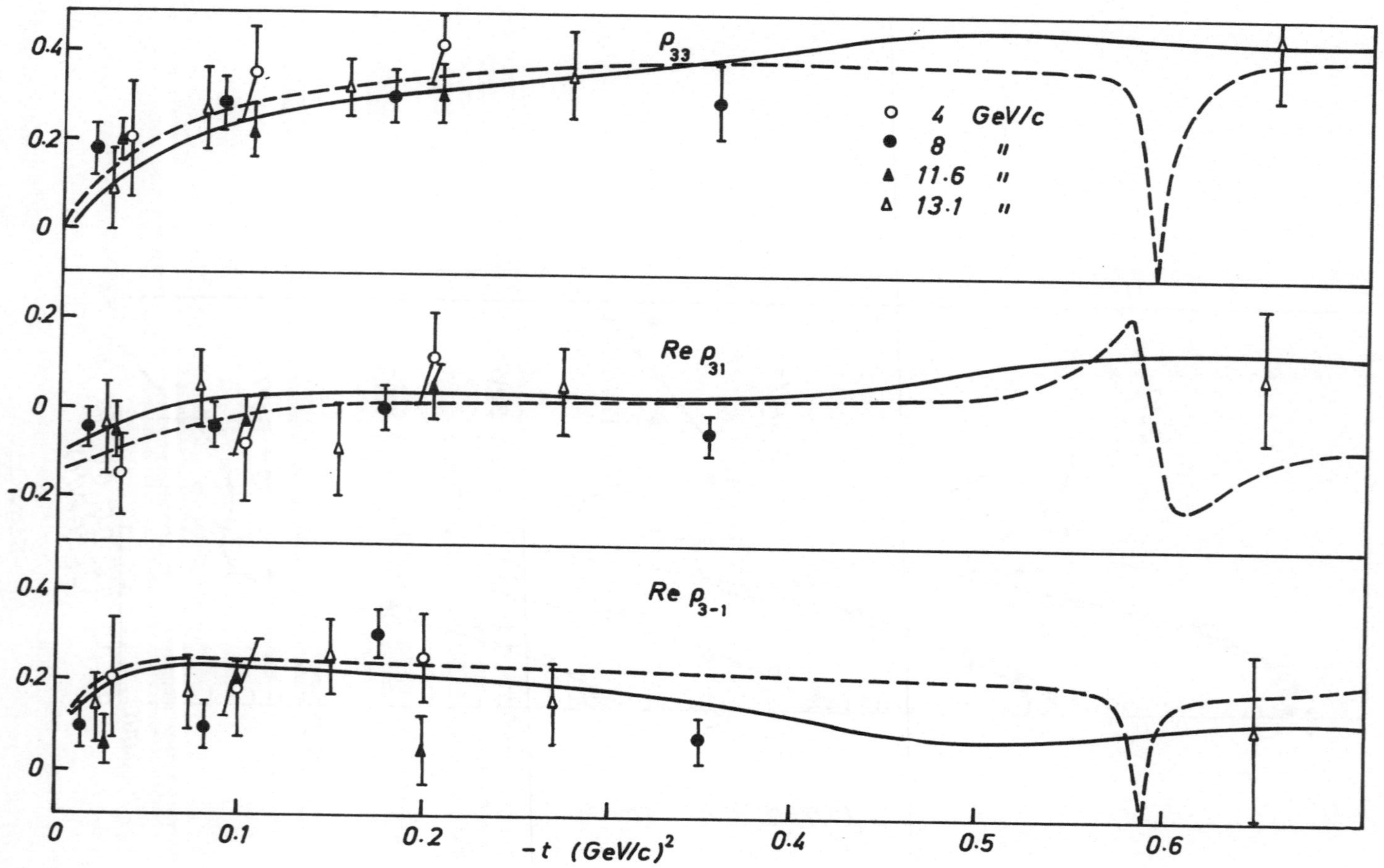

0.4
0.2
0
0.2
0
-0.2
0.4
0.2
0
ρ_{33}
Re ρ_{31}
Re ρ_{3-1}
○ 4 GeV/c
● 8 "
▲ 11.6 "
△ 13.1 "
0
0.1
0.2
0.4
0.5
0.6
$-t\ (GeV/c)^2$

Fig. 12.

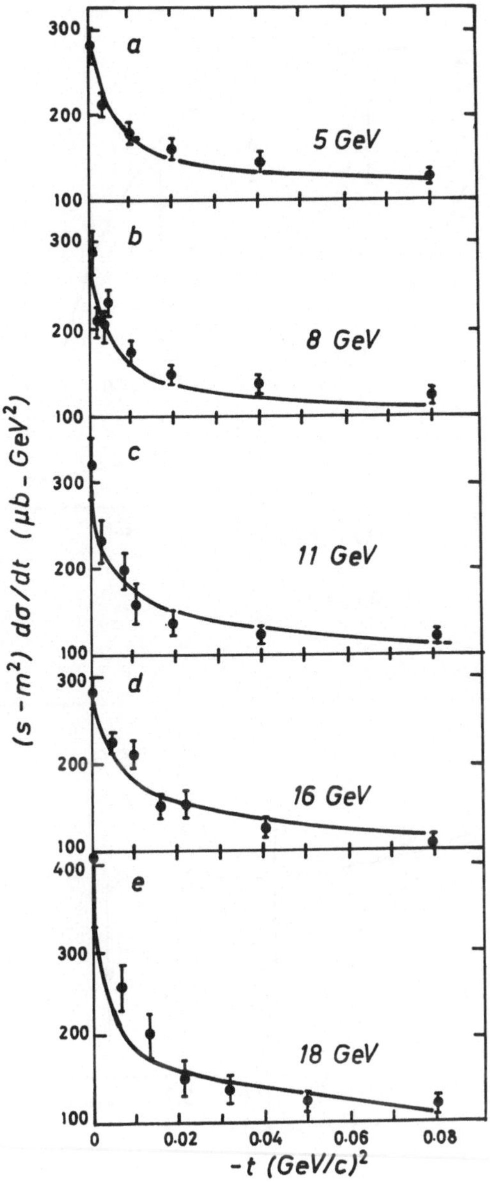
a
5 GeV
b
8 GeV
c
11 GeV
d
16 GeV
e
18 GeV
(s - m²) dσ/dt (μb - GeV²)
-t (GeV/c)²
0
0.02
0.04
0.06
0.08
100
200
300
400

Fig. 13.

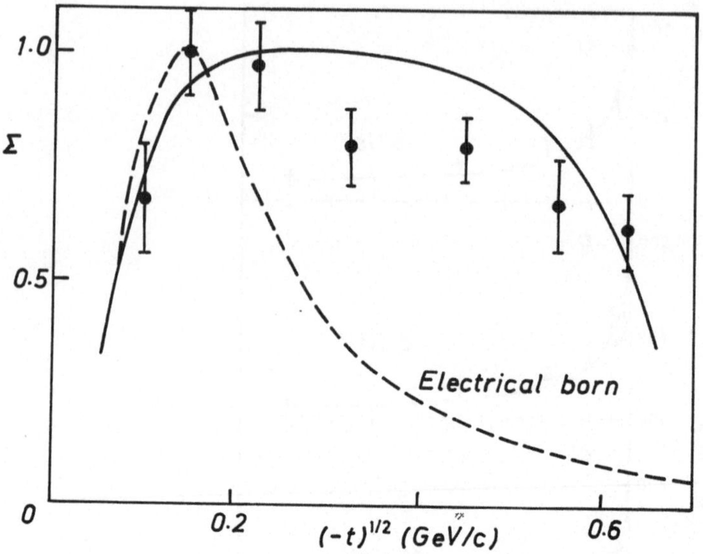
1.0
0.5
0
Σ
0.2
0.6
$(-t)^{1/2}$ (GeV/c)
Electrical born

Fig. 14.

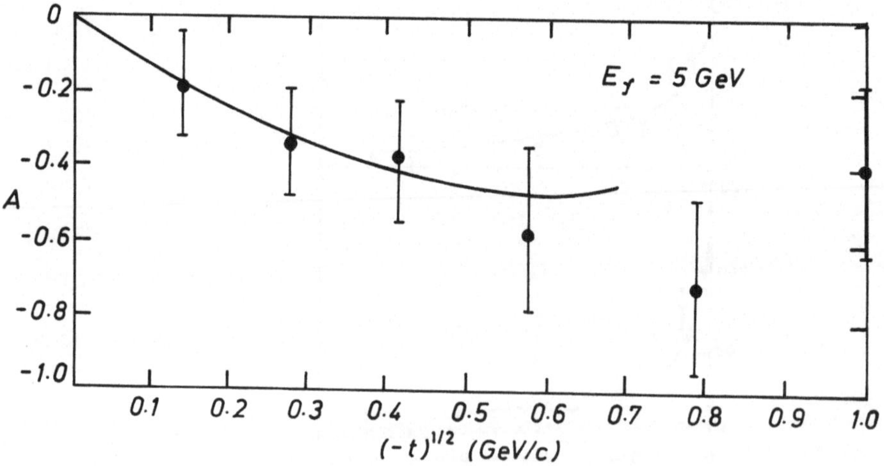
0
-0.2
-0.4
-0.6
-0.8
-1.0
A
E_γ = 5 GeV
0.1
0.2
0.3
0.4
0.5
0.6
0.7
0.8
0.9
1.0
$(-t)^{1/2}$ (GeV/c)

Fig. 15.

Acta Physica Austriaca, Suppl. VIII, 91—120 (1971)

MULTI PARTICLE DUAL MODEL[*]

BY

I. DRUMMOND
Department of Applied Mathematics
and Theoretical Physics
Cambridge

I. INTRODUCTION

The multiparticle dual model is a generalisation of the simple Beta-function model for two-particle scattering introduced by Veneziano [1]. For the elastic scattering of two identical particles, A, of mass m,

$$A(p_1)+A(p_2) \rightarrow A(p_3)+A(p_4)$$

we have as the invariant amplitude

$$T = B(-\alpha(s), -\alpha(t)) + \text{two other crossed terms.}$$

Here
$$\alpha(s) = \alpha_o + \alpha's \tag{1.1}$$

and
$$s = (p_1 + p_2)^2, \; t = (p_3 - p_2)^2 . \tag{1.2}$$

---

[*] Lecture given at X. Internationale Universitätswochen für Kernphysik, Schladming, March 1 - March 13, 1971.

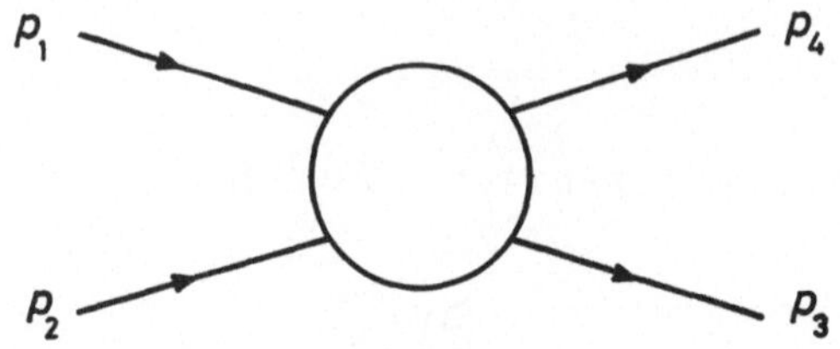

Fig. 1.

The reason for considering this amplitude is that it provides a simple solution to Finite Energy Sum Rules [2] with narrow width resonance saturation of the absorptive part. That is, we have

i) Regge asymptotic behaviour in the whole s-plane cut along the positive real axis,

$$T = \Gamma(-\alpha(s))\Gamma(-\alpha(t))/\Gamma(-\alpha(s)-\alpha(t)) \sim (-\alpha(s))^{\alpha(t)}\Gamma(-\alpha(t)), \qquad s \to \infty, \tag{1.3}$$

ii) Dispersion relations (with determined subtractions)

$$T = \sum_{n=0}^{\infty} \frac{1}{n-\alpha(s)} \frac{\Gamma(\alpha(t)+n+1)}{\Gamma(\alpha(t)+1)} \tag{1.4}$$

We see from (1.1) ad (1.3) that the Regge trajectory is linearly rising, which is not inconsistent with the known level structure of hadrons, and from (1.4) that T has poles in the s-plane with residues polynomial in t. They can therefore be interpreted as representing a superposition of intermediate states with angular momentum J, $0 \leq J \leq n$ where n is the position of the pole.

Of course we can write

$$T = \sum_{n=o}^{\infty} \frac{1}{n-\alpha(t)} \frac{\Gamma(\alpha(s)+n+1)}{\Gamma(\alpha(s)+1)} \qquad (1.5)$$

so T has poles in both s and t but not simultaneously.

This property of the Beta-function, that it can be written either as in (1.4) or (1.5) is referred to as duality and is taken to be a fundamental property of physical scattering amplitudes, at least in a narrow resonance approximation. In these lectures I hope to explain how this principle of duality for strong interaction amplitudes has been investigated in an attempt to gain a deeper understanding of its meaning.

2. THE FIVE POINT FUNCTION AND N-POINT FUNCTION

The "reason" for the duality property of the Beta function can be understood from its integral representation

$$B(-\alpha(s), -\alpha(t)) = \int_0^1 dx\ x^{-\alpha(s)-1}\ (1-x)^{-\alpha(t)-1} \qquad (2.1)$$

i) The large s asymptotic behaviour comes from a neighbourhood of x = 1 and its nature is controlled by $\alpha(t)$. Put

$$x = 1 - \frac{u}{-\alpha(s)},\ \alpha(s) \to -\infty$$

$$x^{-\alpha(s)-1} \sim e^{-u}$$

$$(1-x)^{-\alpha(t)-1} \sim \left(\frac{u}{-\alpha(s)}\right)^{-\alpha(t)-1};$$

$$B(-\alpha(s),-\alpha(t)) \sim \int_0^{\infty} du \cdot e^{-u}\ u^{-\alpha(t)-1}(-\alpha(s))^{\alpha(t)} =$$

$$= \Gamma(-\alpha(t))\,(-\alpha(s))^{\alpha(t)} \qquad (2.2)$$

The applicability of the asymptotic behaviour throughout the whole cut s-plane depends on the simplicity of the analyticity properties of the integrand particularly in the neighbourhood of x=o.

Similar remarks can be made _mutatis mutandis_, about the large t-behaviour.

We shall not discuss the asymptotic behaviour of the subsequent generalisations though once again its nature and region of applicability depend on the simple analytic character of the integrand.

ii) The fact that there are no simultaneous poles in s and in t is due in turn to the fact that the former arise from divergences at x=o and the latter from divergences at x=1.

We can put this in a language that generalizes. Thus, with $\alpha(s)=\alpha_{12}, \alpha(t)=\alpha_{23}$

$$T = \int dI \, u_{12}^{-\alpha_{12}-1} \, u_{23}^{-\alpha_{23}-1} \tag{2.3}$$

where $$u_{12} = 1-u_{23} \tag{2.4}$$

and $$dI = du_{12} = du_{23}$$

The integrand then has a factor $u_{ij}^{-\alpha_{ij}-1}$ for each channel that can have a pole. The pole arises from a divergence at u_{ij}=o. The appearance of simultaneous poles is prevented by the _duality constraint_ eg. (2.4). The integration measure is chosen to make the amplitude cyclically symmetric in a manner consistent with having poles at integer values of α_{ij}.

Bardacki and Ruegg [3] and Goebel and Sakita [4] generalized this result to the Five Point Function.

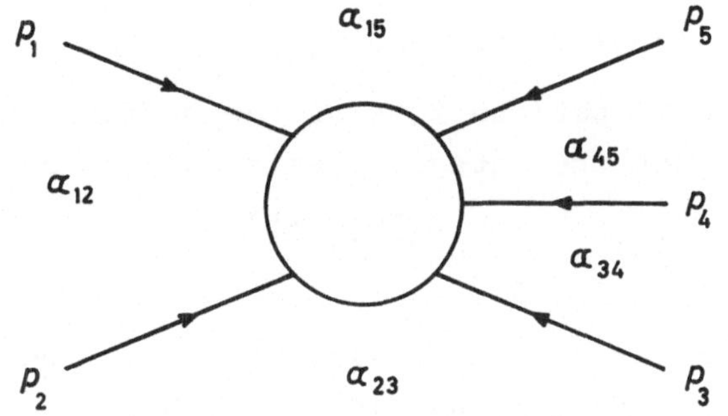

Fig. 2.

They suppose that the amplitude is made up of $\frac{1}{2}(5-1)!=12$ pieces, one for each different ordering of the external particles, just as for the 4-point function. A typical piece is represented by Fig. 2. It is required to have an internal pole structure consistent with any Feynman diagram which can be drawn with the given ordering of the external lines. Fig. 3 shows two typical cases.

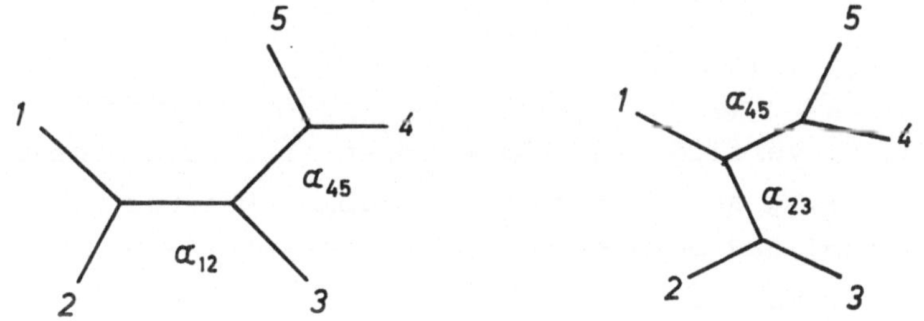

Fig. 3.

Thus simultaneous poles can appear in α_{12} and α_{45} but not in α_{12} and α_{23}. We say that the channels associated with these latter two variables are dual to one another. The list of duality relationships is

α_{12} is dual to α_{23} and α_{15}

α_{23} is dual to α_{12} and α_{34}

and further cyclic permutations of the suffixes.

We construct the integrand as suggested by the B-function.

$$T = \int dI \prod_{(i,j)} u_{ij}^{-\alpha_{ij}-1}$$

with one factor for each channel with poles. The duality constraints must be such that if $u_{12}=o$ then both the dual variables u_{23} and u_{15} must be unity and if either of those dual variables is zero then u_{12} must be unity. It is easy to see that these conditions are satisfied by

$$u_{12} = 1-u_{23}\, u_{15} \tag{2.5}$$

similarly we have

$$\begin{aligned} u_{23} &= 1-u_{12}\, u_{34} \\ u_{34} &= 1-u_{23}\, u_{45} \\ u_{45} &= 1-u_{34}\, u_{15} \\ u_{15} &= 1-u_{45}\, u_{12} \ . \end{aligned} \tag{2.6}$$

Again it can be verfied that these constraints are compatible with choosing as independent variables the two variables associated with the internal lines of any of the allowed Feynman diagrams. They may be used to express the integration measure. Thus

$$dI = \frac{du_{12}\, du_{45}}{1-u_{12}\, u_{45}} = \frac{du_{23}\, du_{45}}{1-u_{23}\, u_{45}} \tag{2.7}$$

where the weight function has been chosen to ensure the

cyclic symmetry necessary for duality.

It is very easy to check that the pole at $\alpha_{12}=o$ is

$$T \sim \frac{1}{\alpha_{12}} \int_0^1 du_{45}\, u_{45}^{-\alpha_{45}-1}\, u_{34}^{-\alpha_{34}-1}$$

where now from (2.5) and (2.6)

$$u_{34} = 1-u_{45}$$

since u_{12} has been put equal to zero and $u_{15}=u_{23}=1$. Thus the residue of the lowest pole yields the correct four body amplitude for us to identify this internal particle with the external one, assuming the internal trajectory has been chosen to have the correct intercept

$$\alpha_o = -\alpha' \, m^2.$$

The generalization to the N-body case was given by Chan [5]. Once again we assume there are 1/2(N-1)! different contributions one for each ordering of the external lines a typical one being given in Fig. 4.

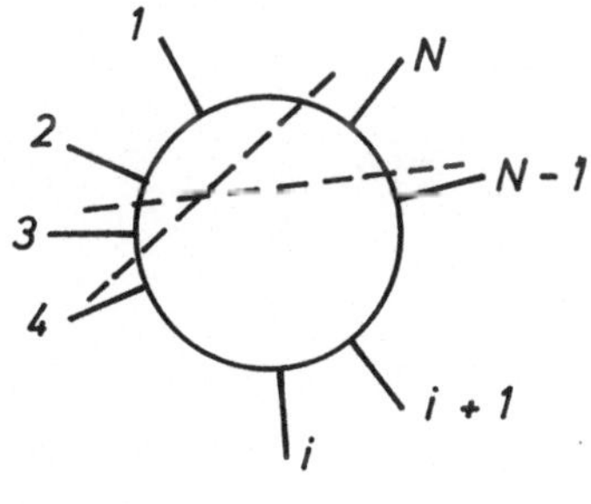

Fig. 4.

The channels which can have poles correspond to partitions of the N lines into two groups. Two partitions are dual if the lines determining them cross. A little thought shows that channels dual in this sense are also dual in the sense of not having simultaneous poles.

The amplitude is now

$$T = \int dI \prod_P u_P^{-\alpha_P - 1} \qquad (2.8)$$

where the product is over all partitions P. The duality constraints are

$$u_P = 1 - \prod_{\bar{P}} u_{\bar{P}} \qquad (2.9)$$

where $\bar{P}$ is any partition dual to P. These constraints leave N-3 variables independent. They may be chosen to be the group of N-3 variables associated with the internal lines of any allowed Feynman diagram. A particularly useful set is that associated with the multiperipheral diagram of Fig. 5.

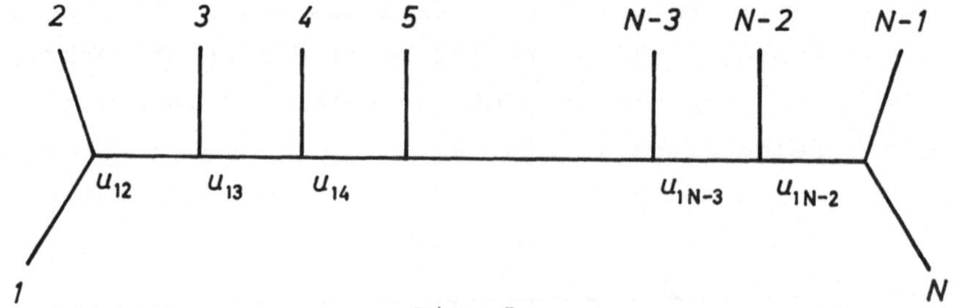

Fig. 5.

If we denote a partition by (ij) where i and j are the end lines in the group that does not contain line N then the variables we are regarding as independent are u_{1i} i=2...N-2.

Cyclic symmetry then requires

$$dI = \prod_{i=2}^{n-2} du_{1i} \Big/ \prod_{i=2}^{n-3} (1-u_{1i} \cdot u_{1i+1}) \qquad (2.10)$$

The constraints (2.9) can be solved by the equations

$$u_{ij} = \frac{(1-x_{ij})(1-x_{i-1j+1})}{(1-x_{i-1j})(1-x_{ij+1})} \qquad x_{ij} = u_{1i}\, u_{1i+1} \cdots u_{1j-1} \qquad (2.11)$$

3. FACTORISATION

If we use eq. (2.11) to eliminate the dependent variables from the integrand in eq. (2.8) we find, assuming all internal trajectories are identical and pass through the external particle,

$$A = \int \prod_{i=2}^{n-2} du_{1i}\, u_{1i}^{-\alpha_{1i}-1} (1-u_{1i})^{\alpha_0-1} \times$$

$$\times \prod_{2\leq i<j\leq n-1} (1-x_{ij})^{2p_i\cdot p_j} . \qquad (3.1)$$

We use the space like metric (+1, +1, +1, -1) and assume the trajectories to have unit slope.

For clarity we introduce a different notation [6] for the integration variables and momenta. It is illustrated in Fig. 6.

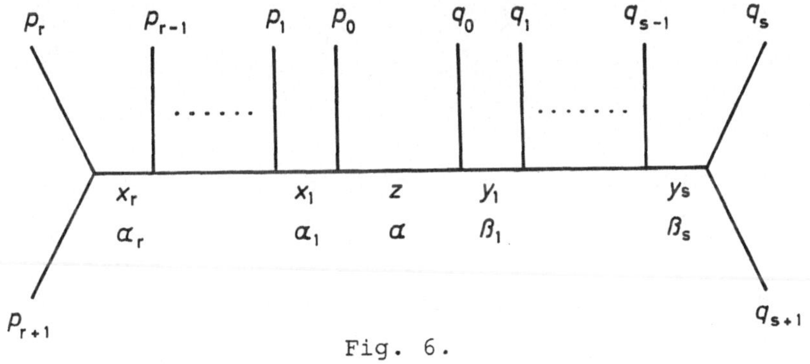

Fig. 6.

We find

$$A_{r+s+4} = \int dx\phi(x,p)\,dy\phi(y,q)\,dz \times$$

$$\times\ z^{-\alpha-1}(1-z)^{\alpha_0-1} \prod_{i=0}^{r}\prod_{j=0}^{s} (1-\rho_i\, z\, \sigma_j)^{2p_i\cdot q_j} \qquad (3.2)$$

where

$$\rho_i = x_1x_2\ldots x_i \qquad \rho_0 = 1$$

$$\sigma_j = y_1y_2\ldots y_j \qquad \sigma_0 = 1$$

$$\phi(x,p) = \prod x_j^{-\alpha_i-1}(1-x_i)^{\alpha_0-1} \times$$

$$\times \prod_{i<j} (1-x_ix_{i+1}\ldots x_j)^{p_{i-1}\cdot p_j} \qquad (3.3)$$

and similarly for $\phi(y,q)$.

We see that $\phi(x,p)$ and $\phi(y,q)$ are the Chan integrands for the subgraphs shown in Fig. 7.

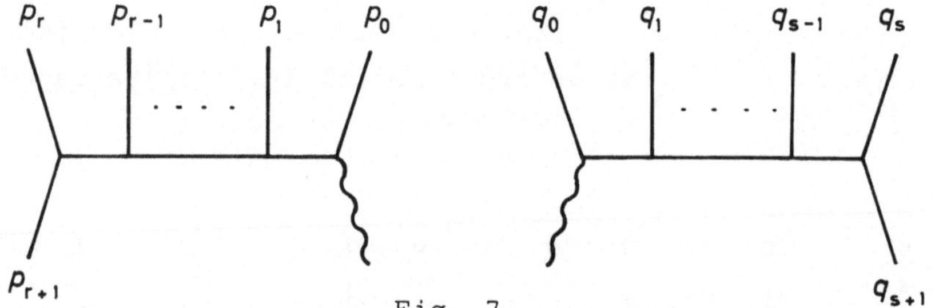

Fig. 7.

The origin of the poles of A in the variable α is in the divergence of the z-integration at z=o. The residues can be obtained by expanding the integrand in powers of z and treating each one separately. However we shall use an indirect but more revealing way of dealing with the problems. Consider the "linking factor"

$$L = \prod_{ij} (1-\rho_i\ z\ \sigma_j)^{2p_i \cdot q_j} =$$

$$= \exp \sum_{ij} 2p_i \cdot q_j \ \ln(1-\rho_i\ z\ \sigma_j) =$$

$$= \exp\{-2\sum_n \frac{z^n}{n} \sum_i \rho_i^n\ p_i \cdot \sum_j \sigma_j^n\ q_j\} = \tag{3.4}$$

$$= e^{-\ 2\sum_n \frac{z^n}{n} P^{(n)}(\rho) \cdot Q^{(n)}(\sigma)} \ .$$

Now introduce the famous harmonic oscillator operators $\{a_{n\mu},\ a^\dagger_{n\nu}\}$ [7] which satisfy

$$[a_{n\mu},\ a^\dagger_{m\nu}] = \delta_{mn}\ g_{\mu\nu} \tag{3.5}$$

and which annihilate a ground state $|o\rangle$

$$a_n|o\rangle = o = \langle o|a^\dagger_m \ . \tag{3.6}$$

We need only a few very simple equations from the theory of the quantum mechanical S. H. O. For an oscillator with a single mode we have $[a,a^\dagger]=1$ and a Hamiltonian $H=a^\dagger a$. The following results are easily proved

$$\begin{aligned} &\text{(i)}\ \ e^{za}|o\rangle = |o\rangle \\ &\text{(ii)}\ \ x^H e^{za^\dagger} x^{-H} = e^{x\ za^\dagger} \\ &\text{(iii)}\ \ e^{za} e^{wa^\dagger} = e^{wa^\dagger} e^{za} e^{wz} \ . \end{aligned} \tag{3.7}$$

In our generalisation to an infinite number of 4-vector modes we have

$$H = \sum_{n=1}^{\infty} n\ a^\dagger_n\ a_n$$

$$\text{(i)}\quad \exp\{\sum \frac{a_n}{\sqrt{n}}\cdot q^{(n)}\}|o\rangle = |o\rangle$$

$$\text{(ii)}\quad x^H[\exp\{\sum \frac{a_n^\dagger}{\sqrt{n}}\cdot q^{(n)}\}]x^{-H} = \exp\{\sum x^n \frac{a_n^\dagger}{\sqrt{n}}\cdot q^{(n)}\}$$

$$\text{(iii)}\quad \exp\{\sum \frac{a_n}{\sqrt{n}}\cdot q^{(n)}\}\exp\{\sum \frac{a_n^\dagger}{\sqrt{n}}\cdot p^{(n)}\} = \tag{3.8}$$

$$= \exp\{\sum \frac{a_n^\dagger}{\sqrt{n}}\cdot p^{(n)}\}\exp\{\sum \frac{a_n}{\sqrt{n}}\cdot q^{(n)}\}\exp\{\sum \frac{p^{(n)}\cdot q^{(n)}}{n}\} \; .$$

We see then that the linking factor can be expressed as

$$L = \langle o|\exp\{i\sum \frac{a_n\cdot q^{(n)}(\sigma)}{\sqrt{n}}\}z^H\exp\{i\sum \frac{a_n^\dagger\cdot p^{(n)}(\rho)}{\sqrt{n}}\}|o\rangle \tag{3.9}$$

so that

$$A = \langle q|\int_0^1 dz\; z^{H-\alpha-1}(1-z)^{\alpha_0-1}|p\rangle \tag{3.10}$$

with

$$|p\rangle = \int dx\phi(x,p)\exp\{i\sum \frac{a_n^\dagger\cdot p^{(n)}(\rho)}{\sqrt{n}}\}|o\rangle \tag{3.11}$$

and similarly for $\langle q|$. Take the case $\alpha_0=1$ for simplicity, then

$$A = \langle q|\frac{1}{H-\alpha-1}|p\rangle =$$

$$= \sum_\lambda \langle q|\lambda\rangle\frac{1}{n_\lambda-\alpha-1}\langle\lambda|p\rangle \tag{3.12}$$

where $|\lambda\rangle$ is an occupation number state

$$|\lambda\rangle = \prod_n \frac{(a_n^\dagger)^{\ell_n}}{\sqrt{\ell_n!}}|o\rangle,\; H|\lambda\rangle = n_\lambda|\lambda\rangle \tag{3.13}$$

so that

$$n_\lambda = \sum_1^\infty n\ell_n \tag{3.14}$$

Thus we have exhibited the poles explicitly, they lie at points $\alpha=0$, 1, 2, ... The pole residue at $\alpha=m$ comprises a sum over a finite subspace of the occupation number space corresponding to the solutions of eq. 3.14 with $n_\lambda=m$.

We can regard the different terms in the residue as representing different intermediate resonant states. Hence $<\lambda|p>$ is the amplitude for the excited state $|\lambda>$ with momentum $(-\Sigma p_i)$, to couple to the scalar particles (o, 1, 2, ... r+1).

This however is only a preliminary analysis of the factorisation. The physical states for reasons which will emerge later lie only in a subspace of the occupation number space associated with a given pole residue. Moreover this latter property gives us the hope of eliminating the negative metric ghost states in $|\lambda>$ created by the presence of time like excitations.

$$<o|a_4\ a_4^\dagger|o> = g_{44} = -1. \qquad (3.15)$$

4. KOBA-NIELSEN VARIABLES

Although it is possible to continue with the approach used so far it is rather inconvenient in that the total number of variables considerably exceeds the number of independent ones. Koba and Nielsen [8] solved this problem by showing that the many Chan [5] variables could be parametrized in terms of n variables $\{z_i\}$ one for each external leg.

These variables are placed round an arbitrary circle in the complex plane in an order which reflects that of the external lines. For the contribution we have been considering the situation is shown in Fig. 8.

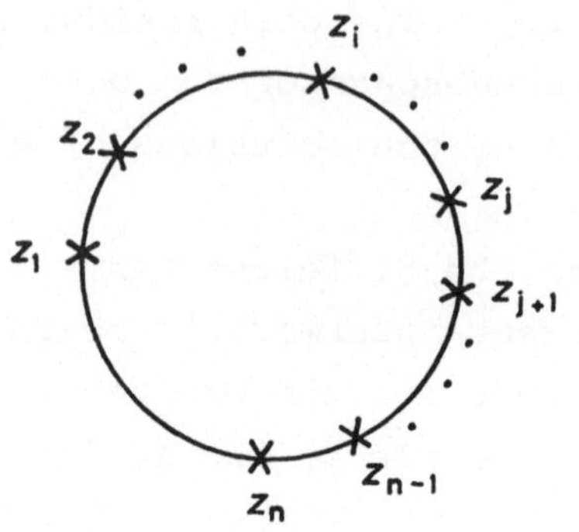

Fig. 8.

The Chan variable for the (ij)-channel is then

$$u_{ij} = \frac{(z_i-z_j)(z_{i-1}-z_{j+1})}{(z_i-z_{j+1})(z_{i-1}-z_j)} = (z_i,z_j,z_{i-1},z_{j+1}) \; . \tag{4.1}$$

Because the cross-ratio is real when the four points lie on a circle u_{ij} is real and if the order of the points is maintained $0 \le u_{ij} \le 1$. Furthermore the duality constraints are satisfied. Let us check this for the 4 pt function (Fig. 1). Here $z_0 = z_4$ and

$$u_{12} = \frac{(z_1-z_2)(z_4-z_3)}{(z_1-z_3)(z_4-z_2)} \; . \tag{4.2}$$

Clearly

$$1-u_{12} = \frac{(z_2-z_3)(z_1-z_4)}{(z_2-z_4)(z_1-z_3)} = u_{23} \; . \tag{4.3}$$

Why is the circle arbitrary? Any two circles can be mapped into one another by a Möbius transformation

$$z \to \xi = \frac{\alpha z+\beta}{\alpha z+\delta} \tag{4.4}$$

and the cross-ratio is invariant under these transformations. This follows from the fact that

$$\xi_i-\xi_j = \frac{(z_i-z_j)(\alpha\delta-\beta\gamma)}{(\gamma z_i+\delta)(\gamma z_j+\delta)} \tag{4.5}$$

which also has the consequence that

$$d\xi = \frac{dz(\alpha\delta-\beta\gamma)}{(\gamma z+\delta)^2} \,. \tag{4.6}$$

Each Möbius transformation corresponds to a matrix

$$\begin{pmatrix} \alpha & \beta \\ \gamma & \delta \end{pmatrix}$$

and the matrix corresponding to the result of successive transformations is the product of the two individual matrices. Since we can fix $\alpha\delta-\beta\gamma=1$ we see that Möbius transformations correspond to the group SL(2C), which has 6 real parameters or 3 complex ones, in the defining representation.

After fixing our circle there is still a three parameter subgroup of transformations which map the circle into itself. This is most easily seen by choosing the real axis as the circle in which case the subgroup is that of <u>real</u> Möbius transformations. For any other circle the matrices are obtained by a similarity transformation on the real Möbius matrices.

The three degrees of freedom represented by these parameters can be removed by fixing arbitrarily any three of the variables at points on the circle. The cross-ratio space can still be explored in its entirety by moving the other variables round the circle. This is a very important point which has a dominant role in what follows. We note in passing that the possibility of choosing the real axis as the K-N circle demonstrates easily that the cross ratio of four points on a circle is real.

Let us consider then the case where the K-N circle is the real axis and let us use the freedom of real Möbius invariance to fix $x_1=0$, $x_{n-1}=1$, $x_n=\infty$. We have then $(i=2,\dots n-2)$

$$u_{1i} = \frac{(x_1-x_i)(x_n-x_{i+1})}{(x_1-x_{i+1})(x_n-x_i)} = \frac{x_i}{x_{i+1}} \tag{4.7}$$

which we can invert to give

$$x_i = u_{1i}\, u_{1i+1}\, u_{1i+2} \cdots u_{1n-2} \,. \tag{4.8}$$

Remembering the definition in eqs. (2.11) we have

$$x_{ij} = x_i/x_j$$

and since

$$u_{ij} = \frac{(x_i-x_j)(x_{i-1}-x_{j+1})}{(x_i-x_{j+1})(x_{i-1}-x_j)} = \frac{(1-\frac{x_i}{x_j})(1-\frac{x_{i-1}}{x_{j+1}})}{(1-\frac{x_i}{x_{j+1}})(1-\frac{x_{j-1}}{x_j})} \tag{4.9}$$

we find again (eq. 2.11)

$$u_{ij} = \frac{(1-x_{ij})(1-x_{i-1j+1})}{(1-x_{ij+1})(1-x_{i-1j})} \,. \tag{4.10}$$

Hence the duality constraints are satisfied. Notice that the number of free K-N variables is N-3 which is also the number of independent Chan variables.

Now we come to the problem of constructing the K-N expression the dual n-point function.

This will have the general form

$$A = \int dI_n(x)F_n(x,p) \tag{4.11}$$

where

$$x = \{x,\dots x_n\}, p = \{p_1\dots p_n\}$$

and both the integrand and measure are invariant under real Möbius transformations that is

$$F_n(Rx,\ p) = F_n(x,\ p)$$
$$dI_n(Rx) = dI_n(x)\ . \tag{4.12}$$

This is because the x-dependence is only through cross-ratios. Now we want to treat all particles symmetrically. A measure which satisfies these requirements is

$$\prod_{i=1}^{n} \frac{dx_i}{|x_i - x_{i+1}|} \quad \text{(remember } n+1=1\text{)}$$

which is both cyclically symmetric and Möbius invariant, we have (see eq. (4.5) and (4.6))

$$\begin{aligned} \prod_i dx_i &\to \frac{\prod dx_i}{(\gamma x_i+\delta)^2} \\ \prod_i \frac{1}{(x_i - x_{i+1})} &\to \frac{\prod_i (\gamma x_i+\delta)(\gamma x_{i+1}+\delta)}{\prod_i (x_i - x_{i+1})}\ . \end{aligned} \tag{4.13}$$

Clearly the measure is not uniquely fixed but it turns out to be the one we want.

It has the defect, however, of involving too many integration variables, n in fact, when we want only n-3. As we shall see it would also give use to an infinite result when used in eq. (4.11). The way to obtain the right number of variables and to remove the infinity is to divide out of this measure the invariant volume element dR of the group of real Möbius transformations {R} which leave the integrand unchanged. That is

$$dI_1(x) = [\prod_i \frac{dx_i}{|x_i - x_{i+1}|}]\ /\ dR. \tag{4.14}$$

Let us explain this in more detail. We can express the

variables $\{x_i\}$ in terms of a basic set $\{\bar{x}_i\}$ of which an arbitrary three $(\bar{x}_a, \bar{x}_b, \bar{x}_c)$ are fixed. We could have them at $(o, 1, \infty)$ say. The general set $\{x_i\}$ is obtained from them by boosting with a real Möbius transformation R which contains the necessary extra three parameters. Thus

$$x_i = R\bar{x}_i \ .$$

These parameters can be chosen in different ways. For example they could be the elements $(\alpha, \beta, \gamma, \delta)$ of the 2×2 matrix, in which case

$$dR = \delta(1-\alpha\beta+\gamma\delta)d\alpha\ d\beta\ d\gamma\ d\delta$$

or they could be chosen as the three points (x_a, x_b, x_c) into which $(\bar{x}_a\ \bar{x}_b\ \bar{x}_c)$ are mapped by R, in which case

$$dR = \frac{dx_a\ dx_b\ dx_c}{|x_a-x_b||x_b-x_c||x_c-x_a|} \ . \tag{4.15}$$

Now if we calculate the Jacobian change of variables $\{x_i\}\to\{R\bar{x}_i\}$ we find the general form

$$\prod \frac{dx_i}{|x_i-x_{i+1}|} = dR\ dI_n(\bar{x})$$

where $dI_n(\bar{x})$ is a measure over the basic set which is determined by the Jacobian. This $dI_n(\bar{x})$ is what we really want. In fact a practical method of calculating it is to operate naively with eqs. (4.14) and (4.15). For example if we want to fix x_1, x_{n-1} and x_n then we write down

$$dI_n(x) = \prod_1^n \frac{dx_i}{|x_i-x_{i+1}|} \frac{|(x_1-x_{n-1})(x_{n-1}-x_n)(x_n-x_1)|}{dx_1\ dx_{n-1}\ dx_n}$$

$$= \frac{\prod_2^{n-2} dx_i}{\prod_1^n |x_i-x_{i+1}|} |(x_1-x_{n-1})(x_{n-1}-x_n)(x_n-x_1)| \ . \tag{4.16}$$

Thus although our measure has a symmetrical definition it has a set of equivalent but unsymmetrical realisations.

To cut a long story short, the integrand we want is

$$F_n(x,p) = \prod_i |x_i - x_{i+1}|^{\alpha_o} \prod_{i<j} |x_i - x_j|^{2p_i \cdot p_j}. \qquad (4.17)$$

The way to check this is to use eqs. (4.16) and (4.17) and then take the limit as $(x_1 x_{n-1} x_n) \to (01\infty)$, then compare the result with eq. (3.1). Remember to use momentum conservation.

From now on I would like to deal with the simplified case where $\alpha_o=1$ and the external scalar particles therefore are tachyons with $P_i^2=\alpha_o=1$. Although it seems some-what artifical this case is much favoured among the cognescenti for reasons which appear later.

Our amplitude can be written simply

$$\int \frac{\Pi dx_i}{dR} \prod_{i<j} |x_i - x_j|^{2p_i \cdot p_j} . \qquad (4.18)$$

However it is convenient also to retain a separately invariant measure and integrand

$$A_n = \int [\prod_i \frac{dx_i}{|x_i - x_{i+1}|}] \; / \; dR.F_n(x,p) . \qquad (4.19)$$

Let us check the Möbius invariance of F_n.

$$F_n(Rx,q) = \prod_i |Rx_i - Rx_{i+1}| \prod_{i<j} |Rx_i - Rx_j|^{2p_i \cdot p_j}$$

$$F_n(Rx,q) = \prod_i \left|\frac{\alpha\delta - \beta\gamma}{(\gamma x_i + \delta)^2}\right|^{p_i \cdot p} F_n(x,p) \qquad (4.20)$$

where $p=\Sigma p_i$. When momentum is conserved p=o and F_n is invariant.

5. FACTORISATION REVISITED

It is convenient, though not essential for this discussion, to group all the (n-1)!/2 terms in the amplitude into one expression. This can be achieved simply by letting the K-N variables range round the whole circle independently except for the three variables held fixed. Their orders must be summed over too however. Now consider the amplitude for n+m tachyons q with two sets of K-N variables $(x_1 \ldots x_n)(t_1 \ldots t_m)$ and momenta $(p_1 \ldots p_n)$ $(k_1 \ldots k_m)$. The integrand can be written ($x_{n+1}=x_1$, $t_{m+1}=t_1$)

$$\int [\prod_i \frac{dx_i}{|x_i - x_{i+1}|} \prod_j \frac{dt_j}{|t_j - t_{j+1}|}] \,/\, dR . F_n(x,p) F_m(t,k) \times$$

$$\times \prod_{ij} |x_i - t_j|^{2p_i \cdot k_j} . \tag{5.1}$$

We can suppose the $\{x_i\}$ obtained from a basic set $\{\bar{x}_i\}$ (with three fixed) by the equations

$$x_i = M\bar{x}_i \tag{5.2}$$

where M is a real Möbius transformation. We have

$$\prod \frac{dx_i}{|x_i - x_{i+1}|} = dM \, dI_n(\bar{x}) . \tag{5.3}$$

Similary if

$$t_j = N\bar{t}_j$$

$$\prod \frac{dt_j}{|t_j - t_{j+1}|} = dN \, dI_m(\bar{t}) . \tag{5.4}$$

So finally removing the bars over the new integration variables we have

$$A_{n+m} = \int dI_n(x)\,dI_m(t)\left(\frac{dM\,dN}{dR}\right) F_n(Mx,p)F_m(Nt,k) \times$$

$$\times \prod |Mx_i - Nt_j|^{2p_i \cdot k_j} \, . \tag{5.5}$$

Now we eliminate the dR as follows. Put

$$\begin{aligned} P &= NM \\ dP\, dN &= dM\, dN \end{aligned} \tag{5.6}$$

and parametrize P as

$$P = B_2^{\dagger}\, \Lambda\, B, \tag{5.7}$$

where   invariant points

$$\begin{aligned} B_1 &= \begin{pmatrix} 1-v_1 & 0 \\ 1 & -v_1 \end{pmatrix} && (0,\ 1) \\ B_2^{\dagger} &= \begin{pmatrix} 1-v_2 & -1 \\ 0 & -v_2 \end{pmatrix} && (1,\ \infty) \\ \Lambda &= \begin{pmatrix} \lambda & o \\ o & 1 \end{pmatrix} && (o,\ \infty)\ . \end{aligned} \tag{5.8}$$

In this case

$$dP = \frac{dv_1}{|v_1(v_1-1)|}\,\frac{dv_2}{|v_2(v_2-1)|}\,\frac{d\lambda}{\lambda^2}\, . \tag{5.9}$$

Now we introduce the redundant integration R by

$$\begin{aligned} N &= B_2^{\dagger} R \\ M &= R^{-1}\, \Lambda\, B\ , \end{aligned} \tag{5.10}$$

then $\qquad dP\, dN = dP\, dR$

and now dR can be divided out.
Afterwards we can put R=1 and get

$$A_{n+m} = \int dI_n(x)\, dI_m(t)\, dB_1 dB_2 \frac{d\lambda}{\lambda^2} F_n(\Lambda B_1 x, p) F_m(B_2^\dagger t, k)$$

$$\prod |\Lambda B_1 x_i - B_2^\dagger t_j|^{2p_i \cdot p_j} .$$

Now

$$F_n(\Lambda B_1 x, p) = |\lambda|^{p^2} F_n(B_1 x, p)$$

so we have

$$A_{n+m} = \int_{-\infty}^{+\infty} d\lambda\, |\lambda|^{p^2-2} \{.........\}$$

which diverges at the origin to create poles at $p^2=1,0,-1,$ -2 ... i.e. just where we want them. This corresponds essentially to the divergences of the z-integration at zero in section 3.

After further manipulations we can put the integrand into the form

$$A_{n+m} = \int dI_n(x)\, dI_m(t)\, dB_1 dB_2 d\lambda\, |\lambda|^{p^2-2}\ F_n(B_1 x, p) F_m(B_2 t, k)$$

$$\prod_{ij} |1+\lambda (B_1 x_i)(B_2 t_j)|^{2p_i \cdot k_j} . \qquad (5.11)$$

Putting

$$\xi_i = B_1 x_i$$

$$\eta_j = B_2 t_j \qquad (5.12)$$

the linking factor can be written in exponential form as

$$\exp\{-\sum_n \frac{(-\lambda)^n}{n}\ p^{(n} \cdot k^{(n)}\}$$

$$p^{(n)}(\xi) = \sum_i p_i\ \xi_i^n \qquad (5.13)$$

$$k^{(n)}(\eta) = \sum_i k_j\ \eta_j^n$$

and then put in the form

$$<o|\exp\{i\sum\frac{a_n}{\sqrt{n}}\cdot p^{(n)}(\xi)\}(-\lambda)^H \exp\{i\sum\frac{a_n^\dagger}{\sqrt{n}}\cdot k^{(n)}(\eta)\}|o> . \qquad (5.14)$$

Finally restricting our attention to the contribution $\int_{-1}^{1} d\lambda$ we obtain for the contribution to A which contains the poles

$$A'_{n+m} = <p|\frac{1+(-1)^H}{H+p^2-1}|k> \qquad (5.15)$$

$$<p| = \int dI_n(x)\,dB_1 F_n(\xi,p)<o|\exp\{i\sum\frac{a_n}{\sqrt{n}}\cdot p^{(n)}(\xi)\} . \qquad (5.16)$$

Again the amplitude for coupling to an intermediate excited state $|\lambda>$ can be obtained as $<p|\lambda>$.

6. GAUGE CONDITIONS

This method of factorising suggests that we need the complete subspace of states $|\lambda>$ which satisfy

$$H|\lambda> = n|\lambda> \qquad (6.1)$$

to factorize the residue at the nth pole. We will now see that this is no so by demonstrating that the physical states satisfy in addition extra conditions to (6.1) [10, 11].

First we introduce the operator

$$L_o = p^2 + H \qquad (6.2)$$

which is closely related to H, on mass-shell the states factorizing the nth pole satisfy

$$L_o|\lambda> = |\lambda> . \qquad (6.3)$$

Then we introduce the operators [10, 11]

$$L_+ = -i\sqrt{2}\, p\cdot a_1^\dagger + \sum \sqrt{n(n+1)}\, a_{n+1}^\dagger \cdot a_n$$

$$L_- = +i\sqrt{2}\, p\cdot a_1 + \sum \sqrt{n(n+1)}\, a_n^\dagger\, a_{n+1} \; . \tag{6.4}$$

We can check that

$$[L_+, L_-] = -2L_0$$

$$[L_0, L_\pm] = \pm L_\pm \tag{6.5}$$

which are just the commutation relations of the generators of the group of real Möbius transformations. This in fact is the reason we are interested in such quantities. We can be more definite. The generator that corresponds to an expansion Λ is L_0 and that which corresponds to a number of the group $\{B_1\}$ which leaves (o,1) invariant is

$$W_+ = L_0 - L_+ \; . \tag{6.6}$$

We shall show that [10, 11]

$$\langle p|W_+ = \langle p|(L_0 - L_+) = o \; . \tag{6.6}$$

This is one of the "gauge conditions" which are the starting point of the analysis of the structure of physical states. We shall derive this result by showing [9]

$$\langle p|(1-v)^{W^+} = \langle p| \tag{6.7}$$

for arbitrary v.

Now one can show from the commutation relations of the generators that

$$\langle o|\exp\{i\sqrt{2} \sum \frac{a_n}{\sqrt{n}}\cdot p^{(n)}(\xi)\}(1-v)^{W^+} = \tag{6.8}$$

$$= (1-v)^{+p^2} \prod_{i=1}^{n} (1-v\xi)^{-2p\cdot p_i} \langle o|\exp\{i\sqrt{2} \sum \frac{a_n}{\sqrt{n}}\cdot p^{(n)}(C\xi)\}$$

where

$$C\xi = \frac{(1-v)\xi}{1-v\xi} \tag{6.9}$$

so that C is a member of the one-dimensional subgroup $\{B_1\}$ of Möbius transformations which leave (o,1) invariant. Moreover the numerical factors in eq. (6.8) can be absorbed into F_n (see eq.(4.20)) in such a way as allows us to derive

$$\langle p|(1-v)^{W_+} = \int dI_n(x)\,dB_1\,F_n(C\xi,p) \times$$

$$\times \langle o|\exp\{i\sqrt{2}\sum \frac{a_n}{\sqrt{n}} \cdot p^{(n)}(C\xi)\} . \tag{6.10}$$

Since

$$\xi' = C\xi = CB_1\xi = B_1'x \tag{6.11}$$

and the invariance of the group measure tells us

$$dB_1 = d(CB_1) = dB_1' \tag{6.12}$$

we find the result eq. (6.7).

7. GENERAL GAUGE CONDITIONS AND THE ELIMINATION OF GHOSTS

It turns out that the states $\langle p|$ in the case $\alpha_o=1$ also satisfy extra Virasoro [12] gauge conditions

$$\langle p|W_\ell = o \qquad \ell = 1, 2, 3, \ldots \tag{7.1}$$

where

$$W_\ell = L_o - L_\ell + \ell - 1$$

$$L_+ = -i\sqrt{2\ell}\; p\cdot a_\ell^\dagger + \sum_1^\infty \sqrt{n(n+\ell)}\; a_{n+\ell}^\dagger\cdot a_n + \tag{7.2}$$

$$+ \frac{1}{2}\sum_{n=1}^{\ell-1}\sqrt{n(\ell-n)}\; a_n^\dagger\cdot a_{\ell-n}^\dagger .$$

We see that $L_1=L_+$ and $W_1=W_+$ so the case $\ell=1$ in eq. (7.1) is the one we derived. We shall not derive the others since time is short. They can be obtained by related techniques - though only when $\alpha_0=1$.

The commutation relations obeyed by the quantities L_ℓ are

$$[L_n, L_m] = (m-n)L_{n+m} \tag{7.3}$$

where we extend the definition so that $L_{-n}=L_n^+$.

Because the bras $\langle p|$ lie in the subspace defined by the Virasoro conditions they do not "see" the complete occupation number space $|\lambda\rangle$ when we form a scalar product $\langle p|\lambda\rangle$. The part they do see in the space of physical states and the orthogonal part is the space of spurious states.

A spurious state [13] in general has the form $|s\rangle = W_\ell|\chi\rangle$ for some $|\chi\rangle$. It will lie on mass shell at the nth pole, when (see eq. 6.3)

$$L_0|s\rangle = |s\rangle \tag{7.4}$$

which is satisfied when

$$(L_0 + \ell-1)|\chi\rangle = o\ . \tag{7.5}$$

That is if $|\chi\rangle$ is an occupation number state with a smaller eigenvalue of R. The spurious states then can be constructed in the form

$$W_\ell|\chi\rangle = (L_0-L_\ell+(\ell-1))|\chi\rangle = -L_\ell|\chi\rangle\ . \tag{7.6}$$

The physical states $|\psi\rangle$ are orthogonal to such states

$$\langle\psi|L_\ell|\chi\rangle = o\ .$$

We have proved this for a $|\chi\rangle$ satisfying eq. (7.5). For other $|\chi\rangle$ it holds trivially hence we have

$$\langle\psi|L_\ell = o$$

or

$$L_{-\ell}|\psi> = o \ . \qquad (7.7)$$

Now because

$$-\ell L_{-\ell-1} = [L_{-1}, L_{-\ell}] \qquad (7.8)$$

we need only require

$$L_{-1}|\psi> = o$$

$$L_{-2}|\psi> = o \qquad (7.9)$$

and for mass shell conditions

$$L_o|\psi> = |\psi>. \qquad (7.10)$$

These are the conditions for a state to be physical [13]. The question is does it imply that the states have positive norm? People feel the answer is yes but they don't know for sure, though extensive checking has been done [14].

Let us look in detail at the first two excited levels [13].

<u>n = 1</u> .

The general state has the form

$$|\psi> = \alpha \cdot a_1^\dagger |o> \ .$$

Now $L_{-2}|\psi>=o$ automatically and

$$L_{-1}|\psi> = \alpha \cdot p|o> = o \qquad (7.12)$$

which implies

$$\alpha \cdot p = o \ . \qquad (7.13)$$

Since $\alpha_o=1$ we have $p^2=o$ at this excitation level. Hence if we choose our frame of reference so that p=(o,o, 1,1) we see that the candidates for physical states are

$$a_{11}^\dagger|o>, \ a_{12}^\dagger|o> \quad \text{and} \quad p \cdot a_1^\dagger |o> \ .$$

The first two states are genuine and have positive norm.

However the latter state is spurious, we have,

$$p\cdot a_1^\dagger|o\rangle = W_1|o\rangle . \qquad (7.14)$$

The reason it slipped through our net is that it has zero norm. Our equations (7.9) (7.10) are deficient then in that we must remove zero norm spuricus states after solving them.

$\underline{n = 2}$

The general state is

$$|\psi\rangle = (a_1^\dagger\cdot\underset{\uparrow}{\underline{\underline{\alpha}}}\cdot a_1^\dagger + \underset{\uparrow}{\beta}\cdot a_2^\dagger)|o\rangle \qquad (7.15)$$

4-tensor 4-vector

the physical state conditions are effectively

$$[ip\cdot a_1 + a_1^\dagger\cdot a_2]|\psi\rangle = o$$

$$[-2ip\cdot a_2 + \frac{1}{2} a_1^2]|\psi\rangle = o \qquad (7.16)$$

we find

$$\beta = -2i\ p\cdot\underline{\underline{\alpha}}$$

$$\mathrm{tr}\ \underline{\underline{\alpha}} - 2i\ p\cdot\beta = o .$$

Choosing $p=(o,o,o,1)$

this gives
$$\beta_\mu = 2i\ \alpha_{4\mu}$$

$$\alpha_{44} = \frac{1}{5}\sum_i \alpha_{ii} .$$

The general state becomes then

$$|\psi\rangle = \sum_{i,j=1}^{3} \alpha_{ij}\ a_i^\dagger\ a_j^\dagger|o\rangle + \frac{1}{5}\sum_{i=1}^{3}\alpha_{ii}(a_{14}^{\dagger 2} - 2i\ a_{24}^\dagger)|o\rangle +$$

$$+ \sum_{i=1}^{3} \beta_i(a_{2i}^\dagger + ia_{1i}^\dagger\cdot a_{14}^\dagger)|o\rangle$$

the last contribution has zero norm. Furthermore it is proportional to $W_1 \cdot a_1^{\dagger} |o\rangle$ hence it is a spurious state. Everything is now satisfactory since the other states have positive norm. Whether on not the same works for all higher levels is an open question.

REFERENCES

1. G. Veneziano, Nuovo Cimento 57A, 190 (1968).
2. R. Dolen, D. Horn and C. Schmid, Phys. Rev. 166, 1768 (1968).
3. K. Bardacki and H. Ruegg, Phys. Letts. 28B, 342 (1968).
4. C. G. Goebel and B. Sakita, Phys. Rev. Letts. 22, 257 (1969).
5. Chan Hong-Mo and Tsuo Sheung Tsun, Phys. Letts. 28B, 425 (1969).
6. S. Fubini and G. Veneziano, Nuovo Cimento 64A, 811 (1969);
   K. Bardacki and S. Mandelstam, Phys. Rev. 184, 1640 (1969).
7. S. Fubini, D. Gordon and G. Veneziano, Phys. Letts. 29B, 679 (1969);
   Y. Nambu, Enrico Fermi Institute preprint COO264-507.
8. Z. Koba and H. G. Nielsen, Nuclear Physics B12, 517 (1969).
9. This particular approach is used in a forthcoming preprint by the author "Dual Amplitudes for Currents".
10. F. Gliozzi, Lett. Nuovo Cimento 2, 864 (1969).
11. C. B. Chiu, S. Matsuda and C. Rebbi, Phys. Rev. Letts. 23, 1526 (1969).

12. M. A. Virasoro, Phys. Rev.
13. E. Del Giudice and P. Di Vacchia, M. I. T. preprint "Characterisation of Physical States in Dual Resonance Models".
14. R. C. Brower and C. B. Thorn, work in progress.

There are several excellent reviews:

1. V. Alessandini, D. Amati, M. Le Bellac and D. Olive Dual Multi Particle Theory, CERN preprint TH. 1160.
2. G. Veneziano, Erice Summer School.
3. S. Mandelstam, Brandeis Lectures.

Acta Physica Austriaca, Suppl. VIII, 121–135 (1971)

NON-LOCAL APPROACH TO WEAK INTERACTIONS* **

BY

H. PIETSCHMANN
Institut für Theoretische Physik
Universität Wien

1. INTRODUCTION

Last year, I had the pleasure to give a Seminar on "Structure Effects in Weak Interactions" at the 9th Schladming Winter School. In order to be able to present continuations of this investigation, let me briefly summarize the underlying model. Details can be found in the literature [1].

Under the assumptions of

a) V-A nature of the weak currents

b) local action of the weak currents

the most general form of the effective weak S-operator for processes without possible real intermediate bosons is

$$S = I - \frac{iG}{\sqrt{2}} \int d^4x d^4y \; J_\lambda(x) K^{\lambda\nu}(x-y) J_\nu^+(y) + O(G^2) \qquad (1)$$

where $K^{\lambda\nu}$ is a c-number structure tensor whose invariant decomposition is given by

---

*Lecture given at X. Internationale Universitätswochen für Kernphysik, Schladming, March 1 - March 13, 1971.

**Supported by "Fonds zur Förderung der wissenschaftlichen

$$K_{\lambda\nu}(x) = M^4\{g_{\lambda\nu}\ F_1(x) + \frac{1}{M^2}\frac{\partial^2}{\partial x^\lambda \partial x^\nu} F_2(x)\} \qquad (2)$$

$F_i(x)$ are dimensionless functions.

$J_\lambda(x)$ is the usual total weak current

$$J_\lambda(x) = \ell_\lambda(x) + \cos\theta\ j_\lambda^\pi(x) + \sin\theta\ j_\lambda^K(x) \qquad (3)$$

with

$$\ell_\lambda(x) = \sum_\ell \psi_{\bar{\ell}}(x)\gamma_\lambda(1+\gamma_5)\psi_{\nu_\ell}(x) \qquad \ell = e,\ \mu \qquad (4)$$

In eq. (2), M is some characteristic parameter of the dimension of a mass but <u>not</u> necessarily the mass of a physical particle such as the intermediate boson. In order to keep also the Fourier transform of the structure functions dimensionless, we define

$$F_i(x) = (2\pi M)^{-4}\int d^4k\ e^{ikx}\ F_i(k) \qquad i = 1,2 \qquad (5)$$

All this has been presented last year and is here only repeated for reference purpose. Let me mention that the basic idea of a weak interaction like eq. (1) goes back to Lee and Yang [2]. In meson theories, similar ideas have been tried even earlier [3]. In these early attempts, local action of the currents was not assumed, however.

In section 2, we shall summarize possible realizations of the structure tensor. Since the influence of the structure tensor on lowest order processes has been discussed in detail in refs. [1], we shall give the calculations for the decay $K_L^0 \to \mu\bar{\mu}$ in section 3. Section 4 is devoted to an investigation of the relation of our model to other attempts at finite weak interactions.

2. REALIZATIONS FOR $K_{\lambda\nu}(x)$

The S-operator (1) is the most general ansatz under the assumptions a) and b) of section 1. Therefore, each existing model of weak interactions consistent with these assumptions must correspond to a particular form for the structure tensor $K_{\lambda\nu}(x)$. In the following we shall list the most prominent models of weak interactions and describe $K_{\lambda\nu}(x)$ belonging to each model.

a) The Current-Current (C.C.) Model.

In the traditional case of direct four fermi interaction, the structure tensor is simply given by

$$K_{\lambda\nu}(x) = g_{\lambda\nu}\ \delta^{(4)}(x) \tag{6}$$

b) The Intermediate Vector Boson (IVB) Model.

Also this model has a more than decade long tradition [4]. If the IVB is a real particle, the S-operator (1) has to be amended by terms describing production and decay of the IVB. Forgetting about these terms (in a low energy approximation), the IVB model can be incorporated in (1) by setting

$$K^{\lambda\nu}(x) = \frac{g^2\sqrt{2}}{G}\ \Delta_F^{\lambda\nu}(x) \tag{7}$$

where g is the semi-weak coupling constant of the IVB to the weak current (3). In this case, M in eq. (2) is to be interpreted as the mass of the IVB, m_W.

c) The "Dipole Boson" Model.

When the "dipole fit" to nucleon form factors [5] was experimentally established, speculations began whether or not rapidly decaying vector mesons behave at least phenomenologically like "dipoles" rather than single poles [6]. Although this path has so far led to nowhere, it is interesting to speculate about this same possibility in weak interactions. In this case, we have

$$F_1(k^2) = F_2(k^2) = (k^2-m_W^2)^{-2} \tag{8}$$

where $F_i(k^2)$ are defined in eq. (5). Also here, we have to identify M in eq. (2) with m_W, the mass of the "dipole boson".

d) The Model of Fivel and Mitter.

Some years ago, Efimov, Fradkin and Volkov proposed and studied a field theoretical model with a non-polynomial interaction Lagrangian [7]. Their method is, in our days, well known and has been presented by A. Salam last year in Schladming [8]. Let us therefore not go into details of the general method but rather directly turn to its application in weak interactions which has been worked out by Fivel and Mitter [9].

Starting point is the following interaction Lagrangian for weak interactions:

$$L_I = g\{J^\lambda(x)U_\lambda^+(x) + h.c.\} \tag{9}$$

where $J^\lambda(x)$ is the current (3) but $U_\lambda(x)$ is the following non-polynomial vector operator

$$U_\lambda(x) = \frac{\phi_\lambda(x)}{(1+f\phi_\mu^+\phi^\mu)^\gamma} \tag{1o}$$

ϕ_λ being the field of the IVB. The Dyson-index [8] of the interaction Lagrangian (9) is

$$D = 5 - 4\gamma \tag{11}$$

Since a theory is renormalizable if D is less or equal 4, Fivel and Mitter choose

$$\gamma > 1/4 \tag{12}$$

Before writing down the structure tensor for this case, let us quote the main properties of this model, as collected by Fivel and Mitter [9]:

"The weak interactions are mediated by a universal, local coupling of a minimal set of vector bosons to the usual weak currents. The interaction Lagrangian, however, is not an entire function of the fields".

"Each order of perturbation theory is finite without the use of cutoffs".

"The low energy (low momentum transfer) predictions agree with those of universal Fermi theory and, in fact, this agreement persists until one considers processes requiring far greater center-of-mass energy than is currently available".

"S-matrix elements calculated with the theory in each order of the semiweak coupling are covariant, crossing-symmetric, and are expected to have analytic structure consistent with unitarity in all orders of perturbation theory".

In this model, the structure tensor (2) takes the following form [9]

$$K_{\lambda\nu}(x) = \frac{1}{[\Gamma(\gamma)]^2} \frac{\Pi}{2^{2\gamma-1}} \int_0^\infty dw\, w^{2\gamma-1} \times$$

$$\times \frac{\sin \pi\gamma\, J_0(w) - \cos \pi\gamma\, N_0(w)}{(1+\lambda_1^2 w^2)^4 (1+\lambda_2^2 w^2)} \{D_1 g_{\lambda\nu} + \frac{4(1-\lambda_1\lambda_2 w^2)}{1+\lambda_2^2 w^2} D_2 x_\lambda x_\nu\} \quad (13)$$

where the following abbreviations have been used

$$\lambda_1 = \frac{1}{2} f\, D_1$$

$$\lambda_2 = \frac{1}{2} f\, (D_1 + 4D_2 x^2) \quad (14)$$

and

$$D_1 = \left(1 - \frac{2}{m_W^2}\frac{d}{dx^2}\right)\Delta_F(x^2, m_W^2)$$

$$D_2 = -\frac{1}{m_W^2}\left(\frac{d}{dx^2}\right)^2 \Delta_F(x^2, m_W^2) \tag{15}$$

so that

$$\Delta_F^{\lambda\nu}(x, m_W^2) = g^{\lambda\nu} D_1 + 4x^\lambda x^\nu D_2 \tag{16}$$

The structure functions $F_i(x^2)$ can easily be extracted from eq. (13) by means of the identity

$$\frac{\partial^2}{\partial x^\mu \partial x^\nu} f(x^2) = 2g_{\mu\nu} f'(x^2) + 4x_\mu x_\nu f''(x^2) \tag{17}$$

where primes denote total derivative with respect to the argument.

3. THE DECAY $K_L^0 \to \mu^+\mu^-$

Before we enter any calculation, let us recall the experimental situation for this decay. The branching ratio is now limited by

$$\frac{\Gamma(K_L^0 \to \mu^+\mu^-)}{\Gamma(K_L^0)} < 6{,}8 \cdot 10^{-9} \tag{18}$$

with 90% confidence level [10]. Of course, this decay can also proceed via a combination of weak and electromagnetic interactions. If one puts an upper limit on second order weak interactions by means of (18), he obtains an upper limit for the cut-off of 29 GeV in the IVB model and of 15 GeV in the CC-model.

Let us now try to calculate this decay in second order weak interactions using the general form of the structure tensor, eq. (2), mediating between the weak currents. The decay is depicted in fig.1, where the round blob stands for the strong interaction part and the two droplet blobs denote the structure tensor of eq. (1). The matrix element is then given by

$$<q_1,q_2|S|k> = \frac{i\,G^2\,m_\mu}{(2\pi)^3\sqrt{q_1^0\,q_2^0}}\,\delta^{(4)}(k-q_1-q_2)\int d^4p\,d^4z\,\times$$

$$\times\,\tilde{K}^{\lambda\tau}(p+q_1)\tilde{K}^{\sigma\rho}(q_2-p)\frac{1}{p^2}\,\bar{u}(q_1)\gamma_\tau\not{p}\gamma_\sigma(1+\gamma_5)v(q_2)\quad\times$$

$$\times\,<o|T\{J_\lambda^+(z)J_\rho^{\cdot}(o)\}k>e^{iz(p+q_1)}\quad. \tag{19}$$

Here, $\tilde{K}^{\lambda\nu}(p)$ denotes the Fourier transform of the structure tensor (2). Whether or not T is the "fancy" time-ordering product introduced last year in Schladming [1] plays little role for our rather crude estimate here. Gaining the advantage of using standard formulae, we take it to be the ordinary T-product of perturbation theory.

It is now necessary to evaluate the matrix element of the hadronic currents which contains all unknown contributions of strong interactions. We cannot simply copy from the standard paper on second-order weak processes by Mohapara, Rao and Marshak [11] because these authors were only interested in the contribution to the most divergent part of the matrix element. Let us therefore make the simplest possible assumption and approximate the matrix-element by its contribution from K- and π-poles

$$<o|T\{J_\lambda^+(z)J_\rho(o)\}|k>_{\pi\text{-pole}} =$$

$$= \frac{1}{(2\pi)^{3/2}} \frac{i}{\sqrt{2k_0}} \int d^4q\, e^{-iqz} \frac{f_\pi \sin\theta \cos\theta}{q^2-\mu^2} q_\lambda \{f_+(\tau)(q+k)_\mu + f_-(\tau)(q-k)_\mu\} \quad (2o)$$

where

$$\tau = (q-k)^2 \quad (21)$$

f_π and $f_\pm$ are the usual constant and form factors occuring in the matrix-element of π- and $K_{\ell 3}$-decay. A similar expression holds for the K-pole contribution.

It is clear, that in the integral over the closed loop of fig. 1 the pole approximation will be very poor for very large values of τ.

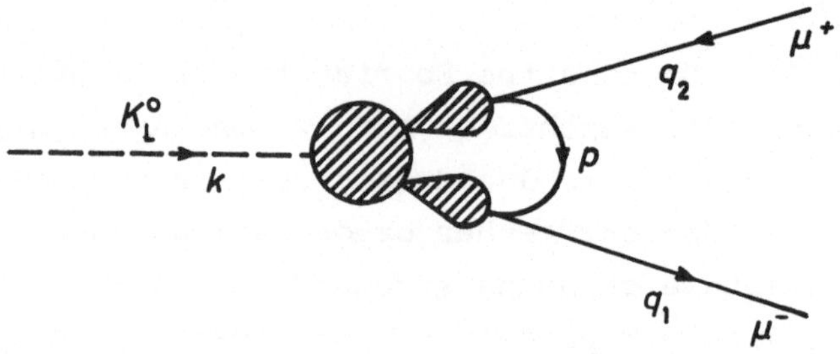

Fig. 1: The decay $K_L^0 \to \mu^+\mu^-$.

This is also seen from the fact that the form factors will damp the matrix element for these values. We therefore take constant form factors, guessing that the integration will give a compensation which suffices for our order of magnitude computation.

It is known from the literature [11] that both the CC-model and the IVB-model give badly divergent results for this decay. The simplest model for the structure tensor is therefore the dipole boson model listed in section 2c. With eq. (8), the calculation can be carried through and yields

$$\Gamma(K_L^o \to \mu^+\mu^-) = \frac{G^4\, m_k^3\, m_\mu^2}{2\pi} \sqrt{1-4\,\frac{m_\mu^2}{m_k^2}} \left[M^2\, \frac{\sin\theta\, \cos\theta\, f_\pi(2f_+ - f_-)}{32\pi^2\, m_k}\right] . \tag{22}$$

Numerical values for M derived for various branching ratios (18) are listed in table 1.

Table 1: Dipole masses derived from eq. (22) for various branching ratios (18).

$\frac{\Gamma(K_L^o \to \mu^+\mu^-)}{\Gamma(K_L^o)}$	M_{Dipole}
10^{-6}	100 GeV
10^{-8}	35 GeV
10^{-10}	10 GeV

That these values are not just random numbers despite our rough approximations is corroborated by a more sophisticated calculation [12] using hard pion techniques [13] which gives roughly the same numerical results.

4. OTHER NON-LOCAL MODELS

Our non-local approach to weak interactions has been based on assumptions a) and b) of section 1. There are several other attempts at finite higher order weak processes by means of non-local interactions which do not assume local action of the currents. Hence the S-operator (1) does not follow from those models. We shall discuss two of these models here.

a) The model of Efimov and Seltser [14]:
This model is based on the observation that we only know the properties of charged leptons whereas neutrinos cannot be studied in detail. Therefore, no experiment forbidds a non-local substitution for the neutrino fields in the following form

$$\psi_\nu(x) \rightarrow \int d^4y \; K(x-y)\psi_\nu(y) \tag{23}$$

With this substitution, one can manufacture a neutrino propagator which damps higher orders to yield a finite number of infinite renormalization constants. The price is, of course, loss of uniformity among leptons. In order to insure that matrix elements for processes with real neutrinos are unchanged, one requests

$$\tilde{K}(q^2=0) = 1 \tag{24}$$

where $\tilde{K}$ is the Fourier transform of the smearing function in eq. (23).

In this model, the decay $K_L^0 \to \mu^+\mu^-$ would be described by a graph as shown in fig. 2.

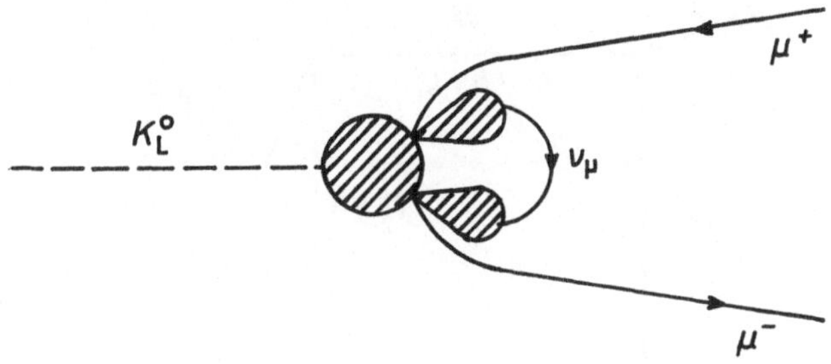

Fig. 2: The decay $K_L^0 \to \mu^+\mu^-$ in the model of Efimov and Seltser.

It is quite clear that it is of different nature than fig. 1. In fact, the leptonic part of the matrix element (19) would be replaced by

$$[\tilde{K}(p)]^2 \frac{1}{p^2} \bar{u}(q_1)\gamma_\lambda \not{p} \gamma_\rho (1+\gamma_5) v(q_2) \tag{25}$$

showing that the structure functions do not depend on the momentum transfer but on neutrino momentum only.

b) The model of Kummer and Segré [15]:
This model has been described in detail during this Winter School by W. Kummer [16]. Let us therefore only point to the essential relations and differences to our approach. A "lowest order" weak process, such as μ-decay, for example, is obtained in this model from a box graph as shown in fig. 3.

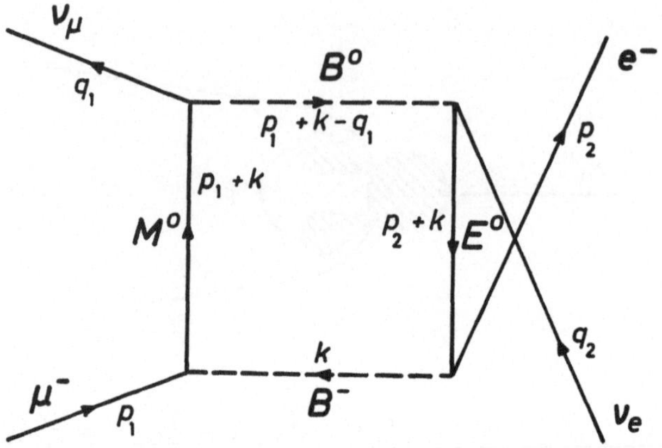

Fig. 3: The process $\mu^-\nu_e \to e^-\nu_\mu$ in the model of Kummer and Segrè.

M^o and E^o are heavy neutral leptons and the B's are scalar intermediate bosons. The matrix element for this process is given by

$$T(\mu^-\nu_e \to e^-\nu_\mu) = \frac{4f^4}{\pi^4}\,\bar{u}(q_1)\gamma_\mu(1+\gamma_5)u(p_1) \times$$

$$\times\ \bar{u}(p_2)\gamma_\nu(1+\gamma_5)u(q_2)J^{\mu\nu} \tag{26}$$

where $\sqrt{4\pi}\,f$ is the coupling constant of the scalar bosons to the scalar-pseudoscalar weak "currents" and

$$J^{\mu\nu} = \int d^4k\,[k^2-M^2]^{-1}[(k+p_1-q_1)^2-M^2]^{-1}[(k+p_2)^2-m_E^2]^{-1} \times$$

$$\times\ [(k+p_1)^2-m_M^2]^{-1}(p_1+k)^\mu(p_2+k)^\nu\ . \tag{27}$$

M is, as always, the boson mass. On a first glance, it looks as if eq. (16) would be of the form derived from the S-operator (1). This is, however, only true in the limit in which eq. (26) takes on the CC-form. From the standard Feynman integral formula

$$\int (1;\ k_\mu;\ k_\mu\ k_\nu) d^4k (k^2-2pk-\Delta)^{-4} =$$

$$= \frac{i\pi^2}{6}\{1;\ p_\mu;\ p_\mu\ p_\nu - \frac{1}{2} g_{\mu\nu}(p^2+\Delta)\}(p^2+\Delta)^{-2} \tag{28}$$

it is immediately seen that the $g_{\mu\nu}$ term carries the smallest number of boson masses in the denominator. (Note that M^2 occurs only in Δ of eq. 28). Hence, if M is taken to be large, one can drop all but the $g_{\mu\nu}$ terms and he obtains

$$T(\mu^-\nu_e \to e^-\nu_\mu) = \frac{4f^4}{\pi^4} \bar{u}(q_1)\gamma_\lambda(1+\gamma_5)u(p_1)\bar{u}(p_2)\gamma^\lambda(1+\gamma_5) \times$$

$$\times u(q_2) I(s,t) + O(\frac{m^2}{M^2}) \tag{29}$$

where I(s,t) is the following Feynman-integral

$$I(s,t) = \int_0^1 x(1-x)dx \int_0^1 dy \int_0^1 dz \{M^2x + m_M^2 z(1-x) +$$

$$+ m_E^2(1-z)(1-x) + t\cdot A + s\cdot B + m_\mu^2 C + m_e^2 C\}^{-1} \tag{30}$$

with

$$A = x^2y^2 + xyz(1-x) + xy(1-z)(1-x) + z(1-z)(1-x)^2$$

$$B = z(1-z)(1-x)^2$$

$$C = z(1-x)(z-xz+xy-1) - xy$$

$$D = (1-x)(1-z)[(1-x)(1-z)+xy-1] \tag{31}$$

Neglecting terms $O(m^2/M^2)$ gives a constant for the integral (3o). Therefore, in the same limit, in which the matrix element is of the CC-form, the structure function becomes a constant. The correction terms, however, are both of different nature (also scalar and pseudoscalar) and of different dependence in the structure function (dependence on t and s).

REFERENCES

1. H. Pietschmann, Springer Tracts in Modern Physics 52, 193 (197o);
   H. Pietschmann, Acta Phys. Austr., Suppl. VII, 588 (197o);
   H. Pietschmann and M. Stremnitzer, Nucl. Phys. B22, 493 (197o).
2. T. D. Lee and C. N. Yang, Phys. Rev. 1o8, 1611 (1957).
3. P. Kristenson and C. Møller, Dan. Mat. Fys—Medd. 27/7 (1952).
4. T. D. Lee and C. N. Yang, Phys. Rev. 119, 141o (196o).
5. W. Bartel, B. Dudelzak, M. Krehbiel, J. M. Mc Elroy, U. Meyer-Berkhout, R. J. Morrison, H. Nguyen Ngoc, W. Schmidt and G. Weber, Phys. Rev. Lett. 17, 6o8 (1966);
   W. Albrecht, H. J. Behrend, F. W. Brasse, W. Flauger, H. Hultschig and K. G. Steffen, Phys. Rev. Lett. 17, 1192 (1966).
6. R. E. Kreps and J. W. Moffat, Phys. Rev. 175, 1942, 1945 (1968).
7. G. V. Efimov, Soviet Phys. JETP 17, 1417 (1963);
   E. S. Fradkin, Nucl. Phys. 49, 624 (1963);
   G. V. Efimov, Nuovo Cim. 32, 1o46 (1964);
   M. K. Volkov and G. V. Efimov, Soviet Phys. JETP 2o, 1213 (1964);

G. V. Efimov, Nucl. Phys. 74, 657 (1965);
G. V. Efimov, Soviet Phys. JETP 21, 395 (1965).

8. A. Salam, Acta Phys. Austr., Suppl. VII, 1 (197o).
9. D. I. Fivel and P. K. Mitter, Phys. Rev. D1, 327o (197o).
10. A. R. Clark, T. Elioff, R. C. Field, H. J. Frisch, R. P. Johnson, L. T. Kerth and W. A. Wenzel, Proceedings of the Kiev Conference 197o.
11. R. N. Mohapatra, J. Subba Rao and R. E. Marshak, Phys. Rev. 171, 15o2 (1968).
12. H. Kühnelt and H. Pietschmann, to be published.
13. I. S. Gerstein and H. J. Schnitzer, Phys. Rev. 175, 1876 (1968).
14. G. V. Efimov and Sh. Z. Seltser, Dubna preprint submitted to Ann. Phys. (N.Y.).
15. W. Kummer and G. Segrè, Nucl. Phys. 64, 585 (1965).
16. W. Kummer, Acta Phys. Austr., Suppl. VIII (1971).

Acta Physica Austriaca, Suppl. VIII, 136–153 (1971)

RENORMALIZABLE "DECEPTION" THEORY OF WEAK INTERACTIONS*

BY

W. KUMMER
Institut für Hochenergiephysik der
Österreichischen Akademie der Wissenschaften
and
Institut für Theoretische Physik
der Technischen Hochschule
Wien

ABSTRACT

We review the very general conditions under which a nonlocal theory of weak interactions with a renormalizable Lagrangian yields the effective V-A-form for the leptons in leptonic and semileptonic matrix elements. The simplest special case is the two-boson exchange model, when both particles have spin zero. After several attempts to elaborate this model so as to avoid qualitative contradictions in the realm of nonleptonic processes by the introduction of several other particles, in its most recent version by Gupta and Patil even less particles are required than in the original box model for leptonic and

---

* Lecture given at X. Internationale Universitätswochen für Kernphysik, Schladming, March 1 - March 13, 1971.

semileptonic processes. We treat here also the last basic open question, namely the incorporation of the CVC-hypothesis, by the assumption of a product development of the hadronic source-operators in the hadronic part and arrive in this way at a satisfactory solution. It seems unreasonable to go any further in the theory at a moment, when our experimental knowledge about the nonlocal structure of weak interactions is still nonexistent.

1. INTRODUCTION

The large bulk of experimental information on weak interactions so far determines only certain constant parameters, essentially effective coupling constants. Up to now any nonlocalities in the well observed weak reactions at low energies can be completely attributed to the effect of participating hadrons. This holds as well for the interpretation of the still scarce data from high energy neutrino scattering. The phenomenological description of leptonic and semileptonic processes in terms of the current-current theory with a V-A-coupling of the leptons has led some time ago to the hypothesis of a charged intermediate massive vector particle (W) [1] as the most obvious means to provide the nonlocality. The main theoretical difficulty connected with this perhaps aesthetically appealing idea is the fact that higher orders cannot be computed in a reliable way due to the nonrenormalizability of the theory [2]. Because no experimental proof for the existence of the W existed in 1965 - a situation which has not changed since then - G. Segrè and the author proposed an analysis of weak interactions based on the possible exchange of more than one intermediate boson which can be scalar and which nevertheless effectively provide a vector coupling theory in low energy reactions [4]. It was found that in fact a very large class of renormalizable field theories can reproduce the V-A-form in leptonic or semi-leptonic processes.

Consider renormalizable interactions [5]

$$L_{int} = \sum_i g_i \bar{\ell}_i^* (1+\gamma_5) \ell \, B_i + h.c. + L'_{int}(B_i, \ell^*) \tag{1.1}$$

where ℓ is one of the observed leptons, ℓ_i^* are new leptons and B_i new scalar bosons, "inside" the weak interactions. The ℓ_i and B_i can be neutral or charged.

(1.1) together with the free Lagrangian is invariant with respect to

$$\ell \rightarrow \gamma_5 \ell, \; m_\ell \rightarrow -m_\ell \tag{1.2}$$

for each "normal" lepton ℓ separately. Among the possible Dirac combinations of e.g. a four-lepton process of "normal" ℓ-s like μ-decay, a term $\bar{\ell}'\ell$, $\bar{\ell}\gamma_5\ell$, or $\bar{\ell}'\sigma_{\mu\nu}\ell$ must necessarily appear multiplied with a factor $m_{\ell'}, m_\ell$ in order to satisfy (1.2). If a sufficient number of the "internal" particles is heavy enough (mass $M \sim O(\text{GeV})$) so that the limit $m_\ell \rightarrow 0$ (or $m_\ell \ll M$) does not introduce mass singularities of the four fermion processes, we remain with a form

$$\bar{\ell}'\gamma^\nu(1+\gamma_5)\ell\bar{\ell}''\gamma^\mu(1+\gamma_5)\ell''(C\, g_{\mu\nu} + D\, P_\mu Q_\nu/M^2) + O\left(\frac{m_\ell^2}{M^2}\right)$$

which is just the observed effective V-A-matrix element, if (in low energy processes) the typical lepton momenta P, Q can be neglected compared to M. Since the same argument can be applied to the leptonic part of a semileptonic process (like neutron-β-decays) as well, again the (V-A) matrix element is selected for the leptons, which from Lorentz invariance requires the hadrons to be coupled also as vectors or axial-vectors only. In the investigation of one of the most simple versions of such a theory one assumes e.g. two light, neutral ℓ-s associated with muon and electron respectively and two scalar very heavy bosons (one scalar, one charged) [4]. Other authors assume heavy, charged ℓ-s and two heavy bosons [6,7] or one heavy charged and one relatively light neutral boson [8,9]. Armed with the knowledge of some simple properties of the box graph in such models from section 2, we show in section 3 and 4 (mostly qualitatively) that all known phenomena of weak interactions fit into such a picture. We also discuss in section 4 how the CVC-hypothesis might be included into the scheme.

2. THE BOX-GRAPH

With an interaction of the form (1.1) the expression for the sum of graphs of Fig. 1 can be written as

$$M_{f,i} \propto \int d^4q\{(q^2-M_1^2)[(p_1-p_3-q)^2-M_2^2][(p_1-q)^2-M_\ell^2]\}^{-1} \times$$
$$\times\ \bar{\ell}'(p_3)(1-\gamma_5)(\not{p}_1-\not{q}+M_\ell)(1+\gamma_5)\ell(p_1)\bar{\ell}''(p_4)(1-\gamma_5) \times$$
$$\times \left\{\frac{\not{p}_2+\not{q}+M_\ell'}{(p_2+q)^2-M_\ell'^2}+\frac{p_4-\not{q}+M_\ell'}{(p_4-q)^2-M_\ell'^2}\right\}(1+\gamma_5)\ell'''(p_2) \tag{2.1}$$

The crossed graph is of course only allowed in certain cases with "neutral currents", i.e. in processes like $e\nu$, or $\nu\nu$-scattering, if the exchanged bosons are both neutral.

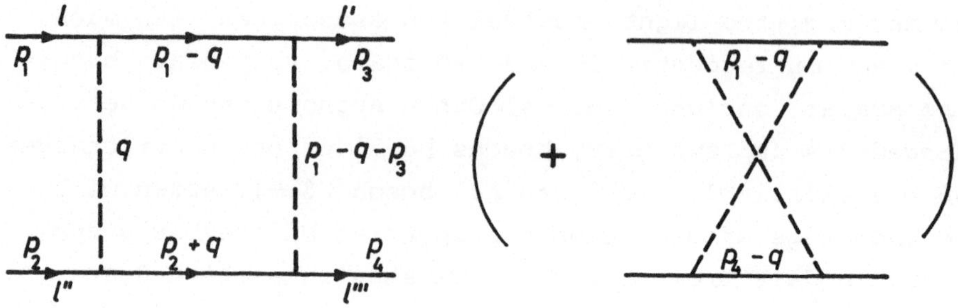

Fig. 1: Direct and crossed box graph.

The following properties of (2.1) are obvious:

a) because of $(1-\gamma_5)(1+\gamma_5)=0$ all terms vanish with M_ℓ or M'_ℓ in the numerator (as expected from our invariance argument).

b) $\not p_1$, $\not p_2$ and $\not p_4$ can be replaced by appropriate external lepton masses, yielding terms of $O(\frac{m_\ell}{M})$ or smaller.

c) in the limit of small external momenta p_i the integral for each graph separately is easily evaluated, if for simplicity all heavy masses are set equal. The result is proportional M^{-2}, the factor depending only on the number of heavy masses. At least two masses have to be large in order to avoid an infrared divergence (or more precisely, a dependence on the "small" masses or momenta in this limit of small p_i, m_ℓ).

d) If the crossed graph is allowed, at low energies only the terms mentioned in b) remain, suppressing further such "neutral current" reactions.

3. LEPTONIC INTERACTIONS

With e.g.

$$L_{int}^{lept} = g\, B^- \sum_{\ell=e,\mu} \bar{L}_\ell^-(1+\gamma_5)\nu_\ell +$$

$$+\, g'\, B^0 \sum_{\ell=e,\mu} \bar{L}_\ell^-(1+\gamma_5)\ell^- + \text{h.c.} \qquad (3.1)$$

one obtains for μ decay just the first graph of Fig. 1, where $\ell=\mu^-$, $\ell'=\nu_\mu$, $\ell''=e^-$, $\ell'''=\bar{\nu}_e$. We have assumed here a version with a heavy charged "internal" lepton L. Taking account of the appropriate factors and putting $M_{B^\pm} = M_{L^-} >> M_{B_o}$, (lepton masses and momenta), we have

$$- \frac{4ig^2g'^2}{(2\pi)^4} \int \frac{d^4q}{[q^2-M^2]^3} = - \frac{2}{M^2}\left(\frac{g^2}{4\pi}\right)\left(\frac{g'^2}{4\pi}\right) = \frac{G}{\sqrt{2}} \qquad (3.2)$$

where $G=10^{-5}m_p^2$ is the usual coupling constant of weak interactions. From (3.1) and the study of the respective box graphs we see by using the arguments of the last section that here e.g. ν_μe-scattering and $\nu_e\mu$-scattering are forbidden to order G, weak μe-scattering is suppressed according to $G\times\left(\frac{m_e m_\mu}{M}\right)^2$ because the crossed graph is allowed, $\nu\nu$-scattering is of O(G). In this model L^- must be of course heavy enough in order not to be produced in pairs in presently attainable high energy electromagnetic interactions. Its contribution to the g-2 of the muon

$$g-2 = \frac{g^2 m_\mu^2}{24\pi^2 M^2} \qquad (3.3)$$

in units of the Bohr magneton is already too small to be noticed in present day experiments, if $M \gtrsim 2$GeV.

4. PROCESSES INVOLVING HADRONS

In order to allow semileptonic processes there must exist also interactions like

$$L_{int}^{hadr} = g_h \bar{N}' (1+\alpha\gamma_5) N B \qquad (4.1)$$

between the "external" hadron N and the internal particles N, B including also a possible change of strangeness between N and N'. From β-decay, where the box contains then two interactions like (4.1) and two interactions of the type (1.1) or (1.3) we have apart from factors

$$(\frac{g^2}{4\pi})(\frac{g_h^2}{4\pi})M^{-2} \sim G/\sqrt{2} \qquad (4.2)$$

Comparison with (3.2) shows that $g^2 \sim g^2_h$ is required to preserve universality. At the same time however (4.1) yields corrections represented by the graphs of Fig. 2 to hadronic processes which violate parity and strangeness to $O(g^2) \sim O(\sqrt{G}M)$ and thus would become too large to remain unobserved. This problem has been solved in ref. [6] and [7] within simple models of the hadronic part of the reaction by introducing further particles, so that such diagrams cancel. The most elegant proposal is the recent one by Gupta and Patil [8] who assume very unequal couplings g_h for the hadronic part. Their hadronic Lagrangian reads

$$L_{int} = ig_+ B^+ [\cos\theta (S_2^1+P_2^1) + \sin\theta (S_3^1+P_3^1] + h.c.$$

$$+ ig_\pi \Pi_B^A P_A^B + \text{other pure hadronic couplings} \qquad (4.3)$$

The neutral boson B_o is simply identified with the combination Π_1^1 of the SU(3)-matrix of pseudoscalar mesons Π_A^B which are coupled strongly ($g_\pi=\sqrt{2}\, g_{\pi NN}$) and parity conserving as usual to the pseudoscalar hadron quark-source but only with g' to the leptons (cf.(3.1)). S_B^A is the scalar hadron density. This immediately eliminates the difficulty of the graphs of Fig. 2:

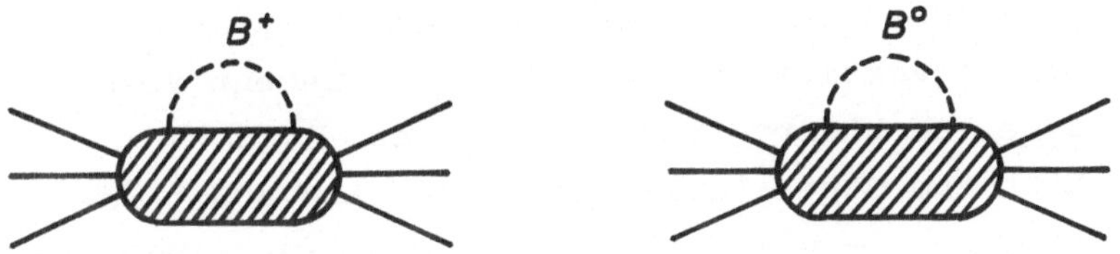

Fig. 2: Possible parity violating or strangeness violating contributions to order g_h^2.

the first one is now $O(g_+^2) \lesssim GM^2$ because β-decay $[gg'g_+g_\pi(16\pi^2M^2)^{-1} \sim G/\sqrt{2}]$ and universality $[gg' \sim g_+g_\pi]$ entail a very small g_+ (remember $g_\pi^2/(4\pi) \sim 15$). The second graph, on the other hand, becomes just the usual selfenergy-contribution from the Π_1^1-component of the hadronic interaction, conserving parity and strangeness. As mentioned in the introduction, the vector character of the leptonic part forces the vector (or axial vector) coupling of the hadrons in a trivial way. More difficult problems present the incorporation of universality (in a more precise way than in the discussion above) and the CVC-hypothesis [10]. The reason is obvious, because the hadronic matrixelement contains here two (scalar) source operators instead of the one vector operator in the usual theory. Only in the limit of small external variables the vector appears "phenomenologically" in the hadronic matrix elements. In order to understand the absence of renormalization of the vector part $V^{\mu}(x)$, the CVC-hypothesis

$$\partial_{\mu} V^{\mu} = 0 \tag{4.4}$$

must hold at least as a weak operator relation in the Hilbert space of hadronic states.

We propose to use here the product expansion of operators, investigated extensively in recent years, and applied to the theory of hadrodynamics [11].

With the Gupta-Patil Lagrangian the graph for $\nu(p_1)$ + hadrons $\rightarrow e^-(p_2)$ + hadrons becomes ($\Delta S=0$, the factor $\cos\theta$ is omitted, $Q=p_2-p_1$, m_o is some average mass of π_o, η, χ; $A(x)=P_1^1(x)$, $B(o)=P_2^1+S_2^1$)

$$\begin{aligned} T_{fi} &= -\frac{2gg'g_+g_\pi}{(2\pi)^4}\,\bar{u}_e(p_2)\gamma^{\mu}(1+\gamma_5)u_{\nu}(p_1) \times \\ &\times \int d^4q\; q_{\mu}\{[q^2-M^2][(q-p_1)^2-M^2][(q+Q)^2-m_o^2]\}^{-1} \times \\ &\times \int d^4x\; e^{i(q+Q)x} <\beta|T(A(x)B(o))|\alpha> \;. \end{aligned} \tag{4.5}$$

The limit of small p_1, p_2, m_e and m_o yields

$$\begin{aligned} T_{fi} &= 2igg'g_+g_\pi\bar{u}_e(p_2)\gamma^{\mu}(1+\gamma_5)u_{\nu}(p_1) \times \\ &\times \int d^4x <\beta|T(A(x)B(o))|\alpha> \frac{\partial J}{\partial x^{\mu}} \end{aligned} \tag{4.6}$$

where

$$J(M^2,x^2) = \frac{1}{(2\pi)^4}\int d^4q\; e^{iqx}\{(q^2-M^2)^2q^2\}^{-1} \;. \tag{4.7}$$

This integral is easily evaluated after a development into partial fractions of $\{\}^{-1}$ and comparison of each term with the known x-space result for the causal Wightman function $\Delta_F(x^2,\mu^2)$. As espected the result vanishes exponentially or by rapid oscillations of Besselfunctions outside

$$x^2 \lesssim \frac{1}{M^2} \tag{4.8}$$

so that for large enough internal masses only the light cone singularity of the T product in (4.6) is important. If $M \gtrsim 10$ GeV all hadron masses and symmetry breakings are assumed to be small. Neglecting possible logarithmic factors for the time being, the C-number coefficient functions $C_n(x)$ in

$$A(x)B(o) \underset{x\sim o}{\to} \sum_n C_n(x) O_n(o) \tag{4.9}$$

are easily determined from the dimension of the local operator O_n alone [11]. Since dim A=dim B=$[m]^3$ one finds in a π-fermion theory

$$P_1^1(x)P_2^1(o) \underset{x\sim o}{\to} \frac{C}{x^6} + \frac{D\Pi\Pi}{x^4} + \frac{E\bar{\psi}\psi}{x^3} + \frac{x^\mu}{x^4}[E'\Pi\partial_\mu\Pi + E''\bar{\psi}\gamma_\mu\psi] + \ldots \tag{4.10}$$

where all SU(3)-indices have been suppressed in the different allowed couplings and x^2 is to be regarded always as the appropriate limit $x^2 - i\varepsilon x_o$ $(\varepsilon > o)$ [11]. Another term $\bar{\psi}\sigma_{\mu\nu}\psi x^\mu x^\nu / x^5$ vanishes from symmetry.

If E' and E" are adjusted appropriately (f-coupling!) the term in square brackets can be just the vector current referred to above apart from a constant. (4.10) is not yet the required bahaviour near the lightcone ($x^2 \sim o$) where also powers of x_α in the numerator are allowed besides the terms in (4.10) [12]

$$\begin{aligned} P_1^1(x)P_2^1(o) \underset{x^2\sim o}{\to} & \text{(terms of 4.10)} + \\ & + x^{-3} \sum_{n=1}^{\infty} x^{\alpha_1} \ldots x^{\alpha_n} O^{(n)}_{\alpha_1 \ldots \alpha_n}(o) + \\ & + x^{-4} \sum_{n=2}^{\infty} x^{\alpha_1} \ldots x^{\alpha_n} \bar{O}^{(n)}_{\alpha_1 \ldots \alpha_n}(o) \quad . \end{aligned} \tag{4.11}$$

We are free to select the O_n such that not $g_{\alpha_i\alpha_j}$ occur because this would lead to a reduction of the light cone singularity. For dimensional reasons $O^{(n)}_{\alpha_1\ldots\alpha_n}$ is then produced just by taking the necessary number of derivatives of the $\bar{\psi}\psi$-term:

$$(\partial_{\alpha_1}\ldots\partial_{\alpha_k}\bar{\psi})(\partial_{\alpha_{k+1}}\ldots\partial_{\alpha_n})\psi \ .$$

The $\bar{O}^n$ could contain also $g_{\alpha_i\alpha_j}$. Then all remaining (also nonsingular) quantities are lumped together in this infinite sum.

The T-product is obtained by changing the limit $x^2-i\varepsilon x_o$ with

$$\theta(x_o)(x^2-i\varepsilon x_o)^{-\rho} + \theta(-x_o)(x^2+i\varepsilon x_o)^{-\rho}$$

into $x^2-i\varepsilon$. Because of the ∂_μ in (4.6), any parts of (4.11) depending on x^2 alone or with an even number of x^{α_1}-factors do not contribute to the integral. This makes the first three terms in (4.10) and all the even n in the sums unimportant. The order of magnitude of the different remaining contributions to (4.6) is easily estimated after introducing in the x-space integral the dimensionless variables

$$x^\mu = M^{-1}\xi^\mu \ .$$

The terms with $O^{(n)}$ ($\bar{O}^{(n)}$) are seen to be reduced at least by a factor $\frac{1}{M}(\frac{1}{M^2})$ with respect to the leading vector term. Clearly integrations term by term in the infinite sums diverge, but we know from the fact that the whole boxgraph is convergent, that the x-space integral of the full (4.11) must be convergent too. If the integral with the vector-part alone in (the constant γ is still undetermined)

$$P_1^1(x)P_2^1(o) = i\gamma\, V^\mu(o)\partial_\mu x^{-2} + W(x) \qquad (4.12)$$

gives a finite result - which we shall see below - the same must be true for the integral over the remainder $W(x)$. The latter is then certainly at most $O(m_n/M)$ for low energy semileptonic reactions and can be neglected.

Since

$$(x^2-i\varepsilon)^{-1} = -\,2\pi^2\ \Delta_F(x,\mu=0)$$

(4.12) can be easily used in the T-product

$$\int d^4x\ e^{iqx}T(\text{eq.}4.12) = -4\pi_i^2\ \frac{[\gamma V^\mu(o)+\lambda A^\mu(o)]q_\mu}{q^2} \tag{4.13}$$

which is now true between all hadronic states. We have already included in (4.13) the result of the similar analysis for the axial vector A_μ from $P_1^1\ S_2^1$. So CVC can be assumed at this point with all its desirable consequences. In the limit of small external fourmomenta for α=neutron, β=proton, (β-decay of the neutron), (4.5) with (4.13) and the analogous purely leptonic processes (μ-decay) differ by the factor

$$-4\pi^2\gamma\langle p|\gamma V_\mu + \lambda A_\mu|n\rangle g_+g_\pi \int d^4q\ q^{-2}(q^2-M^2)^{-2}$$

instead of

$$2\pi\ \gamma_\mu(1+\gamma_5)u\ gg' \int d^4q(q^2-M^2)^{-3}$$

The second q-integral is just 1/2 times the first; thus universality means for the vector current

$$gg' = -4\pi^2\gamma g_+g_\pi \tag{4.14}$$

One may go one step further by assuming the equal time commutators (E.T.C) of the quark model to fix γ and λ [13] as long as no logfactors occur. We find from

$$[P_1^1(x),\ P_2^1(o)+\delta_2^1(o)] = -\delta^3(\vec{x})\,(V(o)_2^1 + A(o)_2^1) \tag{4.15}$$

in the development of the commutator derived from (4.12)

$$E_\mu = x_\mu [x^2 - i\varepsilon x_o)^{-2} - (x^2 + i\varepsilon x_o)^{-2}] = F_\mu (x_o)\,\delta^3(\vec{x}) + F_{\mu i}\partial_i \delta^3(\vec{x}) + \ldots$$

that in fact only the term without derivatives of the δ-function

$$F_\mu (x_o) = \int E_\mu (x)\, d^3x = -2\pi^2 i g_{\mu o} + O(x_o)$$

survives in the limit $x_o \to o$ so that

$$\gamma = \lambda = (4\pi^2)^{-1} \; . \qquad (4.16)$$

As shown in ref. [15], it is also possible to admit factors $\log^n(x^2)$ in the development at short distances (4.10). After transformation into q-space the constants γ and λ in (4.13) are then simply replaced by a dependence on the powers of $\log q^2/\mu^2$ which makes the calculation of (4.14) more complicated. The result is however of the same type with a logarithmical dependence on $\log (M/\mu)$. A crucial role in such models play usually processes with "neutral currents" like ν-hadron scattering, weak leptonic-hadron scattering or decays like $K_o \to \ell^+ \ell^-$ via weak interactions. With the Lagrangian (3.1), (4.3) the respective diagrams are set out in Fig. 3.

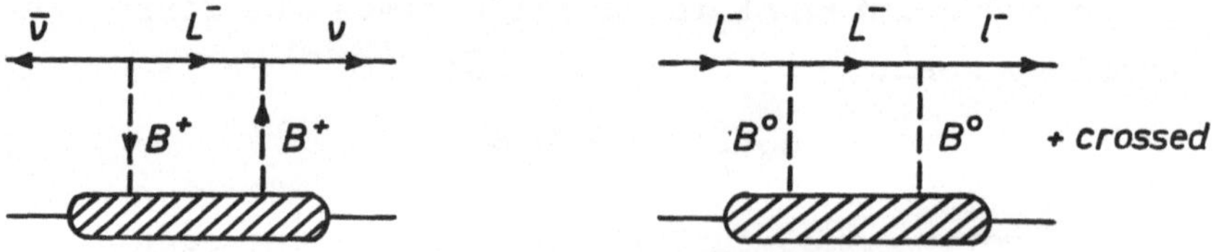

a) $\sim \dfrac{g_+^2 g^2}{M^2} << G$ *b)* $\sim g_\pi^2 g'^2 (\dfrac{m_\ell}{m_N})$

Figs. 3 a and b.

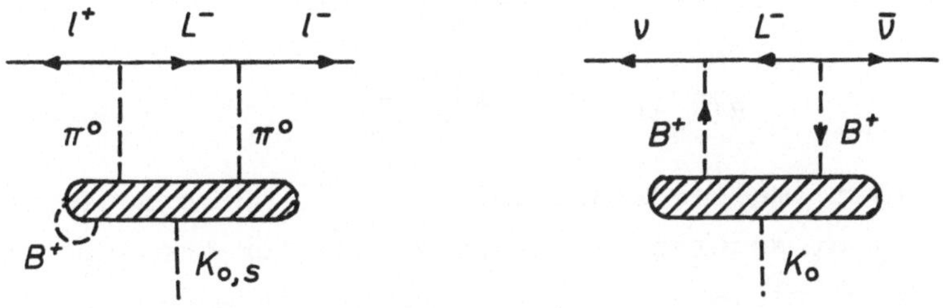

c) $\sim g'^2 \frac{m_\mu m_k}{M^2} \times (\text{amplitude } K_S^O \to 2\pi_0)$ d) $\sim \frac{g_+^2 g^2}{M^2}$

Fig. 3: Graphs for a) ν-scattering, b) weak lepton scattering and the weak decays c) $K_O \to \ell^+\ell^-$ and d) $K_O \to \nu\bar{\nu}$ together with effective couplings.

We see that the reactions involving two neutrinos must be even smaller than the usual weak interactions ($g_+ << g$). In the two other cases despite the relative smallness of the masses of the neutral particle a reduction factor for b) can be estimated because of the partial cancellation of the diagrams to be $(200\ M^2)^{-2}\ \ln M^2/Q^2$ for large momentum transfers (cf. [9]). What happens here is essentially that the propagators (cf. 2.1 with the masses in $\{\}^{-1}$ replaced by m_π, m_N) force the main contributions of the integral to come from low q^2, so that in $\{\}$ we may neglect $(p_2+q)^2$ and $(p_4-q)^2$ as compared to M^2 (in the still converging integral!) which helps sufficiently (together with the factor m_e/m_N) to make this graph a few percent of the electromagnetic one in μ-p-scattering and by another factor m_e/m_μ smaller in the well studied e-p-process.

Estimating $K_{os} \to \mu^+\mu^-$ from the first graph of Fig. 3c) in terms of the $K_{os} \to 2\pi_o$ rate obtains [9]

$$\frac{R(K_s \to \mu^+\mu^-)}{R(K_s \to all)} \gtrsim 10^{-8} \frac{m_N^2}{M^2}$$

and an even smaller limit for $K^+ \to \pi^+\mu^+\mu^-$ Here again the mechanism of section 2d) is essential for the suppression. $K_L \to \mu^+\mu^-$ must violate CP to go through the $2\pi_o$-state. Experimental upper limits for ratios like

$$r = \frac{R(K^+ \to \pi^+ \nu\bar{\nu})}{R(K^+ \to \pi^o e^+ \nu_e)}$$

place an upper limit on

$$g_+^2/g_\pi^2 = \frac{GM^2}{2\sqrt{2}} \; / \; (g_\pi^2/4\pi)^2$$

and hence on M^2. From the experimental result [16] $r < 1.2.10^{-6}$ one finds ($g_\pi = \sqrt{2}\, g_{\pi NN}$, $g_{\pi NN}^2/(4\pi) \sim 15$) that $M \lesssim 16$ GeV. A similar limit can be obtained from $K_{oL} \to \mu^+\mu^-$, taking into account the CP-violation in $K_{oL} \to 2\pi_o$.

The heavy lepton $L^\pm$ and $B^\pm$ could be expected to be produced in cosmic ray events or in proton storage rings ($p+p \to p+p+L^- + \bar{L}^-$) via an intermediate π_o above the respective mass threshold stronger than electromagnetically (L-pairs).

It is interesting to note that e.g. the subsequent fast decay ($\Gamma \; \alpha (g^2/4\pi)(M/ \;) \sim 200$ MeV)

$$L_\mu \to \pi^o + \mu$$

and the production process itself would provide the kind of "instantanous" μ-source required for the isotropic distribution of ultra high energetic muons in the atmospere as observed in the Utah-experiment [17].

The charged $B^{\pm}$ will be produced less copiously, because the rate will be proportional $g_{+}^{2}/4\pi \sim 10^{-4}$. The width of B however will be also smaller (e.g. $\Gamma_{B^{+}\to p+\bar{n}} \sim 1$ MeV) so that it may be easier to see as an enhancement.

5. CONCLUSIONS

What should be taken most seriously in the above considerations is the fact that an infinite manifold of theories produces effectively the V-A-theory of weak interactions at low energies deceiving us to believe in the possible existence of a W-vector-meson. The two boson-model is only the simplest consistent special case of such a theory. At the moment it seems rather futile to go any further, before some experimental knowledge on the actual nonlocalities of weak interactions is available. It may very well be, that a large world of very heavy "internal" particles exist which interact rather strongly among themselves, producing in this way also their large masses (as the hadrons do by means of their interactions).

Thus we cannot exclude the possibility that the field of weak interactions is even richer and more complicated than it appears today. On the other hand this more complicated situation may very well permit a description in terms of a more conventional, i.e. renormalizable, field theory

For the time being the merit of such a model seems to lie mainly in the possibility to "smear out" the weak interactions in a renormalizable way. This may prove fruitful for theoretical investigation of the divergences, encountered in higher orders of the conventional 4-fermion interaction and combined with radiative corrections.

REFERENCES AND FOOTNOTES

1. T. D. Lee and C. N. Yang, Phys. Rev. 119, 141o, (1960).
2. It is not yet clear whether nature in fact can be described by a quantum field theory with indefinite metric [3] where this difficulty may be circumvented.
3. T. D. Lee and G. C. Wick, Nucl. Phys. B9, 2o9 and B10, 1 (1964).
4. W. Kummer and G. Segrè, Nucl. Phys. 64, 585 (1965).
5. We may also allow the exchange of a neutral, massive vector boson, coupled to a conserved current. The latter must be constructed from equal and therefore necessarily "internal" particles like $\bar{\ell}\gamma_\mu\ell$ and can be contained in L'!
6. N. Christ, Phys. Rev. 176, 2o86 (1968).
7. E. P. Shabalin, Yad. Fit. 8, 74 (1968).
7. S. H. Patil and J. S. Vaishya, Nucl. Phys. 19, 338 (1970).
9. V. Gupta and S. H. Patil, Nucl. Phys., to be published.
10. S. S. Gershtein and I. B. Zeldovich, Sov. Phys. JETP, 2, 576 (1956);
R. P. Feynman and M. Gell-Mann, Phys. Rev. 1o9, 193 (1968).
11. K. Wilson, Phys. Rev. 179, 1499 (1969);
R. Brandt, Ann. Phys. (N.Y.) 44, 221 (1967).
12. R. A. Brandt and G. Preparata, CERN-prepr. TH/12o8 (1970).
13. The E. T. C. determine also the Bjorken limit $q_0\to\infty$ ($\vec{q}$ = fixed) [14], which however is not enough to describe the full region $q^2\sim M^2>>m_\ell^2$ in a covariant way. Note also that the light cone development (4.9), amended by log-factors, is true in all renormalizable theories whereas the Bjorken-limit does not hold under all circumstances.

14. cf. the argument of Goldberger, mentioned in ref. [6].
15. W. Kummer, Acta Phys. Austr., to be published.
16. Particle Data Group, "Review of Particle Properties" Phys. Lett. August 1970.
17. H. E. Bergeson et al., Phys. Rev. Lett. 19, 1487 (1967) and 21, 1o89 (1968).

Acta Physica Austriaca, Suppl. VIII, 154–176 (1971)

A NON-LOCAL FIELD THEORY APPROACH TO STRONG INTERACTION DYNAMICS*

BY

J. NILSSON
Institute of Theoretical Physics
Göteborg, Sweden

1. INTRODUCTION

Quantum electrodynamics is beset by some internal consistency problems, but otherwise it has been overwhelmingly successful within its domain of applicability.

Attempts to describe other interactions than the electromagnetic one by means of local field theory cannot claim the same success. Local field theory applied to hadron dynamics for one thing fails miserably. It may be that we lack adequate methods to solve the equations of motion. On the basis of that hope we frequently use local field theory as a guide when we develop more general schemes - to impose analyticity conditions on scattering amplitudes etc. - and I think just about everybody will resort to Feynman graphs when he is in trouble, even if some people may be quick to erase them so as not to fall into disrepute. The reasons for using simple field-theoretic calculations as a guide are many. One obvious reason is

---

* Lecture given at X. Internationale Universitätswochen für Kernphysik, Schladming, March 1 - March 13, 1971.

that local field theory exhibits so many attractive features which we consider physical such as

(i) it provides a unitary S-matrix

(ii) it provides many analyticity properties of physical origin for scattering amplitudes, e.g. poles and threshold behaviour for appropriate channels

(iii) it satisfies crossing symmetry

(iv) it provides simple and very restrictive means to introduce the dynamics of interactions

(v) if the perturbation expansion makes sense and if it "converges" fast enough it yields a practical theory.

Keeping this in mind it is only natural to ask whether we have so far exploited all possibilities of field theory within the conventional approaches. In a certain sense people working in axiomatic field theory try to answer this question within the framework of local field theory. Without specifying the details of the dynamics they make use only of the most general principles to derive results regarding, say, analyticity properties. The difficulties to make real contact with laboratory physics are formidable, however, and only fragmentary results have been obtained so far. One may also question some of the conditions that one conventionally places on field theory such as strict locality. However, in general one has little to substitute for locality or microcausality and partly for that reason one has made rather little progress along these lines.

We shall in these lectures examine one aspect of field theory which in my opinion has not yet been sufficiently explored - the question of what kind of realization one should use for the Poincaré group when one constructs fields. I shall only be able to outline briefly what possibilities other realizations than the conventional Wigner realizations may hold and what price one may have to pay once one goes beyond conventional fields defined with the Minkowski space as the carrier space. For the sake of simplicity and clarity I shall discuss a rather specific field theory model, but much of what I have to say can be expressed in more general terms.

2. THE CHOICE OF CARRIER SPACE

Conventional field theory is based on single particle wave functions which are functions only of the coordiante vector x_μ, that is, of the parameters of the translation group. Admissable wave functions for a free (stable) particle transform irreducibly under restricted Lorentz transformations (no reflections). This means that the manifold of physical wave functions for a free particle of mass $m>o$ and spin j span a representation space for a unitary irreducible representation (UIR) of the restricted Poincaré group P. Following Wigner one may in this way analyze the kinematics of free particles by studying the UIR's of P. In his classical paper Wigner [1] derived and classified all these representations. The explicit realizations given by Wigner make use of representation spaces which are tensor products of a (2j+1)-dimensional linear space (the spin space) and the space of square integrable functions on the mass hyperboloid. This particular choice of realizations greatly restricts the role that spin may play in field theory. For free particles this is not very crucial, but it may be important for systems in interaction [2]. Of course, from the mathematical point of view the choice of realization is only a matter of convenience since different choices provide unitarily equivalent representations. From the point of view of physics it is not that trivial as we shall see. Hence, the choice of carrier space is an important one that requires special attention. In looking for possible alternatives to the Minkowski space one is naturally led to consider the homogeneous spaces of P or more precisely of $\bar{P}$, the covering group of the Poincaré group. To see this we shall next consider the concept of homogeneous spaces in some more detail before proceeding to make a specific choice.

3. HOMOGENEOUS SPACES OF $\bar{P}$

Definitions:

A homogeneous space H of a group G is a topological space on which G acts as a continous transformation group in a transitive fashion. Hence,

(i) if $y \varepsilon H$ and $g \varepsilon G$, then $gy \varepsilon H$ and the mapping satisfies appropriate continuity conditions

(ii) for any y_1, $y_2 \varepsilon H$ there exists a $g \varepsilon G$ such that $y_2 = gy_1$.

The maximal closed subgroup S_o of G is said to be the stabilizer for the point $y_o \varepsilon H$ if it leaves y_o invariant, that is, if for any $s_o \varepsilon S_o$ one has $s_o y_o = y_o$.

Lemma:
The stabilizer subgroups S_1 and S_2 corresponding to two different points y_1 and y_2 of H are related by conjugation.

Proof:

$$y_1, y_2 \;\varepsilon\; H, \quad s_1 \;\varepsilon\; S_1 \;.$$

From transitivity we know that there is a $g \varepsilon G$ such that

$$y_2 = g\, y_1 \ .$$

Hence

$$y_1 = s_1 y_1 = s_1\, g^{-1} y_2 \ .$$

Multiply by g from the left

$$y_2 = g\, y_1 = g\, s_1\, g^{-1} y_2 \ .$$

Hence

$$g\, s_1\, g^{-1} \ \varepsilon\ S_2 \ .$$

Thus we have found that for any element $s_1 \varepsilon S_1$ the element $g\, s_1\, g^{-1}$ belongs to the stabilizer group of y_2. This proves the essential part of the lemma and the remainder is left for an excerise to the reader.

We next proceed to show that we may establish a mapping between H and the coset space G/S_o where S_o is any stabilizer subgroup of G. Different choices of stabilizers within a conjugate class correspond to different choices of origin in H. Consider any element $g \varepsilon G$ and a corresponding stabilizer subgroup S_o of G. Then we may write

$$g = \tilde{g}\, s_o$$

where $s_o \varepsilon S_o$ and $\tilde{g} \varepsilon G/S_o$. Using this decomposition and the transitivity property of H under G we find

$$y_1 = g\, y_o = \tilde{g}\, s_o\, y_o = \tilde{g}\, y_o \ .$$

Hence, using the point y_o as our origin in H we may label any other point in H by an element $\tilde{g}$ of G/S_o. This establishes the desired mapping. In this way we have introduced a coordinate system in H with each point labelled by the parameters of the factor group G/S_o. We shall make use of this later when we consider the special case of the Poincaré group.

It follows from what we have said above that the homogeneous spaces for a group G are classified once one has classified all the conjugate classes of closed subgroups of G. For the case of the Poincaré group, or $\bar{P}$ rather, this has been done [3]. For our special purpose we shall restrict ourselves further by imposing conditions on H to comply with our physical ideas. In particular we shall require that

(i) the Minkowski space shall be a quotient space of H; we can then label the points in H by the space-time coordinate vector χ_μ in addition to whatever other labels that we may need.

(ii) S shall be a subgroup of $\bar{L}$, the covering of the homogeneous Lorentz group, in order that P shall act on the Minkowski space in the usual way.

For the sake of simplicity we shall choose the smallest possible extension of the Minkowski space for which we still gain some additional degrees of freedom. We shall make this vague statement more precise later.

We shall further need to know how the points in H are transformed when we apply a Lorentz transformation just as we in the ordinary case specify the transformation law for the points in Minkowski space under Lorentz transformation. The action of P on H is given by the usual group multiplication law. If we denote a point in H by (x,z), where $\chi \in T_4$, the translation group, and $z \epsilon \bar{L}/S$ and similarly an element of P by (a,Λ) then we may write the desired transformation law in the following symbolic way

$$(x,) \xrightarrow{(a,\Lambda)} (a + \Lambda x, \Lambda z) \ .$$

To get further we shall next give a specific parametrization of $\bar{P}$ so that we may give the transformation law above in explicit form.

4. A PARAMETRIZATION OF $\bar{P}$

Let me first recall the standard commutation rules for the Lie algebra of the Poincaré group

$$[L_{\mu\nu},L_{\rho\sigma}] = i(g_{\nu\rho}\, L_{\mu\sigma} + g_{\mu\sigma}\, L_{\nu\rho} - g_{\mu\rho}\, L_{\nu\sigma} - g_{\nu\sigma}\, L_{\mu\rho})$$

$$[L_{\mu\nu},P_{\rho}] = i(g_{\nu\rho}\, P_{\mu} - g_{\mu\rho}\, P_{\mu})$$

$$[P_{\mu},P_{\nu}] = 0 \ .$$

We shall further make use of the Iwasawa decomposition of the Lie algebra ℓ of the homogeneous Lorentz group

$$\ell = k + a + n$$

where k is the maximal compact subalgebra spanned by L_{12}, L_{23} and L_{31}, a is an abelian subalgebra spanned by L_{03} and n is a nilpotent subalgebra spanned by $(L_{01}+L_{31})$ and $(L_{02}-L_{23})$. As one can easily see, (a+n) is a solvable algebra with n as an ideal. To the Iwasawa decomposition of the algebra there is a corresponding decomposition of the group which we may write

$$\bar{L} = \bar{K}\, A\, N$$

where $\bar{K}$ is the subgroup generated by k etc. Hence, we may write an arbitrary element Λ of $\bar{L}$ in the following way

$$\Lambda = \exp[-i\phi L_{12}]\exp[-i\theta L_{31}]\exp[-i\psi L_{12}] \cdot$$

$$\cdot \exp[isL_{03}]\exp[-it(L_{01}+L_{31})]\exp[-iu(L_{02}-L_{23})] \ .$$

An explicit realization is for example obtained by the identification

$$L_{ij} = \frac{1}{2}\,\varepsilon_{ijk}\,\sigma_k$$

$$L_{ok} = \frac{i}{2}\,\sigma_k$$

where the σ_i's are the usual Pauli matrices. This would provide us with a non-unitary 2×2 matrix representation, which is useful for computations in many cases. We shall later give other realizations appropriate for our purposes. If $\Lambda\varepsilon\bar{L}$ then the parameters satisfy the following conditions

$$o \le \phi + \psi \le 4\pi$$

$$-2\pi \le \phi - \psi \le 2\pi$$

$$o \le \theta \le \pi$$

$$-\infty \le s,t,u < \infty .$$

5. CONSTRUCTION OF UNITARY IRREDUCIBLE REPRESENTATIONS OF $\bar{P}$ WITH H AS A CARRIER SPACE

As before we shall denote a point in H by (x,z). We shall further consider a space F of functions f(x,z) on H. We shall later specify F more precisely. For the moment we only require that all $f\varepsilon F$ must be differentiable. We now define a representation U_g of $\bar{P}$ on F by the relation

$$(U_g f)(x,z) = f(g^{-1}o(x,z))$$

for all $g\varepsilon\bar{P}$. The action of g on the carrier space we have previously given; it is a simple coordinate transformation. That the equation above in fact provides us with a representation is seen by the following composition law

$$(U_{g_2}U_{g_1}f)(x,z) = (U_{g'}f)(g_2^{-1}o(x,z)) =$$

$$= f(g_1^{-1}g_2^{-1}o(x,z)) =$$

$$= f((g_2g_1)^{-1}o(x,z)) =$$

$$= (U_{g_2g_1}f)(x,z)$$

where we have made use of the equation above and the group multiplication law for $\bar{P}$.

Having obtained a representation of the group we may proceed to derive the representation of the corresponding Lie algebra for the infinitesimal generators. One easily finds

$$P_\mu = i \frac{\partial}{\partial x^\mu}$$

$$L_{\mu\nu} = i\left(x_\mu \frac{\partial}{\partial x^\nu} - x_\nu \frac{\partial}{\partial x^\mu}\right) + S_{\mu\nu}$$

where the $S_{\mu\nu}$'s are differential operators in the variable z only. The explicit form for these operators clearly will depend on the choice of H and we shall later return to this question.

In order to make the representations that we have constructed unitary and irreducible we must

(i) introduce a scalar product in F under which the Lie generators are hermitean.

(ii) impose irreducibility conditions on F making the elements of F eigenfunctions of the Casimir operators $P_\mu P^\mu$ and $W_\mu W^\mu$ of P.

This requires that we make a choice of H first. We have already mentioned some conditions which H must satisfy to comply with our physical ideas. In addition we now require

(i) H shall be chosen in such a fashion that the representation defined above in terms of single-component functions shall contain both integral and half-integral spin representations.

(ii) there must exist an invariant measure on H so that we may define Lorentz invariant interactions.

The smallest homogeneous space of $\bar{P}$ satisfying these conditions is $\bar{P}/N$, where N is the nilpotent subgroup of

SL(2,C) which we have previously defined. With our choice of parametrization of $\bar{P}$ this means that the points in H are parametrized by the coordinates $(x_\mu;\phi,\theta,\psi,s)$. With this particular choice of homogeneous space we may now derive the explicit expressions for the operators $S_{\mu\nu}$. The result is given below

$$S_{23} = -i[c\phi \frac{c\theta}{s\theta} \frac{\partial}{\partial\phi} + s\phi \frac{\partial}{\partial\theta} - \frac{c\phi}{s\theta} \frac{\partial}{\partial\psi}]$$

$$S_{31} = -i[s\phi \frac{c\theta}{s\theta} \frac{\partial}{\partial\phi} - c\phi \frac{\partial}{\partial\theta} - \frac{s\phi}{s\theta} \frac{\partial}{\partial\psi}]$$

$$S_{12} = i \frac{\partial}{\partial\phi}$$

$$S_{01} = -i[- \frac{s\phi}{s\theta} \frac{\partial}{\partial\phi} + c\phi\ c\theta \frac{\partial}{\partial\theta} + s\phi \frac{c\theta}{s\theta} \frac{\partial}{\partial\psi} + c\phi\ s\theta \frac{\partial}{\partial s}]$$

$$S_{02} = -i[\frac{c\phi}{s\theta} \frac{\partial}{\partial\phi} + s\phi\ c\theta \frac{\partial}{\partial\theta} - c\phi \frac{c\theta}{s\theta} \frac{\partial}{\partial\psi} + s\phi\ s\theta \frac{\partial}{\partial s}]$$

$$S_{03} = -i[-s\theta \frac{\partial}{\partial\theta} + c\theta \frac{\partial}{\partial s}] \ .$$

Since none of the coefficients in front of the differential operators depend on ψ or s we find that

$$[\frac{\partial}{\partial\psi}, S_{\mu\nu}] = [\frac{\partial}{\partial s}, S_{\mu\nu}] = o.$$

To obtain on irreducible representation of $\bar{P}$ on H we shall, therefore, insist that all elements of F be eigenvectors of the two Casimir operators of P and of $\partial/\partial\psi$ and $\partial/\partial s$. Hence for any $f \varepsilon F$ we have

$$\frac{\partial f}{\partial\psi} = i\ n\ f$$

$$\frac{\partial f}{\partial s} = -(\alpha+i\beta) f \ .$$

The cyclic nature of the variable ψ requires that n be an integer or a halfinteger. The eigenvalue of $\partial/\partial s$ is, however, an arbitrary complex number and we have introduced the two quantum numbers α and β which then are real. Hence, an irreducible representation will be labelled by the quantum numbers $(m,j;n,\alpha,\beta)$, where m and j as usual denote the mass and the spin of representation. We obtain a basis in F by insisting that we find the eigenvectors of a Cartan subalgebra in F. We may choose the momentum p and the helicity λ as a suitable set of labels. Considering eigenvectors with $\vec{p}=o$ one finds

$$f^{(m,j;n,\alpha,\beta)}_{\vec{p}=o,\lambda}(x_\mu,\phi,\theta,\psi,s)=\frac{1}{\sqrt{2(2\pi)^3}}\exp[imx_o]\exp[-(\alpha+i\beta)s] \times$$

$$\times\ D^j_{\lambda n}(\phi,\theta,\psi)$$

where D^j denotes the usual Wigner D-functions of spin j. A state of arbitrary momentum and with helicity λ is obtained by applying an acceleration ε along the z-axis followed by a rotation θ around the y-axis and a rotation ϕ around the z-axis where these parameters are related to the momentum vector p_μ by

$$p_\mu = m(\cosh\varepsilon,\sinh\varepsilon\sin\theta\cos\phi,\sinh\varepsilon\sin\theta\sin\phi,\\ \sinh\varepsilon\cos\theta).$$

One finds

$$f_{\vec{p},\lambda}(x,z) = \frac{1}{\sqrt{2(2\pi)^3}}\exp[ipx]\frac{S^j_{\lambda n}(\vec{p};\phi,\theta,\psi)}{(\frac{pk}{m})^{\alpha+i\beta}}$$

where

$$S^j_{\lambda n}(\vec{p};\phi,\theta,\psi) \equiv D^j_{\lambda n}(\phi',\theta',\psi')$$

and the primed angles are obtained by the action of the transformation from rest to momentum $\vec{p}$ as previously discussed. The four-vector k is defined by

$k_\mu = \exp[s](1, \sin\theta\cos\phi, \sin\theta\sin\phi, \cos\theta)$.

To complete the specification of F we must introduce a scalar product for which the representations become unitary. For convenience we introduce the Fourier transform in Minkowski space by

$$f(x_\mu;\phi,\theta,\psi,s) = \int d^4q\ \theta(q_0)\delta(q^2-m^2)\exp[iqx]\hat{f}(\vec{q};\phi,\theta,\psi,s) \ .$$

A possible choice of scalar product satisfying our conditions is then

$$(f_1,f_2) = \int \frac{d^3q}{q_0} \frac{\sin\theta\ d\theta\ d\phi}{(\frac{qk}{m})^{-2\alpha}(\frac{q\kappa}{m})^2} \hat{f}_1^* \hat{f}_2$$

with $\kappa_\mu = e^{-s}k_\mu$. This yields the following normalization for the basis vectors.

$$(f_{\vec{p},\lambda},f_{\vec{p}',\lambda'}) = \frac{1}{(2\pi)^2} p_0\ \delta^3(\vec{p}-\vec{p}')\ \frac{\delta_{\lambda\lambda'}}{(2j+1)} \ .$$

With this we have accomplished what we set out to do in this section, that is, we have constructed unitary irreducible representations of $\bar{P}$ and further given a basis for the special case of $H=\bar{P}/N$.

6. AN ALTERNATIVE POINT OF VIEW

For later considerations it is useful to note that the elements of F may be looked upon as matrix elements of the appriate UIR of $\bar{P}$. More precisely one can show that

$$f_{n'}^{(m,j,n)}(g) = U_{n'n}^{m,j}(g) \equiv \langle n'|U(g)|n\rangle$$

where $g \varepsilon \bar{P}$ and $|n\rangle$, $|n'\rangle$ are vectors in an irreducible representation space of $\bar{P}$, with n and n' denoting the necessary labels. It is well known that one may use the elements of a matrix column (fixed n!) to define a basis for a representation. By choosing the vector $|n\rangle$ in a suitable fashion one can arrange it so that the functions f(g) are functions on the homogeneous space $\bar{P}/S$ rather than on $\bar{P}$ itself. To this end we choose $|n\rangle$ so that they satisfy

$$L_i|n\rangle = o$$

for all the Lie generators of the stabilizer group S. This implies that $U(s)|n\rangle = |n\rangle$ for any $s \varepsilon S$. Then

$$\begin{aligned} f_{n'}^{(m,j,n)}(gs) &= \langle n'|U(gs)|n\rangle = \\ &= \langle n'|U(g)U(s)|n\rangle = \\ &= \langle n'|U(g)|n\rangle = f_{n'}^{(m,j,n)}(g) \end{aligned}$$

which proves the assertion above.

7. FREE FIELDS DEFINED ON $\bar{P}/N$

In the previous sections we have derived suitable single particle wave functions and we may now in standard fashion construct second-quantized free fields. To do this we define creation and annihilation operators for particles and antiparticles and impose the following commutation rules

$$[c(\vec{p},\lambda), c^+(\vec{p}\,',\lambda')]_\pm = p_o\ \delta^3(\vec{p}-\vec{p}\,')\delta_{\lambda\lambda'}$$

$$[d(\vec{p},\lambda), d^+(\vec{p}\,',\lambda')]_\pm = p_o\ \delta^3(\vec{p}-\vec{p}\,')\delta_{\lambda\lambda'}$$

with all other commutators vanishing. The choice of

commutation or anticommutation rules we shall discuss later. As far as the statistics is concerned commutation rules will as usual yield bosons and anticommutation rules fermions. At this stage there is no relation between spin and statistics, however.

We now define the free field $\Psi(x,z)$ by

$$\Psi(x,z) = \frac{1}{\sqrt{2(2\pi)^3}} \sum_\lambda \int_{p_0=\sqrt{m^2+\vec{p}^2}} \frac{d^3p}{p_0}\{e^{ipx} \frac{S^j_{\lambda n}(\vec{p};\phi,\theta,\psi)}{(pk/m)^{\alpha+i\beta}} c^+(\vec{p},\lambda) +$$

$$+ e^{-ipx} \frac{S^{j*}_{\lambda -n}(\vec{p};\phi,\theta,\psi)}{(pk/m)^{\alpha+i\beta}} d(\vec{p},\lambda)\}$$

and similarly

$$\Psi^+(x,z) = \frac{1}{\sqrt{2(2\pi)^3}} \sum_\lambda \int_{p_0=\sqrt{\vec{p}^2+m^2}} \frac{d^3p}{p_0}\{e^{-ipx} \frac{S^{j*}_{\lambda n}(\vec{p};\phi,\theta,\psi)}{(pk/m)^{\alpha-i\beta}} c(\vec{p},\lambda) +$$

$$+ e^{ipx} \frac{S^j_{\lambda -n}(\vec{p};\phi,\theta,\psi)}{(pk/m)^{\alpha-i\beta}} d^+(\vec{p},\lambda)\} .$$

Alternatively we may write these fields

$$\Phi_{n'}(g) = \frac{1}{\sqrt{2(2\pi)^3}} \sum_\lambda \int \frac{d^3p}{p_0}\{U^{m,j}_{\vec{p},\lambda;n'}(g) c^+(\vec{p},\lambda) +$$

$$+ U^{m,j\,*}_{\vec{p},\lambda;n'}(g) d(\vec{p},\lambda)\}$$

$$\Phi^+_{n'}(g) = \frac{1}{\sqrt{2(2\pi)^3}} \sum_\lambda \int \frac{d^3p}{p_0}\{U^{m,j\,*}_{\vec{p},\lambda;n'}(g) c(\vec{p},\lambda) +$$

$$+ U^{m,j}_{\vec{p},\lambda;n'}(g) d^+(\vec{p},\lambda)\} .$$

It is of ommediate interest to study the commutation relations of these fields. We define

$$\Delta_\pm(x;z,z') = -i[\Psi(x,z),\ \Psi^+(o,z')]_\pm$$

$$\Delta_\pm(g_1,g_2) = -i[\Phi_n(g_1),\ \Phi_n^+(g_2)]_\pm\ .$$

It is straightforward to compute these commutator functions. One finds

$$[\Phi_n(g_1),\ \Phi_n^+(g_2)]_\pm = \frac{1}{(2\pi)^3}\{U_{nn}^{m,j}(g_1^{-1}g_2)\ \pm\ U_{nn}^{m,j}(g_2^{-1}g_1)\}$$

or for the special case previously considered

$$[\Psi(x,z),\Psi^+(o,z')]_\pm = \frac{1}{2(2\pi)^3}\int\frac{d^3p}{p_o}\{\exp(-ipx)\pm\exp(ipx)\}F(p)$$

where

$$F(p) \equiv \sum_\lambda \frac{S_{\lambda n}^j(\vec{p};\phi,\theta,\psi)\ S_{\lambda n}^{j*}(\vec{p}\ ;\phi,\theta,\psi)}{(pk/m)^{\alpha+i\beta}\ \ (pk'/m)^{\alpha-i\beta}}\ .$$

If we first consider the special case one can show that the numerator of F(p) is an even function of p. Hence, if F(p) will itself be an even or odd function of p if $\alpha=n$, respectively $\alpha=n+1/2$, where n is an arbitrary integer. For general α F(p) is neither even nor odd. If we restrict ourselves to α integer or half-integer we find

$$\Delta_-(x;z,z') = F(i\frac{\partial}{\partial x})\Delta(x) \qquad \alpha = n$$

$$\Delta_+(x;z,z') = F(i\frac{\partial}{\partial x})\Delta(x) \qquad \alpha = n + \frac{1}{2}$$

where

$\Delta(x)$ is the usual causal function

$$\Delta(x) = \frac{-i}{2(2\pi)^3}\int_{p_o=\sqrt{m^2+\vec{p}^2}}\frac{d^3p}{p_o}\{\exp(-ipx)-\exp(ipx)\}$$

with support only inside the light cone. The following observations may now be made

(i) if $\alpha=-j$, where j is integer or half-integer, F(p) is a finite polynomial of degree 2j.

(ii) for general α F(p) is not a polynomial and a power expansion does not terminate.

The first case is of interest as one may show that it is equivalent to a description in terms of a regular Wigner realization (conventional field theory). For that case we obtain causal commutator functions provided we choose the right spin-statistics combination (this is the essence of the spin-statistics theorem) since a finite derivative of a causal function is itself a causal function. However, for $\alpha \neq -j$ we will in general get a violation of micro-causality. We obtain in this sense a non-local theory. One can show, that if we integrate the commutator function over the internal variables then we obtain a function which decreases exponentially as we penetrate the light-cone into the space-like region [3].

For the general case G. Fuchs [4] has investigated the causality properties of the commutator functions. From the expression given above one easily establishes

(i) $\Delta_\pm(g,s,g_2 s)=\Delta_\pm(g_1,g_2)$ for any $s \varepsilon S$ since only diagonal elements enter the commutator functions

(ii) $\Delta_\pm(g'g_1,g'g_2)=\Delta_\pm(g_1,g_2)$ for any $g \varepsilon \bar{P}$.

The latter property, invariance under left translation, corresponds to Lorentz invariance for the commutator function. It is now clear, that choosing the case of a commutator (the boson case), one obtains a vanishing result for points related by

$$g_1^{-1} g_2 = s_1 g_2^{-1} g_1 s \quad \text{for any } s_1, s_2 \;\varepsilon\; S \;.$$

For the fermion case we first note that $U^{mj}(\alpha g)=(-1)^{2j}U(g)$ where α is the element of SL(2,C) which in the 2×2 matrix representation takes the form $\begin{pmatrix}-1 & 0\\ 0 & -1\end{pmatrix}$. Using this we find that the anticommutator vanishes for points which satisfy

$$g_1^{-1}g_2 = \alpha\, s_1 g_2^{-1} g_1 s_2 \quad \text{for any } s_1,\, s_2 \in S\ .$$

One may now look for solutions of these equations, in particular for space-like points in Minkowski space. Fuchs has proved the following results

(i) for $\gamma\equiv g_1^{-1}g_2\equiv(x,\Lambda)$ such that $\Lambda=\underline{1}$ and $S\neq SL(2,C)$ there exist space-like x for which the equations have no solutions, and, hence, the commutator functions will not vanish by this mechanism

(ii) for any space-like x and any S there exist a sub-manifold of SL(2,C)/S of lower dimension than SL(2,C)/S itself for which there are solutions to the equations above.

From these results, which only partially describe the support properties of the commutator functions, we learn that by suitable choice of the "internal" variables we may always make the appropriate commutator function vanish for space-like distances. However, in general one cannot by this mechanism hope for complete micro-causality.

Finally, there are clearly other mechanisms whereby the commutator functions may vanish outside the light-cone. We have already seen, that by choosing $\alpha+i\beta=-j$ in our special case we do get micro-causality. To what extent one would obtain the same result for other choices of quantum numbers or whether one may obtain micro-causality for other choices of homogeneous spaces is not yet known and the field is still subject for investigations.

8. THE FREE FIELD PROPAGATOR FUNCTION

We next consider the propagation properties of the generalized fields. To this end let us compute the conventional free field propagator Δ_F, where

$$\Delta_F(x;z,z') = 2\langle 0|T\{\Psi(x,z)\Psi^+(0,z')\}|0\rangle .$$

Depending on the choice of statistics one finds

$$\Delta_F(x;z,z') = \frac{1}{(2\pi)^3}\int \frac{d^3p}{p_o} e^{-ipx} \sum_\lambda \frac{S^j_{\lambda n}(\vec{p};\phi,\theta,\psi)\; S^{j*}_{\lambda n}(\vec{p};\phi',\theta',\psi')}{(\frac{pk}{m})^{\alpha+i\beta}\;(\frac{pk'}{m})^{\alpha-i\beta}} ; x_o>0$$

$$= \mp\frac{1}{(2\pi)^3}\int \frac{d^3p}{p_o} e^{-ipx} \sum_\lambda \frac{S^j_{\lambda n}(\vec{p};\phi,\theta,\psi)\; S^{j*}_{\lambda n}(\vec{p};\phi',\theta',\psi')}{(\frac{pk}{m})^{\alpha+i\beta}\;(\frac{pk'}{m})^{\alpha-i\beta}} ; x_o<0$$

where $p_o=\sqrt{|\vec{p}|^2+m^2}$. Consider for simplicity the case j=o. Then one may write Δ_F

$$\Delta_F(x;z,z') = \frac{i}{(2\pi)^4}\int \frac{d^3p}{p_o} e^{ipx} \int_{-\infty}^{+\infty} dq_o\, e^{-iq_o x_o} \left\{ \frac{f(pk)f^*(pk')}{q_o-p_o+i\varepsilon} \pm \frac{f(\tilde{p}k)f^*(\tilde{p}k')}{q_o+p_o-i\varepsilon} \right\}$$

with $\tilde{p}\equiv(p_o,-\vec{p})$ and $f(pk)=(pk/m)^{-(\alpha+i\beta)}$. Under a Lorentz transformation x,k and k' transform as fourvectors. The action of Λ on x may be transferred onto the integration variables $(q_o,\vec{p})$. It is then seen that Δ_F is not a Lorentz invariant function in general. However, for the special case $\alpha+i\beta=-j$ for which $f\equiv1$ the invariance is maintained as one would expect. The trouble for the more general situation where $\alpha+i\beta\neq-j$ can be traced back to the lack of microcausality. One easily verifies that products of the form $\Theta(x_o)\Delta(x,z)$ are invariant only if the support of $\Delta(x,z)$ in Minkowski space lies entirely within the light cone. Hence the concept of a T-ordered product is not relativistically invariant for the non-local fields. One may try various remedies to recover Lorentz invariance. For one thing we may always redefine the T-ordering for equal times [5]. For non-local theories more extensive changes in the definition of a T-product may be necessary. For the model considered here (j=o) we suggest

$$\tilde{\Delta}_F(x;z,z') \equiv 2\langle o|\tilde{T}\{\Psi(x,z)\Psi^+(o,z')\}|o\rangle$$

where

$$\tilde{\Delta}_F(x;z,z') = \frac{i}{(2\pi)^4}\int d^4q\, \frac{\exp[-iqx]}{q^2-m^2+i\varepsilon}\, f(qk)\, f^*(qk') \; .$$

This function $\tilde{\Delta}_F$ is manifestly Lorentz invariant. Depending on the analytic structure of the function f special prescriptions may be necessary with regard to the path of integration in the q_o-plane. The implications of this choice of propagator function is presently under investigation.

9. A SIMPLE MODEL

To demonstrate some of the new features that a non-local field theory of the type suggested may exhibit we shall examine a simple model of two kinds of uncharged spinless bosons (mesons) M and m with a trilinear interaction [6]. The fields describing the mesons are given by

$$\Psi_M(x,k) = \frac{1}{\sqrt{2(2\pi)^3}} \int \frac{d^3p}{p_o}\{e^{ipx} f_1(pk)a^+(\vec{p})+e^{-ipx} f_1^*(pk)a(\vec{p})\}$$

$$\Psi_m(x,k) = \frac{1}{\sqrt{2(2\pi)^3}} \int \frac{d^3p}{p_o}\{e^{ipx} f_2(pk)b^+(\vec{p})+e^{-ipx} f_2^*(pk)b(\vec{p})\}$$

with

$$f_1(pk) = (\frac{pk}{m})^{-(\alpha_1+i\beta_1)}$$

$$f_2(pk) = (\frac{pk}{m})^{-(\alpha_2+i\beta_2)} .$$

We shall take the interaction hamiltonian density H(x) to be

$$H(x) = \int d\mu(s,\phi,\theta,\psi)\Psi_M(x,k)\Psi_M(x,k)\Psi_m(x,k)$$

where $d\mu$ is the invariant measure on the internal space

$$d\mu \doteq e^{2s}\, ds\, d(\cos\theta)d\phi\, d\psi .$$

In the usual fashion the S-matrix may be written

$$S = \tilde{T}\{\exp[-ig \int d^4x\, H(x)]\}$$

where $\tilde{T}$ is the modified time-ordering operator.

Consider now the scattering of M-mesons, that is the process

$$M + M \rightarrow M + M .$$

In lowest order perturbation theory the process takes place through the exchange of an m-meson as described by the diagram.

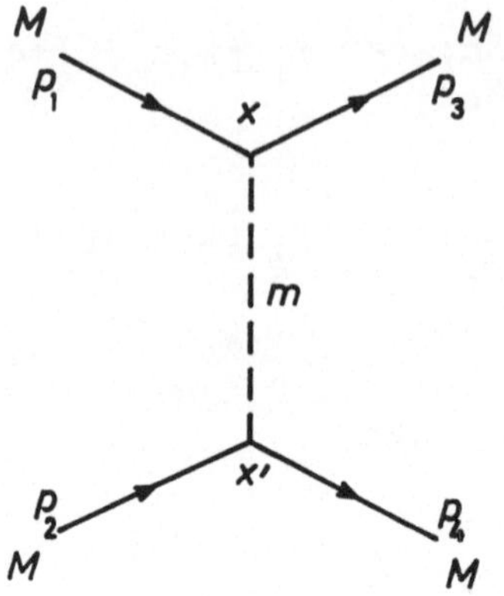

Straightforward application of the model above yields the following expression for the S-matrix element, or rather the T-matrix element

$$T_{fi}=\text{const} \times g^2\int d^4x\ d\ x'\ d\mu\ d\mu'\ \exp[i(p_3-p_1)x]\exp[ip_4-p_2)x'] \times$$

$$\times\ f_1^*(p_1k)f_1(p_3k)\tfrac{1}{2}\tilde{\Delta}_F(x-x';k,k')f_1^*(p_2k')f_1(p_4k') =$$

$$= \text{const} \times g^2\int d^4q\ d\mu\ d\mu'\ \delta^4(p_1-p_3-q)\delta^4(p_2-p_4+q) \times$$

$$\times \frac{f_1^*(p_1k)f_3(p_3k)f_2(qk)f_2^*(qk')f_1^*(p_2k')f_1(p_4k')}{(q^2-m^2+i\varepsilon)} .$$

If we compare this with the result one obtains in conventional field theory it is seen that we have the usual result exept that we have a form factor at each vertex of the type

$$F(p_1,p_3,q) \equiv g\int d\mu\ f_1^*(p_1k)f_1(p_3k)f_2(qk) .$$

The form factor is Lorentz invariant and taking into consideration the δ-functions we see that it only depends on the invariant momentum transfer variable $t=(p_1-p_2)^2$.

To simplify our calculations further we take the M-particles to be conventional scalar mesons so that $f_1=1$. Furthermore, we choose $\alpha=2$ and $\beta=0$ for the m-mesons. Introducing the explicit form for f_2 under these assumption we obtain

$$F(t) = g \int ds\, d(\cos\theta)\, d\phi\, d\psi\, e^{2s} \frac{1}{(\frac{qk}{m})^2} \,.$$

The integral can easily be evaluated with the result

$$F(t) = \text{const.} \times Gm^2 \cdot \frac{1}{t}$$

where G is a renormalized coupling constant incorporating the infinity from the s-integration.

We shall not pursue these model calculations further but make the following observations

(i) the form factors fall off for large momentum transfers as we would like them to do

(ii) the presence of the formfactors enhances the convergence properties of the perturbation expansion and a judicious choice of the quantum numbers α and β may in fact remove the ultraviolet divergences alltogether

(iii) the analytic structure of the form factors one obtains in generalized field theory is quite different from the conventional theory requiring special attention with regard to crossing etc.

1o. CONCLUSIONS

In these lectures I have outlined some of the properties that field theories on homogeneous spaces of the Poincaré group have. We have examined more in detail one specific choice of a homogeneous space. This choice was based on physical considerations but it is, of course, not unique. One may approach these models in different ways. For one thing they provide an opportunity to examine non-local field theory within a frame work that can easily be made rigorous and where the amount of arbitrariness is limited in my opinion. There are indications that non-local field theory may prove to be more adequate for hadron physics than conventional local field theory. In a more ambitious approach to these models one may try to find applications to real physical processes, that is, to find a consistent set of quantum numbers describing the hadrons within this framework. On both levels much work remains to be done but the results so far are encouraging.

REFERENCES

1. E. P. Wigner, Annals of Math., 40, No. 1 (1939).
2. F. Lurcat, Physics 1, 2 (1964) and 1, 95 (1964); A. Kihlberg, Arkiv Fysik 28, 121 (1964) and Nuovo Cimento 53A, 592 (1968).
3. L. Brink, Institute of Theoretical Physics (Göteborg) report .
4. G. Fuchs. Thesis, Faculté des Sciences d'Orsay, 1969, and Ecole Polytechnique preprint, March 1970.
5. K. Hepp, Théorie de la renormalisation, chapter VI, Springer-Verlag, 1969.
6. A. Kihlberg, Institute of Theoretical Physics (Göteborg) report 70-33.

Acta Physica Austriaca, Suppl. VIII, 177—226 (1971)

AUTOMORPHISM GROUPS OF C*-ALGEBRAS, FELL BUNDLES, W*-BIGEBRAS, AND THE DESCRIPTION OF INTERNAL SYMMETRIES IN ALGEBRAIC QUANTUM THEORY*

BY

M.E. MAYER**

Institut des Hautes Etudes Scientifiques
91-Bures-sur-Yvette, France

and

University of California, Irvine, California 92664***

INTRODUCTION

There is relatively little justification for including lectures on the algebraic approach to field theory, particularly lectures devoted to some mathematical aspects, in a course dedicated mainly to hadron physics. My only defense is the fact that I too will be using the word duality in at least two different meanings, and certainly in a sense which has nothing at all to do with the duality concept now fashionable in high energy physics.

---

*Lecture given at X. Internationale Universitätswochen für Kernphysik, Schladming, March 1 - March 13, 1971.

**Based in part on work partially supported by the U.S. National Science Foundation. The hospitality of the IHES is gratefully acknowledged.

***Permanent address. On sabbatical leave 1970/1971.

The concepts of duality which I will mention will be: the duality between groups and certain objects formed from their representations (e.g. the Pontryagin-van Kampen duality theorem for abelian groups, in particular the duality between a compact abelian group -- the gauge group -- and a countable discrete abelian group -- the group of "charges", the Tannaka-Krein duality theorem, which associates to a compact, nonabelian group a Hopf algebra [23], or a Krein algebra [26], and the relatively recent generalizations to locally compact groups and C^*-algebras, by Tatsuuma, Takesaki and others (cf. e.g. [24,25,27] and literature quoted there), the duality discovered by Araki and Dell'Antonio [9] for the von Neumann algebras associated to free fields (cf. eq. (1.5) infra) and which plays an important role in the reconstruction of fields from observables by Doplicher, Haag and Roberts [10], and possibly other concepts of duality. In order to distinguish these homonyms, I shall attach rather arbitrary labels, calling the group-theoretic duality K^2T^3-duality, the field-theoretic duality A-D'A-duality etc.

The purpose of this lecture is in fact twofold: 1° to provide a simple introduction to the mathematical techniques used to describe automorphisms of the algebras associated to infinite quantum systems, and 2° to propose a new way of looking at internal symmetries (or inexact symmetries, since nonabelian internal symmetries seem to be of necessity inexact symmetries) by means of W^*-bigebras (Hopf-von Neumann algebras). The latter concept has only recently (1966-1967) gained some popularity among mathematicians, and will certainly lead to new insights in harmonic analysis, even if it turns out not to be the right tool for us. What W^*-bigebras (or related objects) promise to do for us as physicists is the following:

instead of starting out from a group acting on the states of the system, group which we know from the beginning not to be really a symmetry group, we start out from a collection of "representations of the non-group" forming almost the dual object of a compact group (heuristically we are still guided by a perturbation approach, successively breaking symmetries, until we are left with the Poincaré group and various gauge-groups generating exactly conserved charges). We wish to experiment with such W*-bigebras, abandoning various requirements which guarantee that we have the dual of a compact group, and thus try to infer the symmetries from the algebras of observables.

This program is still in a very initial stage, but encouraged by the success in the abelian case [10], it seems worthwhile pursuing further.

We start with a brief description of the "physical" part, i.e. the concpets of algebra of observables, states, and their representation. This is followed by a discussion of transformations of states and the action of the transformation group on the observable algebra.

The second chapter is devoted to the development (of necessity very brief) of the concept of covariant system, its generalization to Fell-bundles, representations of generalized group algebras, and a discussion of W*-bigebras and the group-theoretical duality theorems. Proofs are mostly omitted, or only sketched.

Chapter III deals with the physical applications, sketching the concept of field algebra, and introducing the *-bigebraic description of internal symmetry.

An appendix collects all the necessary facts from the theory of C*-algebras and von Neumann algebras, for details we refer the reader to Dixmier's fundamental books [5,6].

The bibliography at the end is rather sketchy, and neither in alphabetical, nor in chronological order. I apologize to any author whom I may have omitted, or quoted insufficiently. This is mainly due to the necessity of a rapid submission of the manuscript.

ACKNOWLEDGEMENTS

I would like to take this opportunity to thank L. Motchane and L. Michel for the kind hospitality extended to me at the Institut des Hautes Etudes Scientifiques and to members and visitors of that Institute for fruitful discussions.

I would also like to thank P. Urban for giving me the opportunity to come once again to Schladming.

Finally I would like to thank all those who made my stay at the IHES and at Schladming fruitful and enjoyable.

1. PHYSICAL MOTIVATION. OBSERVABLES, STATES AND THEIR TRANSFORMATIONS

1.1 Description of Observables by C^*-Algebras, Expectations by States

There is almost no need today to motivate the use of a more abstract description of physical entities. Apart from the "pedagogical" advantages, which are perhaps best illustrated in the case of quantum mechanics by the passage from the Schrödinger equation to the operator formulation, transformation theory and ultimately von Neumann's Hilbert space formulation, one sometimes even gains in simplicity of presentation. Thus the occurrence of inequivalent representations of the canonical commutation-anticommutation relations, and the related Haag theorem on the interaction picture, gain considerably in simplicity when one passes to a C*-algebraic formulation; the presente of superselection rules in the "concrete" Hilbert-space formulation of quantum theory, becomes somewhat simpler in the abstract approach -- indeed, under some reasonable additional assumptions, superselection sectors, "unobservable fields" and the Bose-Fermi character of their commutation at spacelike distances can even be derived from a more or less pure algebraic formulation.

Another area where the abstract C*-algebra approach [1] has turned out to be quite useful -- even more useful than in field theory -- is the statistical mechanics of infinite systems [2].

Here the C*-algebra approach allows one to discuss directly infinite systems rather than formulating everything in a box and going to the "thermodynamic limit", it turns out that locally normal states turn out to be what one calls in statistical mechanics normal: they correspond to finite particle densities with well defined entropy densities, and the representations of the C*-algebra are locally Fock representations. Furthermore there are even models for symmetry-breakdown, where the interactions have a higher symmetry than the states involved in the desintegration of equilibrium states. Further, the C*-algebra approach allows one to treat (up to a certain point) classical and quantum statistical mechanics in a unified manner; indeed, the Gel'fand theorem for commutative C*-algebras shows us immediately that in this case any state is represented by a probability measure, and one thus returns to the Gibbs formulation.

Last and not least we should remember that the only nontrivial rigorously solved models of quantum field theory (I have in view the work of Glimme, Jaffe and of Segal and coworkers on the two-dimensional fields theories of polynomial and Yukawa interactions) show that these fields yield indeed representations of the C*-algebras satisfying the Haag-Kastler axioms [3].

We describe here briefly, and incompletely, the C*-algebraic formulation of the quantum theory of infinite systems (we will discuss mainly relativistic field theory, but by replacing Minkowski space by R^3, the Poincaré group by the Euclidian group, and adding some requirement on states to be equilibrium states -- I like best the imposition of the Kubo-Martin-Schwinger boundary condition, cf. e.g. Takesaki [4] -- one can get a formulation of statistical mechanics, cf. e.g. Ruelle [2]).

Before going further, let me make one remark, which may not be acceptable to all workers in this field: there is no absolutely compelling reason to postulate that the observables are elements of a C*-algebra (rather than of a concrete subalgebra of $B(H)$, e.g. a von Neumann algebra of operators, or perhaps a more general "algebra of unbounded operators" (Powers, oral communication). The choice of C*-algebras, is mostly a matter of convenience: most of the assertions which are true algebraically, are automatically true topologically, and of course, one should not forget the existence of Dixmier's excellent books on the subject [5].

If we accept that any self-adjoint operator in a Hilbert space is an observable, it presents no difficulty to consider as observables, abstractly, the hermitian elements of a C*-algebra. The restriction to bounded objects is of no consequence (even for unbounded operators, one actually "measures" only bounded spectral projections), addition and multiplication without restrictions may however give rise to objections, which we sweep under the rug for the time being. The step from hermitian elements to the full C*-algebra also may need some justification. However, since the use of the field of complex numbers is, to say the least, convenient, and since the Gel'fand-Naimark condition, imposed on hermitian elements, almost directly leads to the fact that the algebra is a C*-algebra (cf. e.g. Behncke [6]), we will assume that the observables in a region O of Minkowski space are described by the hermitian elements of a C*-algebra $A(O)$.

Two conditions on the net of C*-algebras $A(O)$ impose themselves immediately, namely:

i) Isotony: if $O_1 \subseteq O_2$ then $A(O_1) \subseteq A(O_2)$ (we avoid here the delicate question whether one may take the space-time open set O to be arbitrarily small!);

We shall return below to the effects of Poincaré transformations of the regions O on the algebras $A(O)$, but we already mention that there will have to be an isomorphism of the algebras associated to regions which are obtained from one another by Poincaré transformations. This, together with the isotony requirement, make it logical to consider the union of the net of all local algebras $A(O)$, and, since we might want to take limits, to consider its closure, the quasilocal algebra

$$\mathrm{A} = \bigcup_{O} A(O). \tag{1.1}$$

Observables in any of the $A(O)$ will be called local observables, quantities in A will be called quasilocal quantities. Global quantities are in the W*-algebra generated by a representation of A.

ii) Locality: if O_1 and O_2 are space-like separated (this is denoted by $O_1 \subset O_2'$ or $O_2 \subset O_1'$) the algebras associated to these regions commute: $A(O_1) \subset A(O_2)^c$. Here $A(O_2)^c$ denotes the commutant of $A(O_2)$ in A, $A^c = A' \cap A$.

The observations carried out on observables yield their expectations, i.e. numerical functions of the observables, which will have to have a few obvious properties:

i) linearity (at least over the reals -- it then extends to the complex field);

ii) positivity (i.e. the expectation value of a positive observable, of the form a^*a, must be nonnegative);

iii) normalization, i.e. the expectation of the identity observable must be one. This leads us immediately to identify expectations with states on the algebra A. We note that since we are dealing with C*algebras, there is no need to even assume continuity, since any state is automatically continuous on a C*-algebra.

A positive linear functional (in particular, a state) is pure if it cannot be written as a convex linear combination of other positive functionals (unless they are proportional to the given functional), i.e. f is pure if

$$f = \lambda f_1 + (1-\lambda) f_2,\ 0 \le \lambda \le 1 \quad \text{implies} \quad f_1 = \alpha f,\ f_2 = \beta f.$$

If we realize the C*-algebra concretely as an algebra of operators on a Hilbert space, i.e. we consider a representation of the abstract C*algebra, the states are (locally) given by traces of density matrices (i.e. normal states of $\pi(A)''$), in particular, the pure states are given by vector states, i.e. usual expectation values in a state described by a normalized vector. Finally, for the commutative case, the states are given by integrals with respect to probability measures, if the functions representing the elements of the algebra, i.e. classical expectation values.

A given C*-algebra may, however, have many inequivalent representations (e.g. G-N-S representations generated by different states, only one of which satisfies the requirement to be the physical vacuum). This might seem to be an argument against the use of abstract algebras, but is in fact the opposite. As has been pointed out by Haag and Kastler [3], it is physically impossible to distinguish two inequivalent representations (as long as they are faithful) of the same algebra, since we are in fact capable only to carry out a finite number of measurements on a given system, i.e. to verify the equality of the numbers given by applying states in the two representations to the operators representing a finite number of observables. It turns out that this criterion for physical equivalence of representations coincides with the concept of weak equivalence proposed by Fell. Two representations $\pi_1(A)$ and $\pi_2(A)$ of the C*-algebra A are called weakly equivalent if the kernels (i.e. the set of elements of A mapped into the zero operator) of the two representations coincide. In particular, if the representations are faithful (proper, nondegenerate), i.e. have no kernel, two representations of the same C*-algebra will always be weakly, hence physically equivalent, i.e. will yield the same expectation values on any finite number of elements of the algebra [3].

1.2 Symmetries and Automorphism Groups. Heuristics.

The analysis of the action of symmetries on the algebra of observables and its states can be carried out in analogy with the analysis of the action of symmetries in Hilbert-space quantum mechanics. One is led to an analogue of Wigner's theorem, which states that any symmetry in the Hilbert-space formulation is implemented by a unitary or antiunitary operator [7]. In the abstract form, the theorem will have the following form: any symmetry group of the physical system described by the C^*-algebra is represented by a subgroup of the automorphism group of the C^*-algebra A (automorphism here always means isometric *-automorphism). If the symmetry group is a topological group, the representation is a strongly continuous function fo the group elements.

We will give here an approximate derivation of this result (which seems to be due to Kadison [8]) by taking the "passive" point of view, i.e. by subjecting the states to symmetry transformations (which have to be at least weak-*-continuous) and then showing that this implies strong continuity of the automorphisms implementing to symmetry, in the C^*-algebra. This kind of reasoning is particularly well suited to the transformations describing "kinematical" degrees of freedom, like Poincaré transformations. On the other hand, so-called internal symmetries, as well as symmetries associated to superselection rules, are of a slightly different nature. They cannot be easily interpreted as transformations of states, and often manifest themselves by the occurrence of certain multiplicities in the representations of the algebras. In the next chapter we will then analyze the covariant systems which seem adequate for the description of such symmetries, and will propose a somewhat unorthodox point of view, allowing one to discuss broken symmetries starting from their representations.

Let us then assume that we have a symmetry transformation of the "measuring device" associated to the state ω, corresponding to a "Schrödinger picture", if the transformation in question is the time-evolution of the system. The set of states $E(A)$ of the C^*-algebra A will be considered with the weak*-topology, i.e. the weakest topology which makes the mappings $\omega \to \omega(a)$ continuous for all $a \in A$. In this topology the set of states $E(A)$ is a convex compact cone, its extremal points being the pure states. If we subject the "apparatus" measuring the state ω to a transformation from the group G (which will always be assumed to be locally compact and separable, or even simpler, a Lie group), the requirement that the state be mapped into a state shows us that in $E(A)$ the group element g must be implemented by an affine transformation α_g (i.e. a map of states, taking convex linear combinations into convex linear combinations), and since we are able to "verify" the invariance only on a finite number of elements of the C^*-algebra, that the mapping $\alpha_g : \omega \to \alpha_g(\omega)$ depends weak* continuously on g; further $\alpha_{g_1 g_2} = \alpha_{g_1} \circ \alpha_{g_2}$.

Once this is accepted, it is a technical matter to prove that a group of weak*-continuous affine transformations of the states (or of a sufficiently large subset of the set of all states, e.g. the vector states of a separating family of irreducible representations, or factor representations of the algebra) induces in the C^*-algebra a strongly continuous group of automorphisms (for the details or the proof, see e.g. Kadison, loc.cit.). The strong continuity of the action of the group G on A means that to each $g \in G$ there exists an automorphism τ_g (the transposed of τ_g is the mapping $\alpha_g{-1}$ in $E(A)$, the g^{-1} being necessary if we want to preserve the order of operations) of A, such that for every $a \in A$ and $\varepsilon > 0$, there is a neighborhood N of the identity $a \in G$, such that for

$g \in N \| \tau_g a - a \| < \varepsilon$.

Thus, the requirement of Poincaré invariance of the theory can be formulated as follows. For any element $L=(\Lambda,a)$ of the Poincaré group (since we are dealing with observables, which are supposedly single-valued, there is no need to go to the covering group) there is an automorphism, τ_L of the quasilocal algebra A such that on a local subalgebra

$$\tau_L A(O) = A(LO). \tag{1.2}$$

In any representation π of the algebra A there is a strongly continuous cyclic representation $L \rightarrow U(L)$ by unitary operators, implementing L

$$\pi(\tau_L a) = U(L)\pi(a)U(L)^{-1} \tag{1.3}$$

and leaving the cyclic vector Ω of the representation invariant. The state ω_o corresponding to this vector is the vacuum state

$$\omega_o(a) = \langle \pi(a)\Omega | \Omega \rangle . \tag{1.4}$$

In addition, in each representation the spectral condition must be satisfied, i.e. the spectrum of the operator implementing the 4-translations must be situated inside the forward light-cone, and the vertex of the cone is an isolated point of the spectrum, with eigenvector Ω.

In addition to the representation $\pi(A)$ of the algebra A, or of any local algebra $A(O)$ it will be important to consider also the weak closures of these representations, i.e. the von-Neumann algebras generated by these representations. In particular, the local algebras associated to closed double cones (a closed double cone is the closed set in Minkowski space having non-empty interior, and being the intersection of a closed forward light-cone with a closed backward light-cone) have particularly simple properties, at least in the case of free fields. A double cone coincides with its "second causal complement", i.e. if we denote the set of points space like to O by O', a double cone has the property $O''=O$. It has been shown by Araki, Dell'Antonio, and others [9], that in the case of free fields, the representations of the local algebras, or more precisely, their associated von Neumann algebras, satisfy the duality requirement.

$$\pi(A(O'))'' = \pi(A(O))' \tag{1.5}$$

(the extra primes on both side guarantee that we are actually dealing with von Neumann algebras). This requirement is a strengthened form of locality, and has played an important role in the recent discussion of superselection sectors by Doplicher, Haag and Roberts [10]. We shall return later to a discussion of its validity or violation in relation to symmetries.

In the next section we analyze the mathematical structures given by a strongly continuous action of a group G on a C*-algebra.

2. AUTOMORPHISM GROUPS, COVARIANT (HEISENBERG) SYSTEMS, FELL BUNDLES AND W*-BIGEBRAS

2.1 Covariant Representations

Let A be a C*-algebra (for simplicity with identity, which is mostly the case in physics, but not for the L_1-algebras of locally compact groups) and let G be a locally compact group "acting" by automorphisms on A. This means that there is a homomorphism τ of G into the automorphism group of A (automorphism always means isometric $*$-automorphism) such that the function

$$\tau : G\times A \to A : (g,x) \to \tau_g(x) \tag{2.1}$$

is strongly (weakly) continuous as a function of g for any $a\varepsilon A$. We will sometimes write briefly $\tau_g(x)=g\,x$. The triplet (A,τ,G) (or briefly, omitting τ, (A,G)) will be called a covariant system or Heisenberg system (the latter term has been used by Dixmier [11]).

Strong continuity means that for every $x\varepsilon A$ and positive ε, there exists a neighborhood N of $e\varepsilon G$ (the identity of the group, represented by the identity automorphism) such that for $g\varepsilon N$, $\|\tau_g(x)-x\|<\varepsilon$.

By transposition, the group G also acts on the dual of the Banach space A, and consequently on the set of states E(A). We denote this transposed action by τ_g^t (this is an "anti-representation", since transposition changes the order):

$$<\omega,\tau_g(x)> = <\tau_g^t(\omega),x> \ \forall\ x\varepsilon A,\ \omega\ \varepsilon\ E(A). \tag{2.2}$$

(It is obvious that τ_g preserves positivity, hence states are transformed into states).

Given the action of τ_g^t on the set of states, not all of the states will make the numerical function $\tau_g^t(\omega)$ norm-continuous. In some cases it will be convenient to restrict one's attention to the convex cone of states which make τ_g^t continuous in the norm-topology on the state space (rather than the weak*-topology, in which the state space is compact for algebras with identity).

The fixed points of the action of G in the set of states are the so-called invariant states: $\tau_g^t\omega=\omega, \forall g\varepsilon G$. Carrying out the Gel-fand-Naimark-Segal construction with an invariant state we are led to a covariant representation* of the covariant system (A,G): let ω be an invariant state of (A,G). Then there exists a unique continuous unitary representation (urep) U(g) of the group G on the Hilbert space H_ω of the G-N-S representation π_ω associated to ω such that

$$\pi_\omega(\tau_g(x)) = U(g)\pi_\omega(x)U(g)^{-1} \tag{2.3}$$

and, if η denotes the canonical mapping of A into H_ω (i.e. the "vector associated to a is $\eta(a)$"), we have

$$\eta(\tau_g(x)) = U(g)\eta(x). \tag{2.4}$$

In particular, the state ω has associated to it the cyclic vector Ω, which is invariant under the representation U(g):

$$U(g)\Omega = \Omega. \tag{2.5}$$

(We all recognize here the situation encountered in field theory, where G is the Poincaré group, or its product with some internal symmetry, Ω is the vacuum state vector $\pi_\omega(A)$ any representation of the algebra of observables).

It is also interesting to consider representations which are not "quite" invariant, the simplest among them being the "quasi-covariant" representations considered by Zeller-Meier [12], Borchers [13] and Guichardet and Kastler [14]. A representation π, U of the covariant system (A,G) is called quasi-invariant if it is quasi-equivalent to a covariant representation, i.e. the von-Neumann algebras generated by these representations are isomorphic, or equivalently, there exist multiples of the two representations which are unitarily equivalent.

A state ω of the algebra A, which is cyclic for a quasi-covariant representation will be called a quasi-covariant state (or quasi-invariant).

---

*Without special mention, all representations to be discussed will be assumed nondegenerate; covariant representations are sometimes called invariant representations because of invariance of the state associated to them.

Until now the states transformed under the action of the group weak-*-continuously. A state $\omega \in E(A)$ is called G-continuous (we denote the subset of G-continuous states by $E_C(A,G)$) if the mapping $g \to \tau_g^t \omega$ is norm-continuous, i.e. $\|\tau_g^t\omega - \omega\|$ tends to zero when g tends to the group identity e. It has been shown in [13,14] that if ω is a normal state on $\pi(A)''$ and U, π is a quasi-covariant representation, then the state $\omega \circ \pi$ is G-continuous. The set $E_c(A,G)$ is convex norm-closed, invariant under τ_g^t, and also under the replacement $\omega \to \omega_a$, where $a \in A$, and $\omega_a(b) = \omega(aba^*)/\omega(a^*a)$.

2.2 Fell-Bundles as Generalized Covariant Systems. Generalized Group Algebras and their Representations

Covariant systems are a special case of a more general equivariant structure based on a locally compact group and a C*-algebra, namely Banach-*-algebraic bundles discussed by Fell [15] (and in special cases by several other authors; for a complete bibliography cf. e.g. the author's review [16]). For simplicity we shall call them Fell-bundles. The motivation for introducing these objects is two fold: first, a generalization of the equivariant structure induced by the group action on the cartesian product $A \times G$ of a covariant system; second, to find objects which in a certain sense contain as limiting cases locally compact groups and C*-algebras, or roughly, two *-Banach algebras, the quotient of which is isomorphic to a group. It turns out that one can carry over a large part of Mackey's theory of induced representations to such objects, and define various generalized group algebras.

Here we give the definition of Fell-bundles for C^*-algebras with identity (the case of algebras without identity, e.g. $L_1(G)$ for a noncompact locally compact, G, is important, but leads to the necessity of introducing the "multiplier algebra, or double centralizer algebra", in order to be able to define unitary elements, cf. [16]). Although not needed, we make the simplifying assumption that G is separable.

Definition. Let G be a locally compact (separable) group with identity e, A a Banach-*-algebra (or C^*-algebra) with identity and let (B,p,G) be a "C^0-vector bundle over G modeled on the Banach space A, with projection p", i.e. B is a Hausdorff space, $p:B\to G$ is an open surjection, and all "fibers" $B_x=p^{-1}(x)$, $x\varepsilon G$ are isometrically isomorphic to A, as Banach spaces. (This means that locally B consists of "products of neighborhoods in G and the Banach-spaces B_x", so that the bundle can be thought of as "glued together" from such products). The group G is assumed to "act" in the bundle space, essentially by combining automorphisms of the algebra A, which is identified with the unit-fiber $A=p^{-1}(e)$, and the Banach-space isomorphisms of fibers. Define a binary operaton $\circ$ in B, and an antilinear unitary operation*, extending to B the product and involution of A, in such a manner that the open surjection p has the following "equivariance" properties:

(i) $p^{-1}(a) = B_e = A$

(ii) $p(s)p(t) = p(s^\circ t)$ (or $B_x B_y \subset B_{xy}$) $s,t \varepsilon B$, $x,y \varepsilon G$

(iii) $p(s^*) = p(s)^{-1}$ (i.e. $B_x^* \subset B_{x^{-1}}$

(iv) each fiber B_x is a Banach-space with norm denoted by $\|\ \|$, and the topology of B relativized to each fiber is the one given by that norm. The operations $s\to\|s\|$ $(B\to \mathbb{R})$, $(s,t)\to s+t$ (in each fiber, i.e.

$p(s)=p(t)$) $(\lambda,s)\to\lambda s$ ($\lambda \in \mathbb{C}$, $s\in B$), and $(s,t)\to s\circ t$; $s, t\in B$, $s\to s^*$ are continuous.

It should be remembered that addition is defined _only within the same fiber_, whereas multiplication is everywhere defined, but the result _does not lie in the same fiber_ as the elements being multiplied, unless they are both in the unit-fiber. That means that B is not quite an algebra (addition is not everywhere defined) but its unit-fiber $B_e=A$ is.

A _cross section_ of B is a map $\sigma:G\to B$ such that $p^o\sigma$ is the identity map on G, i.e. $p(\sigma(x))=x\in G$, and which will always be assumed _continuous_. We say that the cross section _passes through_ s if $s=\sigma(x)$. It will be important to assume or prove that through any $s\in B$ there passes at least one cross section, i.e. that B has enough _corss sections_.

The _trivial case_ is the case when B is the cartesian product $A\times G$. We are then led back to covariant systems (in particular , G may act trivially on A). In that case, cross sections of the bundle are just continuous A-valued functions on G.

Fell managed to prove that the semidirect-product bundle, i.e. the covariant system (A,τ,G) is only a slight specialization, and that the most general Fell-bundle with given unit-fiber-algebra is isomorphic to the one obtained by the following construction (we call it Fell's bundle construction). This construction, for the particular case that A is an algebra of observables and one of the groups involved is a discrete abelian group of charges, is almost identical to the construction of charged superselection sectors by DHR [10].

Consider a Banach-*-algebra A (in particular a C^*-algebra, since we can always take the C^*-envelope of A) with identity $\mathbb{1}$. Let $U(A)$ denote the group of unitary elements in A (i.e. such that $u^*u=uu^*=\mathbb{1}$). It has been shown by Dixmier that this is a polish topological group (complete metric group) in the strong topology. The mapping $a \to uau^*$ is an isometric automorphism of A. Let N be a closed subgroup of $U(A)$ and H a group extension of N, such that H/N is isomorphic to a given locally compact group G -- the group we select as base space for the bundle. Note that the topological group extension problem posed here is by no means trivial (cf. Calvin Moore [17]). However one may consider just abstract extensions, and there may be any solutions to the problem. This means that we have the following exact sequence of group homomorphisms

$$\mathbb{1} \to N \overset{i}{\to} H \overset{j}{\to} G \to e$$

where we identify $N=i(N)$, and $i(N)=\ker j$.

Let there be a homomorphism $\tau: H \to \mathrm{Aut}\, A$ (i.e. a representation of H by automorphisms of A, which on $i(N)=N$ is defined by the "inner" automorphisms $a \to uau^*$, $u \in N$) such that:

(i) $\forall a \in A : h \to \tau_h(a)$ is continuous,

(ii) $\forall u \in N,\ a \in A : \tau_u(a) = uau^*$

(iii) $\forall h \in H,\ u \in N : \tau_h(a) = huh^{-1}$.

The Fell-bundle $(B,p,G,\circ,{}^*,A)$ is then obtained as follows.

Define an equivalence relation $\sim$ on the cartesian product $A \times H$ by setting:

$(a_1,h_1) \sim (a_2,h_2)$ if there is a $u \in N$ such that $a_2=a_1u, h_2=u^{-1}h_1$.

In other words $(a,h) \sim (au,u^{-1}h)$, $u \in N$. If we define the continuous left action of N on $A \times H$ by $u(a,h)=(au^{-1},uh)$, we see that the equivalence class of (a,h) is nothing else but the orbit passing through (a,h) under this action. We denote these equivalence classes by (a,h), and the quotient space under the equivalence by $B=A\times H/\sim$. This is our bundle-space. The bundle projection p is defined as the open surjection

$$p : \underline{(a,h)} \to hN \cong x\in G$$ (identify cosets of N with elements of G).

For fixed $x=hN$, and h the map $a \leftrightarrow \underline{(a,h)}$ is a bijection of $A=B_e$ onto $p^{-1}(x)=B_x$ and we give B_x the Banach-space structure and norm of A via this bijection.

The bundle operations o and * are defined naturally by the "semidirect product" formulas, as operations on orbits of a group action:

$$\underline{(a,h) \circ (b,k) = (a\tau_h(b), hk)},$$

$$\underline{(a,h)^* = (\tau_{h^{-1}}(a^*), h^{-1})}.$$

The only difficult part of the proof is that of continuity, for which we refer to Fell's memoir [15], assuming that the not quite trivial extension problem is solved.

Covariant systems are then obtained as the special case, setting $N=\{1\}$, $H=G$.

Let us sketch briefly the construction of the generalized group algebra $L_1(B,G)$, or its C^*-enveloping algebra $C^*(B,G)$ -- the crossed product (such objects have been studied by Doplicher Kastler and Robinson [18] as covariance algebras, by Takesaki [19], by Leptin [20] as generalized L_1-algebras, and by Behncke [21] and Zeller-Meier [12] as crossed products, in the case of covariant systems).

If the bundle has enough cross sections, one can form the space $K(B,G)$ of cross sections with compact support, and defining as usual the L_p-norms, one can construct the various L_p-completions of that space. In particular, for the cross section f(x), Haar measure dx (left Haar measure, as usual), we define the L_1-norm

$$\|f\|_1 = \int_G \|f(x)\| dx \; .$$

In the space $L_1(B,G)$ we can define a convolution and involution.

We consider the special case of a covariant system (A,τ,G), with the general case following easily from these formulas, by leaving out τ and the cocycle α which can be understood included in the group action.

Let $\alpha(x,y)$ be a mapping from $G\times G$ into the unitaries of the center of A, C_u, where G acts by the same automorphism τ (possibly trivial), and satisfying the usual cocycle identities ($x,y,z\varepsilon G$, $\alpha\ \varepsilon C_u$, $1\varepsilon C_u$):

$$\alpha(e,x) = \alpha(y,e) =$$

$$\tau_x(\alpha(y,z))\alpha(x,yz) = \alpha(xy,z)\alpha(x,y)$$

$$\tau_x(\alpha(x^{-1},x)) = \alpha(x,x^{-1}), \text{ etc.}$$

The coboundaries are of the form

$$\delta\gamma(x,y) = u(x)u(y)u(xy)^{-1} \quad u\varepsilon C_u$$

and two cocycles differing by a coboundary

$$\alpha'(x,y) = \delta\gamma(x,y)\alpha(x,y)$$

are cohomologous, i.e. equivalent.

Then a convolution in $L_1(A,\tau,G,\alpha)$ can be defined by

$$(f \ast g)(x) = \int_G f(y)\cdot\tau_x[g(y^{-1}x)]\alpha(y,y^{-1}x)dy$$

and an involution by

$$[f(x)]^{*} = \Delta(x^{-1})\tau_x[f^{*}(x^{-1})]\alpha(x,x^{-1})$$

where Δ is the modular function of G, and $*$ denotes the involution in A. It is easy to verify that $L_1(A,\tau,G,\alpha)$ thus becomes a $*$-Banach-algebra, and its C^*-completion, to be denoted by $C^*(A,G,\alpha)$ -- a C^*-algebra.

The remarkable thing is, that one can define, as usual, representations for $L_1(A,\tau,G,\alpha)$ or $C^*(A,G,\alpha)$ and that these representations are in one-one correspondence with the covariant α-representations of the covariant system (i.e. α is a multiplier for $U(g)$). (In a certain sense the ones are the integrated form of the others, like for ordinary group algebras). The analogy with ordinary group algebras goes further, one can define positive type functions, induced representations, (this is the reason for considering projective, or α-representations) and even work out a duality theory, as we shall show in the next section.

It can also be shown and will be done, in the Chapter III, on physical applications, that the construction of superselection sectors can be considered as the construction of an appropriate generalized group algebra.

Of independent mathematical interest, with possible physical application, is the study of differentiable group actions in Fell-bundles, study which has been started in [16] and is continuing.

2.3 W^*-Bigebras (Hopf-von Neumann Algebras) and Duality of Groups

In this section we develop the basic elements of the theory of W^*-bigebras (known in the literature under the name of Hopf-von Neumann algebras; the case of C^*-

bigebras can be developed parallel to this, and plays a fundamental role in the duality theory of compact semigroups [22]). We use this concept for a rapid review of duality theory of compact and locally compact groups (due to Tannaka, Krein, Kats, Stinespring, Tatsuuma, Takesaki, cf. [23,24,25,26]; the Krein version involves slightly different algebras called by Hewitt-Ross [26] Krein algebras; it is hard to resist the temptation to denote this duality by the alliterated names of some of these authors: we shall call it K^2T^3-duality). In the next chapter we shall try to make plausible a W*-bigebra approach to the description of internal symmetries.

We first recall the definitions of tensor products for vector spaces, Banach spaces and W*-algebras (for C*-algebras there exist two definitions of tensor products, due to Guichardet [28], and to Takesaki [29] and Wulfsohn [30], which are more complicated, and which we shall avoid for the time being).

2.3.1 Tensor Products

We recall the definition of the (algebraic) tensor product of two vector spaces S, T (over the complex field $\mathbb{C}$; the definition is the same for modules over a ring). Consider the vector space $\mathbb{C}^{(S\times T)}$ of formal linear combinations of elements of the cartesian product $S\times T$ with complex coefficients; this vector space can be thought of in terms of the basis $\{e_i, f_j\}$ of pairs, where $\{e_i\}$ is a basis in S, $\{f_j\}$ a basis in T. This is an abelian group under elementwise addition, which we denote by G. Consider the subgroup H of elements in G on which the bilinear map $S\times T\to G$ vanishes, i.e. the subgroup generated by the elements of the form

$$(s,t+t') - (s,t) - (s,t') \quad (s,t) \in S\times T$$

$$(s+s',t) - (s,t) - (s',t) \quad c \in \mathbb{C}$$

$$(cs, t) - c(s,t), \ (s,ct) - c(s,t).$$

The quotient group $G/H=\mathbb{C}^{(S\times T)}=S \otimes_{\mathbb{C}} T$ is called the tensor product of the spaces S and T over the complex field (since in most cases we consider the complex field, we omit the subscript $\mathbb{C}$ on $\otimes$; we shall use the subscript $\mathbb{R}$ when needed to emphasize that we are dealing with real vector spaces). Another way of defining the tensor product is by considering an additional real vector space V, and a real bilinear map $f:S\times T\to V$, with $f(cs,t)=f(s,ct)$, such that there is a unique vector space isomorphism $f':S \otimes_{\mathbb{C}} T\to V$, with $f'(s \otimes_{\mathbb{C}} T)=f(s,t)$.

If S and T are Banach spaces, any norm on $S\otimes T$ which satisfies

$$\|s\otimes t\| = \|s\| \, \|t\|, \quad s \in S, \ t \in T$$

is called a cross-norm. Completions of $S\otimes T$ under various cross-norms $\| \ \|_j$ are denoted by $S \hat{\otimes}_j T$. In particular, for $x \in S\otimes T$ the greatest cross norm $\| \ \|_g$

$$\|x\|_g = \inf\{ \sum_{i=1}^{n} \|s_i\| \, \|t_i\| : x = \sum_{i=1}^{n} s_i \otimes t_i\}$$

defines the projective tensor product $S \hat{\otimes} T$ (we do not put any subscript on $\hat{\otimes}$). Similarly, the injective tensor product $S \hat{\otimes}_i T$ is defined by completing the algebraic tensor product with the smallest cross-norm

$$\|x\|_s = \sup\{ \sum_{i=1}^{n} \langle f,s_i\rangle\langle g,t_i\rangle : f\in S^*, g\in T^*,$$

$$\|f\| \le 1, \ \|g\| \le 1\}$$

Each cross norm on $S\otimes T$ induces a norm on the dual tensor product $S^*\otimes T^*$, in a natural way. If H and K are Hilbert spaces, the tensor product $H \otimes K$ is uniquely defined, by completing the prehilbert space with the inner product

$$\langle h \otimes k | h' \otimes k' \rangle = \langle h | h' \rangle \langle k | k' \rangle .$$

Now let (M,H) and (N,K) be von Neumann algebras on the Hilbert spaces H, K respectively. Let $X=H \otimes K$ be the tensor product of the Hilbert spaces, and m, n, bounded operators in M, N respectively. The tensor products $m \otimes n$ generate a von Neumann algebra on X called the tensor product of M and N and denoted by $(M,H) \otimes (N,K)$ or simply $M \otimes N$. If the algebra N consists of multiples of the unit operator $N = \mathbb{C}\, 1$, one obtains the algebra $M \otimes I$, called the amplification of M to $H \otimes K$.

It can be proved [4] that the commutant of a tensor product of von Neumann algebras is the tensor product of the commutants:

$$(M \otimes N)' = M' \otimes N' .$$

2.3.2 W*-Bigebras

We can consider the multiplication in a von Neumann algebra A (from now on we abbreviate it W*-algebra) as an operation from A×A to A, i.e. a normal homomorphism m of the W*-algebra A×A into A, such that the following diagram is commutative:

$$\begin{array}{ccc} A \otimes A \otimes A & \xrightarrow{\ m \otimes i\ } & A \otimes A \\ {\scriptstyle i \otimes m}\downarrow & & \downarrow{\scriptstyle m} \\ A \otimes A & \xrightarrow{\ \ m\ \ } & A \end{array}$$

where i is the identity automorphism of A. (Strictly speaking the diagram is unambiguous only if the algebra A is commutative; but we can mentally affix labels to the two copies of A and "preserve order".) The commutative case is obtained when the "multiplication" m commutes with the transposition map $\emptyset : A\otimes A \to A\otimes A : s\otimes t \to t\otimes s$.

This is nothing new, but gives one the idea to try to "dualize" the concept of multiplication, by defining an associative *comultiplication* d as a normal homomorphism of the W*-algebra A into its "tensor square" $A\otimes A : d : A \to A\otimes A$, such that the following diagram is commutative:

$$\begin{array}{ccc} A & \xrightarrow{\quad d \quad} & A\otimes A \\ {\scriptstyle d}\downarrow & & \downarrow{\scriptstyle d\otimes i} \\ A\otimes A & \xrightarrow{\quad i\otimes d \quad} & A\otimes A\otimes A \end{array}$$

where i denotes again the identity automorphism of A. In other words, we have $(d\otimes i)\circ d=(i\otimes d)\circ d$. The notation d has been chosen to remind us that d is essentially a "diagonal map"; thus in the case of function-algebras, (e.g. the algebra A_G which is the von Neumann algebra $L^\infty(G)$ of all essentially bounded functions on a locally compact group) one can actually think of the comultiplication as the diagonal map:

$$d(f)(s,t) = f(st) \quad f \in A_G,\ s,\ t \in G$$

f(st) is in $L^\infty(G\times G)=L^\infty(G)\otimes L^\infty(G)$, and the map is obviously normal.

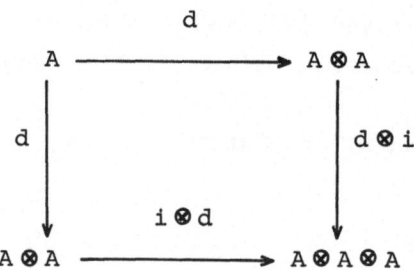

A vector space with such an associative comultiplication is called a cogebra (Hopf algebra) and since A is at the same time a *-algebra (W*-algebra) it corresponds to the standard definition of a bigebra, and we shall call it briefly a W*-bigebra(A,d) (a more current term is Hopf-von Neumann algebra, which is somewhat lengthy; bigebras were also called hyperalgebras, and were studied extensively in Lie-algebraic contexts by Dieudonné and Cartier [31]; the term *-bigebra is becoming more widespread [22]).

If d commutes with the transposition map Ø, the W*-bigebra (A,d) is called cocommutative. If it is both commutative and cocommutative, it is called bicommutative.

A unit in the W*-algebra A can be viewed as a mapping of the number field ($\mathbb{R}$ or $\mathbb{C}$) into the algebra $u=\mathbb{R}\to A$, such that u is the canonical identification of the reals (or complex numbers) with the scalar operators, such that $m\circ(u\otimes i)$ is the identification $\mathbb{R}\otimes A\to A$ (or $\mathbb{C}\otimes A\to A$) and $m\circ(i\otimes u)$ is the identification $A\otimes\mathbb{R}\to A$ (or $A\otimes\mathbb{C}\to A$). Similarly, one can define a counit $c:A\to\mathbb{R}$ (respectively $\mathbb{C}$), such that $(c\otimes i)\circ d=(i\otimes c)\circ d=$identity. In the above example, for compact group G, $c(f)=f(e)$ (we shall see that the existence of a counit is necessary for the W*-bigebra associated to compact groups).

Let (A,d) be a W*-bigebra, and consider the predual A_* of A (this is the set of all ultraweakly continuous linear functionals on A; it is generated by the normal states). Denoting by f,g the elements of A_* (linear functionals, the application of which to elements $x, y\in A$ will be denoted by Ø $\langle f,x\rangle$) we can define a binary operation "convolution" $*$ between these elements, by transposing the comultiplication d to A_*:

$$\langle f*g,x\rangle = \langle f\otimes g, d(x)\rangle \quad x\in A .$$

Note that the right-hand side of this equality is in the dual space of the tensor product $A \otimes A$.
Thus, the predual A_* is transformed into an algebra, and if d is cocommutative, A_* is a commutative (Banach) algebra

A_* can be made into a *-algebra, if the W^*-bigebra A has an involution (or symmetry) (not to be confused with the original *-operation, which is also called an involution) j, i.e. an anti-automorphism (linear, antimultiplicative map of the algebra A into itself) of order 2, i.e. such that

$$j(\alpha x+\beta y) = \alpha j(x)+\beta j(y),\ x,\ y \in A,\ \alpha,\ \beta \in \mathbb{C}$$

$$j(xy) = j(y)j(x)$$

and

$$j \circ j = i, \text{ i.e. } j(j(x)) = x,$$

such that

$$\emptyset \circ d \circ j = (j \otimes j) \circ d.$$

(The last condition means that the two components of the tensor product $A \otimes A$ get "permuted" by the involution.

A W^*-bigebra (A,d) with an involution j will be called an involutive W^*-bigebra (A,d,j).

By dualization to the predual A_*, j defines an involution in the Banach-algebra A_*: denote the restriction to A_* of the transpose ${}^tj$ of j by j_*. Then f^* in A_* is defined by

$$\langle f, jx\rangle = \langle {}^tjf^*, x\rangle = \langle j_*f^*, x\rangle \quad x \in A .$$

The map $f \to j_*(f^*)$ is antilinear, and the properties of involution are easily verified:

i) $j_*(j_*(f^*)^*) = j_*j_*(f^{**}) = f \quad f \in A_*$

ii) $\langle j_*(f*g)^*,\ x\rangle = \langle (f*g)^*,\ j(x)\rangle = \overline{\langle f*g,\ j(x)^*\rangle} =$

$= \langle f*g,\ j(x^*)\rangle = \langle f \otimes g,\ d \circ j(x^*)\rangle =$

$= \langle f \otimes g, \emptyset \circ (j \otimes j) \circ d(x^*) \rangle = \langle g \otimes f, (j \otimes j) \circ d(x^*) \rangle =$

$= \langle j_*(g^*) \otimes j_*(f^*), d(x) \rangle = \langle j_*(g^*) j_*(f^*), x \rangle$.

Following (as we did so far) Takesaki, we denote $j(a) = a^V$, $a \in A$, and $j_*(f) = f^\wedge$ for every $f \in A_*$. (In the case of the group algebra $L^\infty(G)$ $a^V(g) = a(g^{-1})$ -- agreeing with the usual notation).

2.3.3 Invariant Measures and the K^2T^3 Duality Theorems

This subsection contains without proofs statements of the duality theorems for locally compact and compact groups and their associated W*-bigebras. For the proofs we refer to Takesaki's beautiful paper [24]. We must however first introduce a measure on the W*-bigebra, which will distinguish those which are duals of locally compact groups.

We recall that a trace on a W*-algebra A (DN, sec. 6) is a real function ϕ with non-negative values (finite or not) defined on the positive part A^+ of the algebra, which satisfies: i) $\phi(S+T) = \phi(S) + \phi(T)$. $S,T \in A^+$; ii) for $\lambda \geq 0$, $S \in A^+$, $\phi(\lambda S) = \lambda\phi(S)$ (with the convention $0.+\infty = 0$; iii) for $S \in A^+$ and U a unitary in A, $\phi(USU^{-1}) = \phi(S)$. A trace ϕ is faithful if $\phi(S)=0$ implies $S=0$. A trace is finite if $\phi(S) < +\infty$ for all $S \in A^+$. A trace is semi-finite if for all $S \in A^+$ $\phi(S)$ is the sup of the numbers $\phi(T)$ for all $T \in A^+$ such that $T \leq S$, $\phi(T) < +\infty$. A trace is normal if for any increasing net $\{T_i\}_{i \in I} \subset A^+$ with $\sup T_i = S$, $\phi(S) = \sup_{i \in I} \phi(T_i)$. In particular, on $B(H)$ the trace-formula (T a positive operator, e_i an orthonormal base in H)

$$Tr(T) = \sum_{i \in I} \langle Te_i | e_i \rangle$$

is a normal faithful semi-finite trace, which is independent of the choice of orthonormal system.

Traces play a role similar to measures in von Neumann algebras. In particular one can define an L_1-norm by means of a normal, faithful, semifinite trace on a two-sided ideal of a W^*-algebra, by means of the function $T \to \phi(|T|)$ (where $|T|$ denotes the "absolute value" of T in its polar decomposition). The Banach-space obtained by completion of A (or the appropriate ideal of definition) with respect to the norm given by the faithful, normal, semifinite trace μ will be denoted by $L_1(A,\mu)$.

Let μ be such a trace and $L_1(A,\mu)$ be the appropriate space. If for any a, b, c in $L_1(A,\mu) \cap A$ the trace μ satisfies the condition:

$$\mu \otimes \mu((a \otimes b)d(c)) = \mu \otimes \mu(a \otimes c)d(b)) \qquad \text{(L)}$$

or

$$\mu \otimes \mu((a \otimes b)d(c)) = \mu \otimes \mu((c \otimes b)d(a)) \qquad \text{(R)}$$

then μ is called a _left-invariant_, respectively _right-invariant_ measure (gauge) on the involutive W^*-bigebra. If the measure is bi-invariant, the bigebra is called _unimodular_ and the measure _unimodular measure or gauge_. These terms already strongly suggest that there is a close association between involutive W^*-bigebras with measures and locally compact groups. We shall denote an involutive W^*-bigebra with left invariant measure by (A,d,j,μ).

A _morphism_ of two such objects (A_1,d_1,j_1,μ_1) and (A_2,d_2,j_2,μ_2) will be an isomorphism of the von Neumann algebras $\theta: A_1 \to A_2$ which preserves comultiplication, involution and measure:

$$\theta \circ j_1 = j_2 \circ \theta, \ (\theta \otimes \theta) \circ d_1 = d_2 \circ \theta, \ \mu_1 = \mu_2 \circ \theta.$$

We are now in a position to formulate (without proof, for which we refer to the original paper) the

Takesaki [24] version of the duality theorem for locally compact groups. The Tannaka duality theorem for compact groups follows from it, but an independent, and easily accesible proof in terms of Hopf-algebras (involutive bigebras with measure and counit) can be found in Hochschild [23].

We collect the pieces which we gave above as examples. Let G be a locally compact group with left Haar measure dg. Denote by A_G the von Neumann algebra $L_\infty(G)$ consisting of all functions essentially bounded with respect to the measure dg. We recall that the diagonal map:

$$d_G(f)(s,t) = f(st),\ f \in A_G,\ s,t \in G$$

defines a comultiplication in A_G, making (A_G,d_G) into a commutative W^*-bigebra. The predual of A_G is $(A_G)_* = L_1(G)$, and the convolution of its elements is, as expected, ordinary convolution. The symmetry induced by the group-antiautomorphism $s \to s^{-1}$ induces an involution on A_G:

$$j_G(f)(s) = f^V(s) = f(s^{-1}).$$

Finally, the left Haar measure dg on G induces a left invariant faithful normal semi-finite trace on $L_\infty(G)$, namely, the trace associating to any $f \in A_G^+$ (positive essentially bounded function) the value of its Haar integral.

Thus we get the direct theorem, which is relatively trivial:

Theorem 1 [24]. To any locally compact group with left Haar measure corresponds uniquely a commutative, involutive W^*-bigebra with left invariant measure: $(A_G,\ d_G,\ j_G,\ \mu_G)$.

The converse of this theorem, which has a rather involved proof, shows that one can recover the group from the W^*-bigebra. More precisely:

Theorem 2 [24]. Given an involutive W^*-bigebra with left-invariant measure (A,d,j,μ) there exists a unique locally compact group G, with associated W^*-bigebra (A_G, d_G, j_G, μ_G) isomorphic to the given one.

Denoting by M the von Neumann algebra generated by the left regular representation of the convolution algebra $L_1(G)$ on the Hilbert space $L_2(G)$, Takesaki showed that M' is the right regular representation, that M can be made into a cocommutative W^*-bigebra (M,π,κ) with the coproduct π given by a unitary transform of the amplification $a\otimes 1$ (the details are too complicated to reproduce here) and the involution $\kappa:a\to Ca^*C$, where C is complex conjugation, * the involution in L_1. The resulting W^*-bigebra (M,π,κ) will be called the dual of the W^*-bigebra (A,d,j,μ). The predual M_* of the von Neumann algebra becomes a commutative *-Banach algebra with convolution as product and complex conjugation as involution.

The locally compact group G is then reconstructed essentially as the spectrum S of M, i.e. the morphisms of M into the algebra of complex numbers, with composition defined naturally by $s_1 s_2=(s_1\otimes s_2)\circ\pi, s_i \in S$.

2.3.4 The case of compact groups

Since the case of compact groups will be most important for the physical applications in the next chapter, and since it can be more easily understood (the W^*-bigebras being essentially finite-dimensional), we first state the specializations to that case of the Takesaki duality theorem, and then rediscuss the Tannaka theorem, which will facilitate our understanding of what is really involved.

We first notice that the two W^*-bigebras discussed above, namely A and M, were not quite symmetric: A had an invariant measure, whereas nothing of that sort appeared in M. Stinespring and Kats [25] have proved that this symmetry reappears in the case of unimodular locally compact groups, i.e. that in this case the dual W^*-bigebra (M,π,κ) has a unique unimodular (i.e. bi-invariant) measure τ and that conversely, to a given involutive, cocommutative W^*-bigebra (M,π,κ,τ) with unimodular measure τ, corresponds a unique locally compact unimodular group, whose dual W^*-bigebra is isomorphic to the given one.

Two special cases of unimodular groups are of particular importance to us: compact groups and discrete groups. They are related by the following theorem, which in some ways generalizes to the nonabelian case the well known duality between an abelian compact groups and an abelian discrete group:

Theorem (7 of [24]). The following two sets of conditions $(C_1-C_2-C_3)$, and $(D_1-D_2-D_3)$ are respectively equivalent for a locally compact unimodular group G and the two associated W^*-bigebras, the unimodular commutative (A,d,j,μ) and the dual unimodular cocommutative (M,π,κ,τ):

(C_1) G is compact.

(C_2) (A,d,j,μ) is finite, i.e. the measure μ is finite.

(C_3) (M,π,κ,τ) has a counit, which means that M_* has a unit.

(D_1) G is discrete.

(D_2) (A,d,j,μ) has a counit, i.e. A_* has a unit.

(D_3) (M,π,κ,τ) is finite, i.e. τ is a finite measure.

This shows then that the W^*-bigebra of a compact group can be characterized by having a finite measure, similar to the group itself, or the existence of the counit in the dual algebra M (i.e. $f\equiv 1\in L_1(G)$), whereas in the case of discrete groups, the roles are reversed.

Before taking up the compact case again from a slightly different angle, I would like to mention some recent related results on duality obtained by M. Walter [32].

The duality theorem for compact groups can be derived directly [23] in the following way. Owing to the fact that the irreducible unitary representations of compact groups are finite-dimensional, there is no need to consider two W*-bigebras: since the duality theory of finite-dimensional Euclidean spaces allows us to identify M and M_*, it will suffice to find a W*-bigebra with involution, unit and counit, such that the invariant measure on it (gauge) is finite.

Let U be a finite-dimensional representation of a (compact) group G, H_U the representation space (which is a G-module). The space $B(H)$ of all linear transformations of H_U is finite-dimensional, and so is its dual space $B(H)_*$. (Here we use notations differing from [23], in order to be as close as possible to our previous notations). The elements f of $B(H)_*$ are represented by matrices of the same dimensions as U, hence the composites $f \circ U$ will be $\mathbb{R}$-(or $\mathbb{C}$)-valued functions on the group G. These functions are called _representative functions_ associated with the representation U. The set of these functions forms an algebra under pointwise addition and multiplication.

In general, a _representative function_ f (real-or complex-valued) on the topological group G is a continuous function such that _the translates x.f and f.x_ ((x.f)(y) = = f(yx), (f.x)(y)=f(xy), $f^V(x)=f(x^{-1})$ in the sequel) _span a finite-dimensional vector space_. The representative functions associated to a finite-dimensional representation obviously have this property. The Peter-Weyl theorem on the completeness of the system of irreducible representations of a compact group can be recast in the following succint form:

Peter-Weyl-Theorem. Denoting by R the algebra of all representative functions (under pointwise addition and multiplication) on a compact group G, R is uniformly dense in the algebra $C(G)$ of all continuous functions on the group [23].

Now consider the real algebra $R(G)$ of all real-valued representative functions on the compact group G. (This is a von Neumann algebra on $L_2(G)$). The canonical isomorphism of $R(G) \otimes R(G)$ into $R(G\times G)$ given by $u \otimes v(s,y) \longleftrightarrow u(x)v(y)$ is a normal morphism of von Neumann algebras (finite-dimensional algebras, proof by introducing bases). The map sending a pair (x,y) onto f(xy) can then be identified with an element of $R(G) \otimes R(G)$, and the map $f \to df$ realizing the homomorphism $R(G) \to R(G) \otimes R(G)$ is clearly a comultiplication, thus making $R(G)$ into a bigebra. The counit of this object is given by the map: $c(f)=f(e)$, i.e. evaluation of each function at the neutral group element. It is easy to verify that $(c \otimes i) \circ d = (i \otimes c) \circ d = i$. The algebra also has a unit u, the identification of real numbers with scalar operators. An involution in the bigebra $R(G)$ is given by the map $j(f)=f^{V}$ (as defined above). To recover the group from the bigebra we associate to every group element x the algebra homomorphism (character) $\hat{x}: R(G) \to \mathbb{R}$ such that $\hat{x}(f)=f(x)$. These objects can be multiplied according to the rule $(xy)^{\wedge} = (\hat{x} \otimes \hat{y}) \circ d$.

Finally, since G is compact, the Haar functional J gives us an R-module homomorphism (i.e. linear positive functional, or trace) $J: R(G) \to \mathbb{R}$: $f \to J(f) = \int_G f\, dg$, $J(f^2) \geq 0$, and J satisfies $(J \otimes i) \circ d = u \circ d$.

This functional is a finite measure (gauge) on the bigebra.

We can now start from a W*-bigebra A over the reals, having all the above properties, i.e. an involution and a finite measure (in addition to the unit and counit given by the reals in the algebra). Then the group S of algebra-homomorphisms $A \to \mathbb{R}$ with the composition laws given above is a compact topological group, and the reduction to the W*-bigebra (A^o, d^o, j^o, J^o) of the W*-bigebra (A,d,j,J) mediated by the "Gel'fand map" $\hat{f}^o(s)=s(f)$, $f \in A$, $s \in S$, which is an algebra morphism $A \to \mathcal{R}(S)$, is the W*-bigebra associated to the group S.

In other words, we get the following version of the Tannaka duality theorem [23]:

For a compact group G, let $\mathcal{H}(G)$ be the bigebra attached to G in the above canonical fashion. Then $\mathcal{H}(G)$ is a reduced bigebra (i.e. coincides with its reduction, or in other words, is such that the characters, i.e. homomorphisms into the base field, separate the elements of $\mathcal{H}(G)$), having involution and finite measure (gauge). Given a reduced W*-bigebra H with involution and finite measure, let $\mathcal{G}(H)$ denote the topological group of "characters" as described above, Then $\mathcal{G}(H)$ is compact, and the canonical maps $G \to \mathcal{G}(H)$ and $H \to \mathcal{H}(G)$ are isomorphisms.

In the next chapter we take up the problem of how these concepts may be used in quantum field theory.

3. INTERNAL SYMMETRIES AND W*-BIGEBRAS

3.1 A Brief Review of the Abelian Case

The case of abelian superselection sectors has been treated very elegantly in [10], and the only thing one can do to improve on these discussions is to relate the DHR-construction of "charged sectors from observables", by means of localized automorphisms of the observable algebra A, to the Fell-bundle construction described in Section 2.2. (For more details, see a forthcoming article by the author: "Fell-Bundles and Superselection Sectors", IHES, June 1971).

Let $E_o(A)$ denote the set of pure states of the observable algebra A (cf. Sec. 1.1) satisfying the condition that asymptotically (i.e. outside a sufficiently large double cone) they coincide with the vacuum, in a representation which satisfies "duality" (1.5). Two such states ω_1, ω_2 should be obtainable from one another by means of the action of an observable $a \varepsilon A$, i.e.

$$\omega_2 = \omega_{1a} \quad (\text{i.e. } \omega_2(b) = \omega_1(a^* ba)). \tag{3.1}$$

Now we split the set E_o into "sectors", i.e. classes within which the superposition principle for the vectors describing pure vector states holds without restriction:

$$E_o = \bigcup_q E_q \tag{3.2}$$

and such that there is no superposition for states in different sectors E_q. The labels q play the role of charges, i.e. in general form a (discrete) additive abelian group $\hat{G}$ (this is proved in detail in DHR I; an example of continuous "charges" has been discussed by Streater and Wilde [10]). The fact that the charges form a countable discrete group $\hat{G}$ is fundamental in the discussion -- its dual is a compact abelian group G (Pontryagin duality, for a change), and will form the basis for our attempts at generalizations to the nonabelian case.

The family of states $E_o = \bigcup_q E_q$ gives rise to a family representations π_q (in general distinct from the vacuum representation π_o) such that outside some double cone

O we have unitary equivalence (one says that the representations are strongly locally equivalent)

$$\pi|_{A(O')} \cong \pi_o|_{A(O')}$$

and such that the representation induced by the state $\omega \in E_q$ is a faithful covariant representation (π,U) of $(A,\tilde{P}_+^\uparrow)$ ($P_+^\uparrow$ denotes the proper orthochronous Poincaré group) satisfying the spectrum condition). The unitary operator V which intertwines the two representations, $V\,H=H_o$ is such that the vacuum vector Ω is mapped into the cyclic vector Φ of the representation π: $V^{-1}\Phi=\Phi$, and the state $\omega=\omega\circ\pi$ is "strictly localized" in O. The converse also holds, i.e. any strictly localized state induces a local perturbation from the vacuum, and such states can be obtained by the action of unitary "field operators" V. If in addition to the properties outside O we want V to be an automorphism inside O, one has to impose some kind of duality (1.5), which implies, after a relatively complicated argument, that

$$V\,\pi(A(O_\lambda))V^{-1} = \pi_o(A(O_\lambda)),\ O \subset O_\lambda. \tag{3.3}$$

Such automorphisms have been called by Borchers and DHR localized automorphisms. The group of those automorphisms which are covariant with respect to the Poincaré group will be denoted by $L(O)$, its unitary subgroup (i.e. those implemented by unitary elements of $A(O)$) will be denoted $J(O)$, the unitaries in $A(O)$ form a polish topological group, containing $J(O)$. Similarly, one can define the "global", U, L, J. There is a natural injection of U into J, thus J is a normal subgroup of L. It can be shown that the quotient group L/J -- the group of superselection sectors -- is an abelian group $\hat{G}$, which we will assume discrete, i.e. the dual of a compact abelian group G, the gauge group. One is thus led to a typical group-extension problem: find the "field group" F, of unitary operators which are extensions of L by $\hat{G}$. This problem was solved as an abstract group extension problem by DHR II, where it has been shown that one can indeed reconstruct from these ingredients the field group F, its net of "local algebras $F(O)$" (recall that a von Neumann algebra is generated by its unitaries, hence, by abuse of notation, one can call F the field algebra). Moreover, a rather lengthy analysis shows that the commutation structure of the algebras $F(O)$ is such that they obey either Bose-Einstein commutation, or Fermi-Dirac anticommutation at space like separations, and that the spectrum condition is satisfied in each sector.

Let us now recall the Fell construction of a bundle from a C^*-algebra A, a subgroup N of its unitary subalgebra U, and an extension H of the latter, such that H/N is isomorphic to a given locally compact group G. If G is the discrete abelian group of charges $\hat{G}$, we recognize the closeness of the field algebra and some Fell-bundle derived from A as the unit-fiber subalgebra. Indeed, since for abelian groups we have complete duality, in the sense that the dual object of a group is again a group, we can go to the dual picture, replacing $\hat{G}$ by its dual gauge group, and then ordinary Fourier analysis shows us that the unit-fiber algebra A can be obtained by taking the mean over the gauge group. (For more details on this topic, which unfortunately we have no time to discuss in detail here, see the mentioned paper by the author).

As a last remark, let us note that since there is a one-one correspondence between representations of Fell-bundles and their L_1-algebras, one will gain some insights by realizing the field algebra as such an $L_1(A,G,\alpha)$ algebra (with cocycles, because of the group extension problem). This is done in Appendix II of DHR II (in terms of a "covariance algebra" ℓ_1 -- the same as the generalized group algebras discussed in Ch. II), and in a simplified form, in the author's already mentioned paper.

3.2 Nonabelian Exact Symmetries. W*-Bigebraic Description

In the present section we attempt to reformulate the theory of field algebras (as defined in DHR I) in such a manner as to bring out into the open the dual object of the gauge group, destined to play here a role similar to the additive group of charges in the abelian case. It is not yet clear whether the W*-bigebra is the most suitable dual object, or whether one should not turn to Krein algebras [26], or possibily a Fourier, or Fourier-Stieltjes algebra [32] for this purpose. In view of the exploratory nature of this approach, we discuss one of these versions first, hoping to return on a future occasion to a fuller discussion of all available options.

First a few general remarks. In distinction from abelian superselection-sector generating symmetries, where the symmetry manifests itself exactly through its superselection aspect, or the conservation law of some integer-valued charge, the non-abelian symmetries observed in nature manifest themselves only through their breakdown indeed, an absolutely exact nonabelian symmetry could never be discovered (unless it led to some form of parastatistics, cf. DHR III [10]), since we would have no way of distinguishing the members of a multiplet, and would thus just introduce redundant variables into the theory. On the contrary, the approximate nonabelian internal symmetry groups always manifest themselves indirectly, via the existence of finite-dimensional multiplets of states with almost exactly the same properties in particular, belonging to "almost the same" representation of the Poincaré group, i.e. having the same spin and slightly different masses (which we blame on the "symmetry-breaking interactions" in a typical perturbation-theoretic philosophy). It thus seems worthwhile to investigate how far one can go in describing the internal symmetry in terms of a collection of unitary, finite-dimensional representations (either of states or of fields) or the various algebraic objects one can form with these, objects like Hopf-algebras (W^*-bigebras of representative functions, and Krein algebras). The next step should consist in analyzing the breakdown of the symmetry for these objects, so that the duality theorem itself breaks down. Thus in a certain sense, one is led to the concept of a "collection of representations in search for a group".

In order to be able to modify the axioms for field algebras, we restate them briefly, following up to a point the notations and terminology introduced in DHR I.

Let K denote the collection of closed double cones in Minkowski space. All algebras below will be concrete operator algebras on a Hilbert space H, and unless expressly mentioned, we will always take weak closures, so that they are von Neumann algebras (or W^*-algebras, if we prefer to work with equivalence classes of such objects -- we will not be too pedantic in this respect).

1. There is a correspondence

$$0 \to F(0) \quad 0 \in K \tag{3.4}$$

between double cones and W^*-algebras, satisfying the isotony requirement (condition i in Sec. 1.1). The total field algebra $F=\overline{\bigcup_{O\in K} F(0)}$, where the bar means norm-closure, and F is assumed weakly dense in $B(H)$. No locality is assumed, but the algebra $F(0')$ is the C^*-subalgebra of F generated by all the $F(O_1)$, with $O_1 \subset O'$.

2. The covering (N.B.!) group $\tilde{P}$ of the Poincaré group P is represented by automorphisms α_L of F, $\alpha_L F(O)$ $L \epsilon \tilde{P}$, and α_L is implemented by a strongly continuous representation $U(L)$ of $\tilde{P}$, such that the self-representation of F, and U form a covariant representation in the sense of Sec. 2.1, satisfying the spectrum condition: the spectrum of translations is in the forward light-cone and the vertex, 0 is an isolated, nondegenerate eigenvalue, corresponding to the vacuum state ω_o, which is represented in the Hilbert space by the cyclic vacuum vector Ω.

3. Gauge group (this is the "axiom" which we shall attempt to replace by an equivalent one, in the case of exact symmetries). There exists a compact group G, the gauge group, and a faithful strongly continuous representation of G by unitaries in $B(H)$, such that $(F, U(G))$ form a covariant representation (i.e. $U(g)FU(g^{-1}) = \alpha_g(F)$). The "gauge-automorphisms" α_g are strictly local

$$\alpha_g F(O) = F(O), \tag{3.5}$$

and commute with the representation U_L of the Poincaré group P:

$$U(P) \subset U(G)', \text{ thus } U(g)\Omega = \Omega . \tag{3.6}$$

4. The local observables are obtained from the local field algebra by imposing invariance with respect to G, i.e.

$$A(O) = F(O) \cap U(G)' \quad O \epsilon K . \tag{3.7}$$

$A(O)$ already satisfies the locality requirement of Sec. 1.1, and the representation of P (which is a projective representation of P) becomes a true representation of P when restricted to the observable algebra A. Another way of passing from the field algebra F to the observable algebra A (and the respective local nets $F(O)$),

$A(O)$ is by averaging over G, with respect to the Haar measure:

$$m(F) = \int_G \alpha_g(F)dg \qquad F \in \mathcal{F} \tag{3.8}$$

defines a normal positive linear mapping from $B(H)$ onto $U(G)'$, reducing to the identity on $U(G)'$, and such that for any double cone O

$$\begin{aligned} A(O) &= m(\mathcal{F}(O)) = \mathcal{F}(O) \cap U(G)', \\ A(O') &= m(\mathcal{F}(O')) = \mathcal{F}(O') \cap U(G)', \\ A &= m(\mathcal{F}) = \mathcal{F} \cap U(G)', \\ A'' &= m(\mathcal{F}'') = U(G). \end{aligned} \tag{3.9}$$

(For proofs cf. DHR I). We will use these properties as a starting point in the W*-bigebra approach, since there we have an invariant measure (gauge).

Let us list several other assumptions about field algebras, which we will not be needing here.

5. The clustering property: for $\underline{x}$ a space-like translation, $F_i \in \mathcal{F}$

$$\lim_{\underline{x}\to\infty} \langle\Omega|F_1\alpha_{\underline{x}}(F_2)F_3\Omega\rangle = \langle\Omega|F_1F_3\Omega\rangle\langle\Omega|F_2\Omega\rangle . \tag{3.10}$$

6. The Reeh-Schlieder property of analytic vectors: a vector $\psi \in H$ is analytic for the translations (i.e. the series $\sum_{n=0}^{\infty}(z^n/n!)\,\|p^n\psi\|$ has a nonzero radius of convergence in the z-plane) if the function $x \to U(x)\psi$ is holomorphic in the neighborhood of the identity. Analytic vectors are cyclic and separating for each $\mathcal{F}(O)$, and this implies (as shown by Borchers) that "weak additivity" holds, i.e. the translates $\mathcal{F}(O+x)$ and $\mathcal{F}(O+x)'$ generate $B(H)$.

7. The field algebra in a double cone O must commute with the observables of the space-like separated regions:

$$F(O) \subset A(O')' . \tag{3.11}$$

In addition various duality requirements may be imposed on the field algebra. In particular, a duality requirement which is satisfied by abelian sectors is

$$A(O')'' = F(O)' \cap U(G)' , \tag{3.12}$$

which leads to the localized automorphisms discussed briefly in the preceding section (here we have dropped the π, since A is a concrete algebra).

A few words are in order regarding the compactness of the gauge group G. This restriction, which at first seems rather strong, is in fact only a reflection of the fact that all observed particle multiplets are finite (hence transform, in the limit of exact symmetry, under finite-dimensional unitary representations) together with the requirement imposed by scattering theory, that the one-particle states span asymptotically a dense set in Hilbert space. The group of all unitary transformations commuting with Poincaré transformations contains G as a subgroup, and being isomorphic (cf. [10] I, Sec. II) to the product of all unitary one-particle state transformations, is compact, by Tikhonov's theorem. Thus, even if there is an infinite number of finite dimensional multiplets, the gauge group will be compact. We can use the same argument, in showing that the bigebra to be introduced in place of G is the one corresponding to a compact group, i.e. must have finite invariant measure, and involution.

We are now ready to try to modify the axioms 3. - 4. by replacing the group G by its dual object, i.e. the "complete spectrum", as defined by DHR. A close reading of theorem 3.6 in DHR I shows indeed one is dealing with an algebra of representative functions $R(G)$, and that several other theorems of DHR should take a more natural form if the dual object of the gauge group were used directly.

This is what we are going to attempt in the remainder of this section. Thus let $R(G)$ be the algebra of representative functions of a compact group, which we make into a W^*-bigebra with involution and finite measure in the manner described in Section 2.3.4. What we wish to do is to embed this algebra in F, or, to define in F directly a partial trace, singling out A.

Now let us suppose that we do not know that the "label space" Σ is the dual object of a compact group, but that we are given a field algebra, with a partial normal faithful "finite" trace m which maps the field algebra F onto A and each local field algebra $F(O)$ onto $A(O)$, such that the identity is mapped onto the identity, and such that m is idempotent. The only term needing careful explanation is the term finite, since finiteness of the trace (combined with bi-invariance in the bigebra and the involution of the latter) will lead to the identification of the gauge group as the dual of the part of the bigebra which the invariant trace will help identify.

Thus, let m be a linear positive normal map of von Neumann algebras, with the following properties:

i) $m \circ m = m$ (i.e. $m(m(x)) = m(x)$)

ii) $m(x^*x) = a^*a$ (positivity)

iii) $m(x) = o$ implies $x = o$, (faithfulness)

iv) let $\{x_i\}_{i\varepsilon I}$ be an increasing net of positive elements of F such that $\sup_{i\varepsilon I}\{x_i\}=x$, then $m(x) = \sup_{i\varepsilon I} m(x_i)$ (normalcy

v) $[m(F_1), m(F_2)]=o$ if F_1 and F_2 belong to field algebras of spacelike separated double cones.

vi) m is Poincaré invariant.

Now associate to the field algebra F its tensor square $F \hat{\otimes} F$, and define the comultiplication d suggested by W^*-bigebras of covariant systems: denote by A the result of averaging with m.

$$A = m(F)$$

$$A(O) = m(F(O)) \ .$$

Then on the elements of A (denoted by a, b etc.)

$$da = a \otimes I \ .$$

On the unitary group (or the group of invertible elements) of F (denote the "averaged unitary group by W, it should contain "observable" symmetries, like Poincaré transformations, etc.) we have the following action of the comultiplication:

a) on W: $ds = I \otimes I$

b) on $U(F)/W$: $ds = s \otimes s$.

Thus, the field bigebra (F, d) can be considered as generated by this prescription (this argument is heuristic and needs some refinements in the case where F is not separable; U is a polish group, hence separable in the strong topology).

The invariance properties of the partial trace m are expressed by imposing simultaneously the requirements (L) and (R) of Sec. 2.3.3, since we wish to simulate the behavior of a unimodular symmetry group. The compactness, tantamount to finite-dimensional multiplets, has to be expressed as a finiteness property of the trace. This can be done tentatively by requiring that any normal state applied to the result of applying m to an element of F yield a finite result; or, since m is itself normal, the "averaged states" be finite.

Finally, we must define an involution, in order to regain completely the group property. The involution takes the place of the inverse, hence we can again proceed on generators: (C is the "complex-conjugation" required)

$$j(a) = Ca^*C \otimes I \qquad a \in A$$

$$j(s) = Cs^{-1}C \qquad s \in \mathcal{U}(\mathcal{F})/\mathcal{W} .$$

The group structure is then recovered by considering multiplicative linear functionals on ($\mathcal{F}$,d,j) with the composition law of "characters" defined by

$$s_1 s_2 = s_1 \otimes s_2 \circ d .$$

REFERENCES

1. I. E. Segal, Postulates for general quantum mechanics, Ann. of Math. 48, 930 (1947).
2. Cf. e. g. D. Ruelle, Statistical Mechanics, Benjamin, N. Y. 1969.
3. R. Haag and D. Kastler, An algebraic approach to quantum field theory, J. Math. Phys. 5, 848-661 (1964).
4. M. Takesaki, Tomita's theory of Modular Hilbert Algebras and its Applications, Lecture Notes in Math. n°128, Springer, Berlin-Heidelberg-New York, 1970.
5. J. Dixmier, Les algébres d'opérateurs dans l'espace Hilbertien (Algébres de von Neumann), 2e Ed. Gauthier-Villars, Paris 1969, quoted as DN.
J. Dixmier, Les C*-algébres et leurs representations, 2e Ed. Gauthier-Villars, Paris, 1969, quoted as DC*.
6. see e.g. H. Behncke, A remark on C*-Algebras, Commun. Math. Phys. 12, 142 (1969); A note on the Gel'fand-Naimark conjecture, Commun. Pure Appl. Math. 23, 189 (1970).

7. V. Bargmann, Note on Wigner's theorem on symmetry operations, J. Math. Phys. 5, 862 (1964).
8. R. Kadison, Transformation of states in operator theory and dynamics, Topology, 3, Suppl. 2, 177--198 (1965).
9. H. Araki, Von Neumann algebras of local observables for the free scalar field, J. Math. Phys. 5, 1 (1964); G. F. dell'Antonio, Structure of the algebras of some free systems, Commun. Math. Phys. 9, 81--117 (1968). Cf. also (10).
10. S. Doplicher, R. Haag and J. E. Roberts, Fields observables and gauge transformations, I, II, Commun. Math. Phys. 13, 1 (1969), 15, 173 (1969), quoted as DHR I, II; K. Druhl, R. Haag and J. E. Roberts, On parastatistics, Commun. Math. Phys. 18, 2o4--226 (1970). Cf. also the recent example worked out by R. F. Streater and I. F. Wilde, Fermion states of a boson field, Nucl. Phys. B24, 561--575 (1970).
11. e.g. J. Dixmier, Varenna Lectures, 1968 Academic Press, 1969.
12. G. Zeller-Meier, Produits croisés d'une C*-algébre par un groupe d'automorphismes, J. Math. Pures et Appl. 47, 1o1-239 (1968).
13. H. Borchers, On the implementability of automorphism groups, Commun. Math. Phys. 14, 305 (1969), also Cargése Lectures 1969, D. Kastler, Ed. Gordon & Breach, N. Y. 1970; C*-algebras and locally compact groups of automorphisms, Marseille Preprint 70/P. 320.
14. A. Guichardet and D. Kastler, Désintégration des états quasi-invariants des C*-algébres, J. math. Pures et Appl. 49, 349--380 (1970).
15. J. M. G. Fell, An extension of Mackey's method to Banach-*-algebraic bundles, Mem. A. M. S. No 90, 1969.

16. M. E. Mayer, Differentiable cross sections in Banach-*-algebraic bundles, Cargèse Lectures 1969, D. Kastler, Ed. Gordon & Breach, N. Y. 1970).
17. e.g. L. Auslander and C. C. Moore, Unitary representations of solvable Lie groups, Mem. A. M. S. no. 62, 1966.
18. S. Dolplicher, D. Kastler and D. W. Robinson, Covariance algebras in field theory and statistical mechanics, Commun. Math. Phys. 3, 1--28 (1966).
19. M. Takesaki, Covariant representations of C*-algebras and their locally compact automorphism groups, Acta Math. 119, 273-303 (1967).
20. H. Leptin, Verallgemeinerte L^1-Algebren und projektive Darstellungen lokal kompakter Gruppen, I, II, Inventiones Math. 3, 257--281 (1967), 4, 68--86 (1967); Darstellungen verallgemeinerter L^1-Algebren, Inventiones Math. 5, 192--215 (1968), and to be published.
21. H. Behncke, Automorphisms of crossed products, Ph. D. Thesis, Indiana University, June 1968.
22. K. H. Hofmann, The duality of compact semigroups and C*-bigebras, Lecture Notes in Math. No. 129, Springer, Berlin-Heidelberg-New York, 1970.
23. G. Hochschild, The structure of Lie groups, Sec. II.3, Holden-Day, 1965.
24. M. Takesaki, A characterization of group algebras as a converse of the Tannaka-Stinespring-Tatsuuma duality theorem, Amer. J. Math. 91, 529--564 (1969).
25. W. F. Stinespring, Integration theorems for gages and duality for unimodular groups, Trans. Amer. Math. Soc. 90, 15--56 (1969); G. I. Kats (Kac), Kol'tsevye gruppy i printsip dvoistvennosti I, II, (Ring-groups and the duality principle), Trudy Mosk. Mat. Obsch. 12, 259--301 (1963), 13, 84--113 (1965). (English translation available).

26. E. Hewitt and K. Ross, Abstract Harmonic analysis, vol. II, Springer, 1969.

27. J. Ernest, The enveloping algebra of a covariant system, Commun. Math. Phys. 17, 61--74 (1970); A duality theorem for the automorphism Group of a covariant system, Commun. Math. Phys. 17, 75--90) (1970).

28. A Guichardet, Tensornye proizvedeniya C*-algebr (Tensor products of C*-algebras) Dokl. Akad. Nauk SSSR 60, 986--989 (1965) (Sov. Math. Dokl. 6, 210--213 (1965)).

29. M. Takesaki, A note on the direct product of operator algebras, DC* ref. 158, also DC*, ref. 207.

30. A. Wulfsohn, Produits tensoriels de C*-algébres, Bull. Soc. Math. 87, 13--27 (1963); The primitive spectrum of a tensor product of C*-algebras, Proc. Amer. Math. Soc. 19, 1094--1096 (1968); Tensor products of Hilbert spaces and C*-algebras, Ph. D. Thesis, University of California, Irvine, 1969.

31. J. Dieudonne, Hyperalgébres et groupes formels, Mimeographed Noted, IHES, 1958, P. Cartier, Hyperalgébres et groupes de Lie formels, Séminaire Sophus Lie, 2éme année, Paris, 1957.

32. M. Walter, Group duality and isomorphisms of Fourier and Fourier-Stieltjes algebras form a W*-algebra point of view. Bull. Am. Math. Soc. 76, 1321--1325 (1970); W*-algebras and Nonabelian Harmonic Analysis, PH. D. Thesis, University of California, Irvine, 1970.

Acta Physica Austriaca, Suppl. VIII, 227–276 (1971)

ULTRALOCAL QUANTUM FIELD THEORY*

BY

J. R. KLAUDER
Bell Telephone Laboratories, Incorporated
Murray Hill, New Jersey

ABSTRACT

These notes cover lecture material presented at two 1971 Winter Schools.

A special class of model field theories, differing from relativistic theories by the absence of the spatial gradient in the Hamiltonian, is discussed. Simple operator solutions are constructed which circumvent the infinite mass renormalization, infinite coupling constant renormalization and infinite field strength renormalization inherent in a perturbation approach. A number of properties of these solutions are analyzed including the spectrum of the Hamiltonian. Compared to free field theories, the ultralocal models discussed herein may, conceivably, provide an alternate starting point to attack covariant field theories.

---

*Lecture given at VIII. Annual Winter School for Theoretical Physics, Karpacz, February 15 - 27, 1971; and at X. Internationale Universitätswochen für Kernphysik, Schladming, March 1 - March 13, 1971.

1. INTRODUCTION

Soluble quantum field theory models are a rare commodity. An infinite number of degrees of freedom and noncompact invariance groups have a nasty habit of exploding in the model-makers' face. Nevertheless, important progress has recently been made in the class of superrenormalizable relativistic theories, such as a self-interacting boson in a two-dimensional space time [1]. These results have been obtained starting with the free field and adding the interaction in a carefully controlled way. Yet, the models successfully studied in this way do not have an infinite field strength renormalization, which, at least according to perturbation theory, should appear for realistic relativistic models in four-dimensional space time.

Motivation and the Model

The ultralocal scalar field theories discussed in these lecture notes are likewise motivated by relativistic theories but are based on a different approximation. This approximation formally amounts to dropping the spatial gradient term from the Hamiltonian rather than the nonlinear interaction. For a self-interacting boson field in a space-time of (s+1) dimensions (s≥1), the classical ultralocal model Hamiltonian reads

$$H = \int \{\tfrac{1}{2}\pi_{cl}^2(\underset{\sim}{x}) + \tfrac{1}{2}m_o^2\phi_{cl}^2(\underset{\sim}{x}) + V[\phi_{cl}(\underset{\sim}{x})]\}d\underset{\sim}{x} \ . \qquad (1\text{-}1)$$

The quantum theory of this model is the subject of the present paper. This model differs formally from a relativistic theory by the term $\int\frac{1}{2}[\underset{\sim}{\nabla}\phi_{cl}(\underset{\sim}{x})]^2\, d\underset{\sim}{x}$ which, it is hoped, can, in one or another way, be added as a perturbation in the quantum theory. However, that still remains a problem for the future, and we confine our remarks to a careful study of the "unperturbed" model (1-1).*

General Features

The most striking feature of the model in (1-1) is the implied independence for-all-times of the field at distinct spatial points. The light cone has been collapsed to a single "vertical" line passing through the spatial point in question; for this reason we call these models ultralocal. This independence does not mean that all topological features of the configuration space R^s are dispensed with. On the contrary, we still assume the existence of a space translation generator classically given by

$$\underset{\sim}{P} = \int \pi_{cl}(\underset{\sim}{x}) \underset{\sim}{\nabla} \phi_{cl}(\underset{\sim}{x}) d\underset{\sim}{x}; \tag{1-2}$$

and we need the topological properties if ever we are to restore the spatial gradient to the Hamiltonian. One would not expect the ultralocal models to exhibit scattering, either classically or quantum mechanically, since the localized excitations can never separate from one another spatially. The ultralocal models, like the free theory, represent an initial theory to which scattering perturbations are added. It is the differences in these initial theories which may permit the scattering to be added for one initial theory and not for the other. Classically, for instance, it is not unrealistic to regard the ultralocal models as a valid long wavelength limit of the relativistic models, something which the free theory cannot be if $V \neq 0$.

---

*The present notes are an elaboration and extension of the author's paper [2]. Other studies of these models have also been made [3].

Just as the relativistic theory with interaction is different than the relativistic free (Fock) theory, so too is the ultralocal theory with interaction different from *its* free field, the ultralocal, free Fock theory. This difference is apparent if one attempts to pass from the noninteracting to the interacting ultralocal models by a standard perturbation treatment. Divergences enter such a calculation, e.g., in the self-energy bubble in a quartic interaction model where summation over spatial momenta leads to infinite multiplicative factors; indeed, divergences arise even in a *compact* configuration space. Thus the quantum problem is far from being well posed starting from the Fock representation, and must be approached by alternative techniques.

One striking feature of the exact solutions for all the interacting ultralocal models is the *absence* of canonical field and momentum operators obeying canonical commutation relations. Instead the two-point spectral weight function has a divergent integral signifying, from the usual standpoint, an infinite field strength renormalization. The energy levels of the quantum Hamiltonian are all discrete and apart from the ground state, are integral multiples of various levels which can be understood as "single particle" or "multiparticle collective" energies.

Another striking feature of the exact results is the comparatively simple form and conceptual transparency permitted by the operator solutions. These operator solutions can be presented in a form understandable by anyone familiar with the free field and with elementary Fock space methods. Therefore we have chosen to divide our discussion into two parts: The first part, Sec. 2, deals with the operator solutions to the model and to a physically-oriented discussion of their basic properties; the second part, Sec. 3, deals with the justification for the assumed operator solutions, and a number of associated mathematical questions stated without proof in Sec. 2.

It may be noted that our solutions cover a very broad class of potentials $V[\phi_{cl}]$. Subject to the assumption that V is even and the ultimate quantum Hamiltonian positive, we can deal with general polynomial or *non*-polynomial potentials in an arbitrary number of space dimensions $s \geq 1$.

2. HIGHLIGHTS OF OPERATOR FORMULATION

In this section the basic operators of the theory, such as the field operator and the Hamiltonian, are constructed and discussed. Since the properties of the operators themselves are comparatively simple to analyze, we reserve the justification for our construction to the more deductive analysis of Sec. 3.

2.1 Preliminaries

Basic Hilbert Space

Let us introduce H, the Hilbert space of interest to us, in a very conventional way. Let $A(\underset{\sim}{x},\lambda)$ and $A^{\dagger}(\underset{\sim}{x},\lambda)$ denote a usual Fock representation of annihilation and creation operators defined for $\underset{\sim}{x} \varepsilon R^s$ and for an additional variable λ (having nothing to do with $\underset{\sim}{x}$) such that $-\infty<\lambda<\infty$. The real variable λ is simply an auxiliary variable used in the constructions to follow. The only nonvanishing commutation relation is

$$[A(\underset{\sim}{x},\lambda),A^{\dagger}(\underset{\sim}{x}',\lambda')] = \delta(\underset{\sim}{x}-\underset{\sim}{x}')\delta(\lambda-\lambda'),$$

and the vacuum vector $|0\rangle \varepsilon H$ obeys

$$A(\underset{\sim}{x},\lambda)|0\rangle = 0 \ .$$

We suppose that H is spanned by repeated action of the creation operator on the vacuum, or equivalently by vectors of the form*

$$\exp[\int\int \psi(\underset{\sim}{x},\lambda)A^{\dagger}(\underset{\sim}{x},\lambda)d\underset{\sim}{x}d\lambda]|0\rangle$$

for all square integrable $\psi(\underset{\sim}{x},\lambda)$. As a consequence, the vector $|0\rangle$ is the only vector annihilated by all the $A(\underset{\sim}{x},\lambda)$, and we are dealing with a very conventional Fock representation, albeit one with an additional variable λ.

Fundamental Model Function

We next introduce a fundamental c-number function $c(\lambda)$ which effectively labels the various models and will eventually be related to the model potential. This "model function" is chosen real, even, nowhere vanishing and twice differentiable. Indeed, apart from one anomaly $c(\lambda)$ is generally a very regular function. Specifically, we require that

$$\int (\frac{\lambda^2}{1+\lambda^2}) c^2(\lambda)d\lambda < \infty$$

while we must have

$$\int c^2(\lambda)d\lambda = \infty \qquad (2\text{-}1)$$

Thus $c(\lambda)$ has a singularity at the origin; for example, of the form $c(\lambda)\propto|\lambda|^{-1/2}$ in the vicinity of λ=o.

For the most part it suffices to assume $c(\lambda)$ falls off strongly at infinity, and we shall generally assume that

---

* Such states are unnormed coherent states, $|\psi\rangle$ (see Sec. 3).

$$\int |\lambda|^p c^2(\lambda)\,d\lambda < \infty \,; \qquad p = 1, 2, 3, \dots \,.$$

$$|\lambda| \geq 1 \,.$$

Suitable examples of the model function $c(\lambda)$ are thus

$$c(\lambda) = \frac{1}{|\lambda|^{1/2}}\, e^{-y(\lambda)} ,$$

where

i) $$y(\lambda) = \sum_{n=0}^{M} y_{2n}\, \lambda^{2n}; \qquad y_{2n} > 0$$

ii) $$y(\lambda) = \lambda^2(1+a\lambda^2)/(b+c\lambda^2); \qquad a, b, c > 0$$

iii) $$y(\lambda) = (a^2+\lambda^2)^{1/2}; \qquad a > 0$$

etc. Stronger singularities are also possible such as

$$c(\lambda) = \frac{1}{|\lambda|^\gamma}\, e^{-y(\lambda)} , \; \frac{1}{2} \leq \gamma < \frac{3}{2}$$

and we shall discuss especially the simple case

$$c(\lambda) = \frac{1}{|\lambda|^\gamma}\, e^{-1/2\mu\lambda^2} \,.$$

The parameter γ is reserved for the power of the singularity of the model function. [Generalizations to the above choices for $c(\lambda)$ will occasionally be made; the only sacred growth condition is (2-1).]

Translated Fock Operators

We define the operators

$$B(\underset{\sim}{x},\lambda) = A(\underset{\sim}{x},\lambda) + c(\lambda)$$
$$B^{\dagger}(\underset{\sim}{x},\lambda) = A^{\dagger}(\underset{\sim}{x},\lambda) + c(\lambda)$$

(recall c is real) which differ from the Fock operators by an additive multiple of the identity operator. Evidently

$$[B(\underset{\sim}{x},\lambda), B^{\dagger}(\underset{\sim}{x}',\lambda')] = \delta(\underset{\sim}{x}-\underset{\sim}{x}')\delta(\lambda-\lambda'),$$

while we have

$$B(\underset{\sim}{x},\lambda)|o\rangle = c(\lambda)|o\rangle .$$

Indeed, there is no vector which is annihilated by all the $B(\underset{\sim}{x},\lambda)$, and, moreover, since $\int c^2(\lambda)d\lambda=\infty$ there is no vector which is annihilated by all the $B(\underset{\sim}{x},\lambda)$ with $\underset{\sim}{x} \in O$, where O is some bounded region of R^s. Thus $B(\underset{\sim}{x},\lambda)$ is not unitarily equivalent to $A(\underset{\sim}{x},\lambda)$ either globally or locally.

This inequivalence has important consequences for the bilinear forms constructed from $B^{\dagger}$ and B as compared with those constructed from $A^{\dagger}$ and A. With the latter,

$$\int A^{\dagger}(\underset{\sim}{x},\lambda)w\, A(\underset{\sim}{x},\lambda)d\lambda$$

defines a local self-adjoint operator for any self-adjoint operator w acting on the variables λ, e.g., a differential operator. However

$$\int B^{\dagger}(\underset{\sim}{x},\lambda)wB(\underset{\sim}{x},\lambda)d\lambda$$

is suitable for only certain W. For example, w=1 leads to

$$\int B^{\dagger}(x,\lambda)B(\underset{\sim}{x},\lambda)d\lambda = \int c^2(\lambda)d\lambda + \dots$$

which is divergent. The general requirements for suitable w are given in Theorem 3.1. For present purposes we note: If $w=w(\lambda)$, then it is sufficient that

$$\int w(\lambda)c^2(\lambda)d\lambda, \quad \int w^2(\lambda)c^2(\lambda)d\lambda \tag{2-2}$$

exist, the former as a principal value integral; if $e^{isw}c(\lambda)$ is a function then a necessary condition is the existence, for $s\neq 0$, of $\int|(e^{isw}-1)c(\lambda)|^2 d\lambda$.

2.2 Basic Field Operators

The field operator appropriate to the ultralocal models is given by

$$\phi(\underset{\sim}{x}) \equiv \int B^{\dagger}(\underset{\sim}{x},\lambda)\lambda B(\underset{\sim}{x},\lambda)d\lambda \ . \tag{2-3}$$

This expression is of the form above with $w=\lambda$, which in virtue of our basic assumptions regarding $c(\lambda)$, defines a valid, local self-adjoint operator*. Indeed, as we show in Sec. 3, in the favorable cases where $\int\lambda^2 c^2(\lambda)d\lambda<\infty$,

$$\phi(f) \equiv \int f(\underset{\sim}{x})\phi(\underset{\sim}{x})d\underset{\sim}{x}$$

is a self-adjoint operator for all real $f(\underset{\sim}{x})\varepsilon L^2(R^s)$.

Evidently the field operator $\phi(\underset{\sim}{x})$ depends on the model function $c(\lambda)$. This dependence is displayed in the vacuum expectation functional

$$E(f) \equiv \langle o|e^{i\phi(f)}|o\rangle = e^{-\int d\underset{\sim}{x}\int[1-e^{i\lambda f(\underset{\sim}{x})}]c^2(\lambda)d\lambda} \tag{2-4}$$

as evaluated by elementary properties of the operators $B^{\dagger}$ and B. Two field operators based on different model functions are unitarily equivalent if and only if the model functions are equal. Indeed, as shown in Sec. 3, the field operators are locally equivalent (in some compact region O) if and only if the difference in model functions is square integrable.

---

*The c-number $\int\lambda c^2(\lambda)d\lambda$ in (2-3) is defined by a principal value integral and is zero for any $c(\lambda)$. This ensures that $\langle o|\phi(\underset{\sim}{x})|o\rangle=\int\lambda c^2(\lambda)d\lambda=o$.

Functions of the Field

Besides the operator $\phi(\underset{\sim}{x})$ it is also useful to consider the related operators

$$\phi_r^p(\underset{\sim}{x}) \equiv \int B^\dagger(\underset{\sim}{x},\lambda)\lambda^p B(\underset{\sim}{x},\lambda)d\lambda$$

which for p=1,2,3, ... are each well defined local operators ($\phi_r^1 \equiv \phi$). Indeed, we can imagine

$$\phi_r^{(w)}(\underset{\sim}{x}) \equiv \int B^\dagger(\underset{\sim}{x},\lambda)w(\lambda)B(\underset{\sim}{x},\lambda)d\lambda$$

where $w(\lambda)$ is a function fulfilling (2-2). As is easy to see, these operators all commute among themselves for various choices of $w(\lambda)$.

An essential point is that these various operators are all "functions of $\phi(\underset{\sim}{x})$". This will be made precise in Sec. 3, but here we give the heuristic idea. Consider the identity

$$\phi(\underset{\sim}{x})\phi(\underset{\sim}{y}) = \delta(\underset{\sim}{x}-\underset{\sim}{y})\int B^\dagger(\underset{\sim}{x},\lambda)\lambda^2 B(\underset{\sim}{x},\lambda)d\lambda + !\phi(\underset{\sim}{x})\phi(\underset{\sim}{y})!$$

where ! ! denotes normal ordering with respect to B and $B^\dagger$. By using a test function sequence

$$h_n(\underset{\sim}{x},\underset{\sim}{y}) \equiv e_n(\underset{\sim}{x}-\underset{\sim}{z})e_n(\underset{\sim}{y}-\underset{\sim}{z})$$

with the property that $e_n^2(\underset{\sim}{x}) \to \delta(\underset{\sim}{x})$, we obtain

$$\int h_n(\underset{\sim}{x},\underset{\sim}{y})\phi(\underset{\sim}{x})\phi(y)d\underset{\sim}{x}d\underset{\sim}{y}$$

$$\to \phi_r^2(z) = \int B^\dagger(\underset{\sim}{z},\lambda)\lambda^2 B(\underset{\sim}{z},\lambda)d\lambda \ .$$

In effect this is a renormalized multiplication of the field at a point (hence the subscript r). Heuristically, it is useful to write $\phi_r^2(\underset{\sim}{x}) = Z\phi^2(\underset{\sim}{x})$, where $Z^{-1} = \delta(0)$, which puts into symbols the operator multiplication and renormalization factor. A similar line of reasoning suggests that we set

$$\phi_r^{(w)}(\underset{\sim}{x}) = Z^{-1}w(Z\phi(\underset{\sim}{x})),$$

a notation which we shall make precise in Sec. 3.

An important fact regarding the field operator $\phi(\underset{\sim}{x})$ is that polynomials in the smeared field when acting on the vacuum span the Hilbert space H, the same space spanned by polynomials in the (smeared) creation operator $A^{\dagger}(\underset{\sim}{x},\lambda)$ acting on the vacuum. This is demonstrated carefully in Sec. 3, but is already plausible from the foregoing since the renormalized powers of the field $\phi(\underset{\sim}{x})$ can effectively modify the dependence on the variables λ.

2.3 Energy and Momentum Operators

Hamiltonian

The Hamiltonian for the models is given by

$$H = \int d\underset{\sim}{x} \int A^{\dagger}(\underset{\sim}{x},\lambda)\, h\, A(\underset{\sim}{x},\lambda)\, d\lambda$$

where h is a differential operator in λ and independent of $\underset{\sim}{x}$. This basic form is a direct consequence of the ultralocal properties of the models and is established in Sec. 3. It holds plausibly since the Hamiltonian density $H(\underset{\sim}{x})$ must be a functional of $A^{\dagger}(\underset{\sim}{x},\cdot)$ and $A(\underset{\sim}{x},\cdot)$, and the only possible translationally covariant expression is bilinear as above.

The differential operator h pertains to the various models and is intimately connected with the model function. Specifically, we choose

$$h = -\frac{1}{2}\frac{\partial^2}{\partial\lambda^2} + v(\lambda) \equiv -\frac{1}{2}\frac{\partial^2}{\partial\lambda^2} + \frac{1}{2}\frac{c''(\lambda)}{c(\lambda)}$$

which makes use of the differentiability and nonvanishing

of $c(\lambda)$. This form takes its cue from the canonical foundations of the theory. Note, as a differential equation, that

$$\hbar c(\lambda) = o .$$

If we adopt the form

$$c(\lambda) = |\lambda|^{-\gamma} e^{-y(\lambda)} ,$$

then it follows that

$$\hbar = -\frac{1}{2}\frac{\partial^2}{\partial\lambda^2} + \frac{\gamma(\gamma+1)}{2\lambda^2} + e + v_o(\lambda)$$

where

$$v_o(\lambda) \equiv \gamma\lambda^{-1}y'(\lambda) + \frac{1}{2}y'^2(\lambda) - \frac{1}{2}y''(\lambda) - e$$

and e is a constant chosen so that $v_o(o)=o$, i.e.,

$$e \equiv (\gamma - \frac{1}{2})y''(o) .$$

Suppose for the moment that $y(\lambda)$ is a polynomial (even), then so is $v_o(\lambda)$,

$$v_o(\lambda) = \sum_{n=1}^{N} v_{2n}\,\lambda^{2n},$$

which has the gross qualitative features of an attractive well. On the other hand, the term $\frac{1}{2}\gamma(\gamma+1)\lambda^{-2}$ acts as a repulsive potential. The combined potential v has the qualitative shape of the letter "W", i.e., two wells. Such a potential has only discrete eigenstates; but in virtue of the strong repulsive term, the eigenstates can be localized in the positive or negative half of λ space (see Sec. 3). The symmetry of the potential then leads to a two-fold level degeneracy, and we may adopt the convenient, integral operator realization

$$\hbar = \sum_{\ell=o}^{\infty} \mu_\ell u_\ell(\lambda)u_\ell^*(\lambda') ,$$

where the eigenstates $u_\ell(\lambda)$ are real, orthonormal, have the parity of ℓ, i.e.,

$$u_\ell(-\lambda) = (-1)^\ell u_\ell(\lambda),$$

and the energy levels are positive, $\mu_\ell > 0$, two-fold degenerate,

$$\mu_{2\ell} = \mu_{2\ell+1} ,$$

and increasing, $\mu_{2\ell+2} \geq \mu_{2\ell}$; $\mu_{2\ell} \to \infty$. These are basic properties of such Schrödinger operators.

These properties also hold for a nonpolynomial $y(\lambda)$ such that $y(\lambda) \geq p(\lambda)$ where $p(\lambda)$ is a polynomial (even), which is nonconstant and bounded below. For example, if

$$y(\lambda) = \frac{1}{2}\lambda^2 P(\lambda)/Q(\lambda)$$

where $P(\lambda)$ and $Q(\lambda)$ are even polynomials of the same degree [with $Q(\lambda)>0$, $P(\lambda)$ bounded below], then

$$v_0(\lambda) = \frac{1}{2}\lambda^2 \bar{P}(\lambda)/\bar{Q}(\lambda) \tag{2-5}$$

where $\bar{P}(\lambda)$ and $\bar{Q}(\lambda)$ are analogous polynomials of the same degree (which is different than that of P). The qualitative behavior in this case is more or less that for which $P=Q=\bar{P}=\bar{Q}=1$. [On the other hand, if $v_0(\lambda)$ remains bounded at infinity, e.g., when $y(\lambda)=(a^2+\lambda^2)^{1/2}$, then we obtain continuous spectrum for h coupled with some bound states. While this is an acceptable set of models our interest primarily lies with the cases where $y(\lambda)$ is a polynomial or bounded by one as discussed above leading to a purely discrete spectrum for h].

The diagonalization of h permits us to give a more transparent form to H. If we introduce

$$A_\ell(\underset{\sim}{x}) \equiv \int u_\ell^*(\lambda) A(\underset{\sim}{x},\lambda)\, d\lambda \ ,$$

$$N_\ell(\underset{\sim}{x}) \equiv A_\ell^\dagger(\underset{\sim}{x}) A_\ell(\underset{\sim}{x}) \ ,$$

then we may write

$$H = \Sigma \int \mu_\ell N_\ell(\underset{\sim}{x})\, d\underset{\sim}{x} = \Sigma \mu_\ell N_\ell$$

where N_ℓ is the nunber operator for excitations of type ℓ. A physical interpretation of this relation will be given in the next subsection.

A useful alternative expression for h is given by

$$h = b^\dagger b \tag{2-6}$$

where

$$b \equiv \frac{1}{\sqrt{2}}\, c(\lambda) \frac{\partial}{\partial \lambda}\, c^{-1}(\lambda).$$

With this form for h we may write

$$H = \int d\underset{\sim}{x} \int [bA(\underset{\sim}{x},\lambda)]^\dagger [bA(\underset{\sim}{x},\lambda)]\, d\lambda \ . \tag{2-7}$$

Since b involves division by $c(\lambda)$ followed by differentiation we can freely add $c(\lambda)$ to $A(\underset{\sim}{x},\lambda)$ above leading to

$$H = \int d\underset{\sim}{x} \int [bB(\underset{\sim}{x},\lambda)]^\dagger [bB(\underset{\sim}{x},\lambda)]\, d\lambda$$

$$= \int d\underset{\sim}{x} \int B^\dagger(\underset{\sim}{x},\lambda)\, h B(\underset{\sim}{x},\lambda)\, d\lambda.$$

Equation (2-6) makes evident the positivity of h, while (2-7) makes evident the positivity of H. Note that the same Hamiltonian h (and therefore H) arises from model functions which are proportional to one another.

Space Translation Generator

The space translation generator $\underset{\sim}{P}$ for these theories is a straightforward generalization of that for usual free theories. In the present case

$$\underset{\sim}{P} = \int d\underset{\sim}{x} \int A^{\dagger}(\underset{\sim}{x},\lambda)(-i\underset{\sim}{\nabla})A(\underset{\sim}{x},\lambda)\,d\lambda$$

an operator which fulfills the basic requirement

$$-i[\underset{\sim}{P},\phi(\underset{\sim}{x})] = \underset{\sim}{\nabla}\phi(\underset{\sim}{x})$$

placed on such an operator. In addition, it is clear that

$$[\underset{\sim}{P},H] = 0$$

and that

$$\underset{\sim}{P}|0\rangle = 0,$$

and moreover that $|0\rangle$ is a nondegenerate eigenstate of $\underset{\sim}{P}$.

2.4 Interpretation and Basic Properties

Interpretation: Preliminary Remarks

Heuristically, we expect that the Hamiltonian density should split as

$$H(\underset{\sim}{x}) = K(\underset{\sim}{x}) + V(\underset{\sim}{x}),$$

where $K(\underset{\sim}{x})$ denotes the kinetic energy ("$\frac{1}{2}\pi^2$") term while the potential $V(\underset{\sim}{x})$ is a function of the field. Let us set

$$K(\underset{\sim}{x}) \equiv \int B^{\dagger}(\underset{\sim}{x},\lambda)[-\tfrac{1}{2}\partial^2/\partial\lambda^2 + \tfrac{1}{2}\gamma(\gamma+1)\lambda^{-2} + e]B(\underset{\sim}{x},\lambda)\,d\lambda$$

$$V(\underset{\sim}{x}) \equiv \int B^{\dagger}(\underset{\sim}{x},\lambda)v_o(\lambda)B(\underset{\sim}{x},\lambda)\,d\lambda\ .$$

Then $V(\underset{\sim}{x})$ is a well defined local operator and is formally given in terms of the field by

$$V(\underset{\sim}{x}) = Z^{-1}v_o(Z\phi(\underset{\sim}{x})).$$

We identify this as the local potential operator and claim that it is related to the entire classical potential term*

---

*Thus if v_o is nonpolynomial, as in (2-5), then we are dealing with a "nonpolynomial Lagrangian".

$$\tfrac{1}{2}m_o^2\phi_{cl}^2(\underset{\sim}{x}) + V[\phi_{cl}(\underset{\sim}{x})] \equiv v_o(\phi_{cl}(\underset{\sim}{x})).$$

On this basis, the Z factors arise as necessary "renormalizations" in forming $V(\underset{\sim}{x})$. This interpretation suggests that $v_o=\frac{1}{2}\mu^2\lambda^2$ (i.e., $V\equiv 0$) should have something to do with the usual free theory. We will discuss this case in Sec. 3, but we may anticipate somewhat by noting, in this case, that

$$\mu_{2\ell} = \mu_{2\ell+1} = \mu(2\ell+2\gamma+1),$$

an equal spacing rule characteristic of a free theory.

With $V(\underset{\sim}{x})$ meaningful so too is $K(\underset{\sim}{x})$ which we interpret as the "renormalized" kinetic energy term. In effect, the latter two parts provide a renormalization to the differential term making the whole operator well defined.

We now have two expressions for the Hamiltonian. The expression with $B^\dagger$ and B pertains, in the usual picture, to an expression for H in terms of the interacting field containing, for example, a potential which is (say) a polynomial in the field $\phi(\underset{\sim}{x})$. The expression in terms of $A^\dagger$ and A pertains, in the usual picture, to an expression for H in terms of the various asymptotic fields. (Because S=1 past and future asymptotic fields are alike). Since $H=\Sigma\mu_\ell N_\ell$, we may say, in the standard language, that the field $\phi(\underset{\sim}{x})$ can create "particles" of the ℓth species with "mass" μ_ℓ, $\ell=0,1,2,\ldots$. (Why these levels are two-fold degenerate is presently not clearly understood). In the absence of dispersion the various levels μ_ℓ arise from persistent localized interactions since the "particles" cannot escape from one another. If, in effect, there is no localized interaction, then μ_ℓ should be linear in ℓ (apart from degeneracies) which is indeed the case. Departures from these values are suggestive of positive or negative, multiparticle interaction energies. Presumably, if the dispersion could be restored, nearly all of the energy levels would spread into a continuum while the lowest and perhaps a few higher ones with negative interaction energies would remain discrete, the latter corresponding to bound states.

Absence of Canonical Commutation Relations: Infinite Field Strength Renormalization

Without regard to domain questions let us compute $\dot{\phi}(\underset{\sim}{x})$. From the general relation

$$[\int B^{\dagger}(\underset{\sim}{x},\lambda)wB(\underset{\sim}{x},\lambda)d\lambda,\ \int B^{\dagger}(\underset{\sim}{x}',\lambda')w'B(\underset{\sim}{x}',\lambda')d\lambda']$$

$$= \delta(\underset{\sim}{x}-\underset{\sim}{x}')\int B^{\dagger}(\underset{\sim}{x},\lambda)[w,w']B(\underset{\sim}{x},\lambda)d\lambda$$

we readily see that

$$\dot{\phi}(\underset{\sim}{x}) = i[H,\phi(\underset{\sim}{x})] = \int B^{\dagger}(\underset{\sim}{x},\lambda)(-i\frac{\partial}{\partial\lambda})B(\underset{\sim}{x},\lambda)d\lambda.$$

However this expression does not lead to an operator (rather it is a form). For when $w=-i\partial/\partial\lambda$ and $s\neq 0$

$$\int |e^{isw}-1)c(\lambda)|^2 d\lambda = \int |c(\lambda+s)-c(\lambda)|^2 d\lambda = \infty$$

invalidating the requirement for $\dot{\phi}(\underset{\sim}{x})$ to be a local operator (see also Sec. 3). Nevertheless, if we proceeded formally, we would obtain

$$[\phi(\underset{\sim}{x}),\ \dot{\phi}(\underset{\sim}{y}) = i\delta(\underset{\sim}{x}-\underset{\sim}{y})\int B^{\dagger}(\underset{\sim}{x},\lambda)B(\underset{\sim}{x},\lambda)d\lambda$$

which has a leading c-number singularity

$$i\delta(\underset{\sim}{x}-\underset{\sim}{y})\int c^2(\lambda)d\lambda. \tag{2-8}$$

Conventionally, such a divergence is generally interpreted as introducing an infinite field strength renormalization, $\int c^2(\lambda)d\lambda=\infty$. Thus these models do not possess a canonical field and momentum pair obeying traditional canonical commutation relations. Nevertheless, the fact that the dominant contribution to the commutation is a c-number (albeit infinite) lends further support to the initial choice made for the Hamiltonian.

Two-Point Function and Spectral Representation

The basic expression

$$e^{iHt}A(\underset{\sim}{x},\lambda)e^{-iHt} = e^{-iht}A(\underset{\sim}{x},\lambda) \tag{2-9}$$

enables us to readily compute the two-point function. In particular

$$<o|\phi(\underset{\sim}{x},t)\phi(\underset{\sim}{y})|o> = <o|\phi(\underset{\sim}{x})e^{-iHt}\phi(\underset{\sim}{y})|o>$$

$$= <o|(\int c(\lambda)\lambda A(\underset{\sim}{x},\lambda)d\lambda)e^{-iHt}(\int A^{\dagger}(\underset{\sim}{y},\lambda')\lambda'c(\lambda')d\lambda')|o>$$

$$= \delta(\underset{\sim}{x}-\underset{\sim}{y})\int c(\lambda)\lambda e^{-iht}\lambda c(\lambda)d\lambda \ .$$

The positivity of h, $h>0$, suggests, as usual, that we preferably may study

$$\bar{J}(t) \equiv \int c(\lambda)\lambda e^{-ht}\lambda c(\lambda)d\lambda \ .$$

The spectral resolution for h ensures that we may write

$$\bar{J}(t) \equiv \frac{1}{2}\int_0^\infty e^{-\omega t}\rho(\omega)d\omega/\omega$$

where $\rho(\omega)$ -- a sum of δ-functions in the discrete cases -- is the analog of the usual spectral weight function. Evidently

$$\bar{J}(0) = \int\lambda^2 c^2(\lambda)d\lambda = \frac{1}{2}\int_0^\infty \rho(\omega)d\omega/\omega<\infty,$$

while

$$\bar{J}'(0) = \frac{1}{2}\int c^2(\lambda)d\lambda = \frac{1}{2}\int_0^\infty \rho(\omega)d\omega = \infty \ .$$

The latter relation implies that

$$\langle o|\phi(f)H\phi(f)|o\rangle = ||H^{1/2}\phi(f)|o\rangle||^2$$
$$= \frac{1}{2}\int f^2(\underset{\sim}{x})d\underset{\sim}{x} \int c^2(\lambda)d\lambda = \infty ,$$

and thus $\phi(f)|o\rangle$ is not in the domain of $H^{1/2}$, and, a fortiori, not in the domain of H^{ξ}, $\xi \geq 1/2$.

The behavior of such matrix elements for $\xi < 1/2$ may be plausibly estimated by the following hand-waving type argument. Consider

$$h\lambda^{\theta}c = -\frac{1}{2}c^{-1}\frac{\partial}{\partial\lambda}c^2\frac{\partial}{\partial\lambda}c^{-1}\cdot\lambda^{\theta}c$$
$$\sim \lambda^{-2}\cdot\lambda^{\theta}c$$

where the latter behavior is meant to reflect only the leading singularity structure. Similarly,

$$h^2\lambda^{\theta}c \sim h\lambda^{\theta-2}c \sim \lambda^{-4}\cdot\lambda^{\theta}c,$$

and thus

$$h^{\xi}\lambda^{\theta}c \sim \lambda^{-2\xi}\cdot\lambda^{\theta}c .$$

Finally

$$\int \lambda^{\theta}ch^{\xi}\lambda^{\theta}cd\lambda \sim \int \lambda^{2(\theta-\xi-\gamma)}e^{-2y}d\lambda$$
$$\sim [2(\theta-\xi-\gamma) + 1]^{-1}$$

whenever the quantity in brackets approaches zero from above. In the case $\theta=1$, $\xi=\frac{3}{2}-\gamma-\varepsilon\,(\varepsilon>0)$,

$$\int \lambda ch^{3/2-\gamma-\varepsilon}\lambda cd\lambda \sim \varepsilon^{-1},$$

which suggests that $\phi(f)|o\rangle$ is in the domain of $H^{3/4-1/2\gamma-\varepsilon}$, $\varepsilon>0$, and not otherwise.

This estimate is useful in another way. By introducing the eigenstates of h it follows that

$$\Sigma\ \mu_\ell^{3/2-\gamma-\varepsilon} \left| \int u_\ell^* \lambda c d\lambda \right|^2 \sim \varepsilon^{-1} .$$

This means, roughly speaking, that

$$k_\ell \equiv \int u_\ell^* \lambda c(\lambda) d\lambda \sim \ell^{-1/2} \mu_\ell^{-3/4+1/2\gamma}$$

asymptotically. Now, if 2N is the largest power in the potential $v_o(\lambda)$, then crudely (Bohr-Sommerfeld)

$$\mu_\ell \sim \ell^{\frac{2N}{1+N}}$$

which gives an estimate for k_ℓ in terms of γ and N. When N=1, $\mu_\ell \sim \ell$ and $k_\ell \sim \ell^{-5/4+1/2\gamma}$.

The eigenstates of h permit us to write

$$\bar{J}(t) = \Sigma\ e^{-\mu_\ell t} |k_\ell|^2$$

and to identify

$$\begin{aligned} \rho(\omega) &= 2\omega\ \Sigma\ \delta(\omega-\mu_\ell)\, |k_\ell|^2 \\ &= 2\ \Sigma\ \delta(\omega-\mu_\ell)\mu_\ell |k_\ell|^2 \\ &\sim 2\ \Sigma\ \delta(\omega-\mu_\ell)(\mu_\ell/\ell)\mu_\ell^{\gamma-3/2} . \end{aligned} \tag{2-10a}$$

For N=1, our previous estimate then suggests that

$$\rho(\omega) \sim 2\ \Sigma\ \delta(\omega-\ell)\ell^{\gamma-3/2} . \tag{2-10b}$$

While it is difficult to prove this result, we can make it plausible with the idealized example in which $c(\lambda)=|\lambda|^{-\gamma}$, $\gamma>1/2$. In this case

$$h = -\frac{1}{2}\frac{\partial^2}{\partial\lambda^2} + \frac{\gamma(\gamma+1)}{2\lambda^2}$$

which is homogeneous in λ^{-2}. Since $\|\phi(f)|o>\| = \infty$ here, we study $\Delta\bar{J}(t) \equiv \bar{J}(t)-\bar{J}(0)$ and by a change of variables find

$$\Delta\bar{J}(t) = \int \lambda|\lambda|^{-\gamma}(e^{-ht}-1)\lambda|\lambda|^{-\gamma}d\lambda = t^{3/2-\gamma}\Delta\bar{J}(1).$$

It follows, therefore, that

$$\rho(\omega) = \omega^{\gamma-3/2}\rho(1) \qquad (2\text{-}11)$$

not unlike the dependence implied by (2-10).

Space-Time Smearing

At this point it is appropriate to comment on those cases where $\gamma \geq 3/2$. The model function $c(\lambda)$ is then too singular for $\phi(\underset{\sim}{x})$ to be an operator with only spatial smearing [since condition (2-2) is violated]. However the previous definition of the Hamiltonian carries through without difficulty, and we may define the field operator with the help of space-time smearing. Evidently the time smearing $g(t)$ should at least fulfill

$$\int |\tilde{g}(\omega)|^2 \rho(\omega)\, d\omega/\omega < \infty \; .$$

These, too, may be considered ultralocal models; however, we shall not discuss them in any extent*.

2.5 Higher Order Dynamics

Time-Dependent Field Operator

Combining (2-9) with the definition of $\phi(\underset{\sim}{x})$, we find that

$$\begin{aligned}\phi(\underset{\sim}{x},t) &= e^{iHt}\phi(\underset{\sim}{x})e^{-iHt}\\ &= \int A^{\dagger}(\underset{\sim}{x},\lambda)e^{iht}\lambda e^{-iht}A(\underset{\sim}{x},\lambda)d\lambda\\ &+ \int c(\lambda)\lambda e^{-iht}A(\underset{\sim}{x},\lambda)d\lambda\\ &+ \int A^{\dagger}(\underset{\sim}{x},\lambda)e^{iht}\lambda c(\lambda)d\lambda \; . \qquad (2\text{-}12)\end{aligned}$$

Further if we employ the relation

*The role of γ in controlling the high energy behavior of $\rho(\omega)$, apparent in (2-10) and (2-11), makes it an especially interesting parameter. Model functions with stronger types of singularities [e.g., $c(\lambda)=\exp(\lambda^{-2}-\lambda^2)$] may lead to spectral weights $\rho(\omega)$ which grow faster than any power.

$$e^{-i\hbar t}A(\underset{\sim}{x},\lambda) = \Sigma\, u_\ell(\lambda)e^{-i\mu_\ell t}\, A_\ell(\underset{\sim}{x})$$

then

$$\phi(\underset{\sim}{x},t) = \sum_{\ell,m} A_\ell^\dagger(\underset{\sim}{x})\lambda_{\ell m}\, e^{i(\mu_\ell-\mu_m)t} A_m(\underset{\sim}{x})$$

$$+ \sum_\ell k_\ell^* e^{-i\mu_\ell t} A_\ell(\underset{\sim}{x})$$

$$+ \sum_\ell A_\ell^\dagger(\underset{\sim}{x})e^{i\mu_\ell t} k_\ell,$$

where

$$\lambda_{\ell m} \equiv \int u_\ell^* \lambda u_m d\lambda,$$

$$k_\ell \equiv \int u_\ell^* \lambda c(\lambda)d\lambda.$$

We may also make use of the relation

$$e^{-i\hbar t}c(\lambda) = c(\lambda) \tag{2-13}$$

(which lies outside of Hilbert space) to write

$$\phi(\underset{\sim}{x},t) = \int[e^{-i\hbar t}B(\underset{\sim}{x},\lambda)]^\dagger\lambda\; e^{-i\hbar t}B(x,\lambda)d\lambda\;.$$

For the most part it is appropriate to re-express this as

$$\phi(\underset{\sim}{x},t) = \int B^\dagger(\underset{\sim}{x},\lambda)\lambda(t)B(\underset{\sim}{x},\lambda)d\lambda, \tag{2-14}$$

where

$$\lambda(t) \equiv e^{i\hbar t}\;\lambda\; e^{-i\hbar t},$$

so long as we treat $c(\lambda)$ as a formal "left" eigenstate of zero energy. For instance, in writing

$$\langle o|\phi(\underset{\sim}{x},t)\phi(\underset{\sim}{y})|o\rangle$$

$$= \delta(\underset{\sim}{x}-\underset{\sim}{y})\int c(\lambda)\lambda(t)\lambda\; c(\lambda)d\lambda$$

we mean to have

$$\delta(\underset{\sim}{x}-\underset{\sim}{y})\int c(\lambda)\lambda e^{-i\hbar t}\;\lambda c(\lambda)d\lambda$$

since, according to (2-12), this is evidently the intent.

Indeed, computation in the alternate form $(\phi(\underset{\sim}{x},t)|o\rangle,\ \phi(\underset{\sim}{y})|o\rangle)$ using (2-13) directly leads to the right formula.

Truncated Vacuum Expectation Values

Let us consider the expression

$$W_n^T \equiv \langle o|\phi(\underset{\sim}{x}_1,t_1)\phi(\underset{\sim}{x}_2,t_2)\cdots\phi(\underset{\sim}{x}_n,t_n)|o\rangle^T$$

for the truncated n^{th} order vacuum expectation value. It is straightforward to see that this is given by

$$W_n^T = \delta(\underset{\sim}{x}_1-\underset{\sim}{x}_2)\delta(\underset{\sim}{x}_2-\underset{\sim}{x}_3)\cdots\delta(\underset{\sim}{x}_{n-1}-\underset{\sim}{x}_n) \qquad (2\text{-}15)$$

$$\times \int c(\lambda)\lambda(t_1)\lambda(t_2)\cdots\lambda(t_n)c(\lambda)d\lambda$$

with the understanding that $c(\lambda)$ acts as a formal left and right eigenvector of zero energy. If $t_1=t_n=o$ then the formula is literally true as it stands. It is instructive to derive (2-15) using both (2-12) and (2-14) to see the origin of the contributing terms.

An equation such as (2-15) makes abundantly clear the importance of a study of the dynamical evolution in λ space, i.e.

$$\lambda(t) = e^{i\hbar t}\lambda e^{-i\hbar t}.$$

Techniques for the study of this relation, and solutions in a particular case are discussed in Sec. 3.

"Turning Off the Coupling Constant"

It is plausible that an interacting theory passes to a free theory as the coupling constant goes to zero in some sense. In the present case, the free theory is the free Fock theory of mass (say) μ. In terms of the usual annihilation and creation operators,

$$\phi_F(\underset{\sim}{x}) = \frac{1}{\sqrt{2\mu}}\,[A_F(\underset{\sim}{x}) + A_F^{\dagger}(\underset{\sim}{x})],$$

$$\mathcal{H}_F = \mu \int A_F^\dagger(\underset{\sim}{x}) A_F(\underset{\sim}{x}) d\underset{\sim}{x},$$

$$\underset{\sim}{P}_F = \int A_F^\dagger(\underset{\sim}{x}) (-i\underset{\sim}{\nabla}) A_F(\underset{\sim}{x}) d\underset{\sim}{x}.$$

The states $|f\rangle_F \equiv \exp[i\phi_F(f)]|o\rangle$ are coherent states, and fulfill $A_F(\underset{\sim}{x})|f\rangle_F = i(2\mu)^{-1/2} f(\underset{\sim}{x})|f\rangle_F$. Thus

$${}_F\langle f|\mathcal{H}_F|f'\rangle_F = \frac{1}{2} \int f(\underset{\sim}{x}) f'(\underset{\sim}{x}) d\underset{\sim}{x}\, {}_F\langle f|f'\rangle_F,$$

where $E_F(f'-f) = {}_F\langle f|f'\rangle_F$ and

$$E_F(f) = \langle o|e^{i\phi_F(f)}|o\rangle = e^{-\frac{1}{4\mu}\int f^2(\underset{\sim}{x}) d\underset{\sim}{x}} .$$

These functionals determine both $\phi_F(\underset{\sim}{x})$ and $\mathcal{H}_F$ up to unitary equivalence.

The expectation functional of the free theory is obtained [cf. (2-4)] through a sequence of models, or model functions $c_\eta(\lambda)$, such that, as $\eta \to \infty$,

$$\int [1-\cos(\lambda f)] c_\eta^2(\lambda) d\lambda \to \frac{1}{4\mu} f^2.$$

One can easily arrange for such a sequence to yield <u>any</u> pregiven mass $\mu > o$. For example, we may choose $c_\eta^2(\lambda) \equiv \eta^3 c^2(\eta\lambda)$ where $c(\lambda)$ is an interacting model function scaled so that $\int \lambda^2 c^2(\lambda) d\lambda = (2\mu)^{-1}$. Such a sequence effectively turns off the coupling constant as far as the field is concerned, and in the sense described $\phi(\underset{\sim}{x}) \to \phi_F(\underset{\sim}{x})$.

For the Hamiltonian, additional features must be considered first. On the one hand, $e^{i\phi(f)}|o\rangle$ is not in the domain of $\mathcal{H}$; and on the other hand, there is an infinite field strength renormalization, $\int c^2(\lambda) d\lambda$, as illustrated in (2-8). To win the free Hamiltonian we first consider the modified and scaled sequence of Hamiltonians [cf. (2-6) and (2-7)]

$$H_\varepsilon \equiv \int d\underset{\sim}{x} \int A^\dagger(\underset{\sim}{x},\lambda)(b^\dagger M_\varepsilon b)A(\underset{\sim}{x},\lambda)d\lambda ,$$

where $b \equiv 2^{-1/2} c(\partial/\partial\lambda)c^{-1}$ and

$$M_\varepsilon \equiv \frac{e^{-\varepsilon^2/\lambda^2}}{\int e^{-\varepsilon^2/\lambda^2} c^2(\lambda)d\lambda} .$$

The numerator of M_ε fixes the domain problem, while the denominator accounts for the field strength renormalization. As $\varepsilon \to o$, the sequence of operators H_ε converges, not as an operator but as a form. Consider the class of coherent state matrix elements with wave functions of the form $\psi(\underset{\sim}{x},\lambda) = \lambda c(\lambda)h(\underset{\sim}{x},\lambda)$, where h is C^∞ and has compact support in R^s. For such matrix elements,

$$\lim_{\varepsilon \to o} \langle\psi|H_\varepsilon|\psi'\rangle = \frac{1}{2}\int h^*(\underset{\sim}{x},o)h'(\underset{\sim}{x},o)d\underset{\sim}{x} \langle\psi|\psi'\rangle .$$

The states $|f\rangle \equiv \exp[i\phi(f)]|o\rangle$ are, in fact, coherent states with

$$\psi(\underset{\sim}{x},\lambda) = \{e^{i\lambda f(\underset{\sim}{x})} - 1\}c(\lambda) \equiv \lambda c(\lambda)h(\underset{\sim}{x},\lambda) ,$$

and therefore $h(\underset{\sim}{x},o) = if(\underset{\sim}{x})$. Combining, this relation with the limit $\eta \to \infty$, we determine that

$$\lim_{\eta\to\infty} \lim_{\varepsilon\to o} \langle f|H_\varepsilon|f'\rangle = \frac{1}{2}\int f(\underset{\sim}{x})f'(\underset{\sim}{x})d\underset{\sim}{x} \; {}_F\langle f|f'\rangle_F$$

which coincides with the expression for ${}_F\langle f|H_F|f'\rangle_F$. This means, in the sense described, that $H \to H_F$. The limiting form of the momentum operator $\underset{\sim}{P}$ is easily found as $\eta \to \infty$ since

$$\langle f|e^{i\underset{\sim}{a}\cdot\underset{\sim}{P}}|f'\rangle \equiv \langle f|f'_a\rangle \to {}_F\langle f|f'_a\rangle_F$$
$$= {}_F\langle f|e^{i\underset{\sim}{a}\cdot\underset{\sim}{P}_F}|f'\rangle_F$$

where $f'_a(\underset{\sim}{x}) \equiv f'(\underset{\sim}{x}+\underset{\sim}{a})$. In this sense $\underset{\sim}{P} \to \underset{\sim}{P}_F$.

The preceding discussion applies for model functions with singularities γ such that $1/2 \le \gamma < 3/2$. If $\gamma \ge 3/2$, or for more general model function singularities, one may choose suitable renormalized powers of the field and proceed analogously.

3. FUNDAMENTAL OPERATOR PROPERTIES

3.1 Operators and Limits of Operators

Coherent States

In the Hilbert space notation of Sec. 2.1, the states

$$|\psi\rangle = N \exp \{\iint \psi(\underset{\sim}{x},\lambda)A^{\dagger}(\underset{\sim}{x},\lambda)d\underset{\sim}{x}d\lambda\}|o\rangle ,$$

where

$$N = \exp \{-\tfrac{1}{2}\iint |\psi(\underset{\sim}{x},\lambda)|^2 d\underset{\sim}{x}d\lambda\},$$

are the coherent states and are defined for each element $\psi(\underset{\sim}{x},\lambda)\varepsilon\, L^2(R^s,R)$. The matrix element of two such states,

$$\langle\psi|\psi'\rangle = NN' \exp \{\iint\psi^*(\underset{\sim}{x},\lambda)\psi'(\underset{\sim}{x},\lambda)d\underset{\sim}{x}d\lambda\},$$

never vanishes, but these states possess several properties which make them useful. First, these states form a total set, i.e., they span the Hilbert space H. Any state $|X\rangle\varepsilon H$ for which $\langle\psi|X\rangle=0$ for all $\psi\ \varepsilon L^2$ is the zero state, $|X\rangle=0$. In fact, if ψ is restricted to lie in a dense set in L^2 the so restricted set of coherent states is still total. The second important property is

$$A(\underset{\sim}{x},\lambda)|\psi\rangle = \psi(\underset{\sim}{x},\lambda)|\psi\rangle$$

which asserts that $|\psi\rangle$ is an eigenstate of $A(\underset{\sim}{x},\lambda)$ with eigenvalue $\psi(\underset{\sim}{x},\lambda)$. Thus normal-ordered operator expressions formally have easily computed coherent state matrix elements, namely

$$\langle\psi|:O(A^\dagger,A):|\psi'\rangle = O(\psi^*,\psi')\langle\psi|\psi'\rangle$$

where the colons denote normal ordering (all $A^\dagger$ to the left of all A).

Bilinear Operators in Translated Fock Representation

We wish to study the properties of objects of the form

$$W(f) = \int d\underset{\sim}{x}\; f(\underset{\sim}{x}) \int B^\dagger(\underset{\sim}{x},\lambda)\, wB(\underset{\sim}{x},\lambda)\, d\lambda ,$$

for real $f(\underset{\sim}{x})$ and self-adjoint w, and determine when we are justified in treating such objects as self adjoint operators. For the present we assume that $f(\underset{\sim}{x}) \varepsilon C_0^\infty$ and concern ourselves with making the local operator well defined. This we do by introducing a family of approximating operators of the form

$$W_\varepsilon(f) \equiv \int d\underset{\sim}{x} f(\underset{\sim}{x}) \int B_\varepsilon^\dagger(\underset{\sim}{x},\lambda)\, wB_\varepsilon(\underset{\sim}{x},\lambda)\, d\lambda$$

where $\varepsilon > 0$ and

$$B_\varepsilon(\underset{\sim}{x},\lambda) \equiv A(\underset{\sim}{x},\lambda) + c_\varepsilon(\lambda)$$

$$B_\varepsilon^\dagger(\underset{\sim}{x},\lambda) \equiv A(\underset{\sim}{x},\lambda) + c_\varepsilon^*(\lambda).$$

Here $c_\varepsilon(\lambda)$ (possibly complex) is a smoothed-out model function which is chosen such that as $\varepsilon \to o$, $c_\varepsilon(\lambda) \to c(\lambda)$ in some sense. More specifically, we choose

$$\int |c_\varepsilon(\lambda)|^2 d\lambda < \infty , \qquad \int |wc_\varepsilon(\lambda)|^2 d\lambda < \infty \tag{3-1}$$

for all $\varepsilon > o$ which means that besides being square integrable the sequence is adapted to the specific self-adjoint operator w in question. For example, if w denotes multiplication or differentation and we are concerned only about the singularity of $c(\lambda)$ at $\lambda = o$, then we might use

$$c_\varepsilon(\lambda) = \exp[-(1-|\lambda|^{-\varepsilon})^2]c(\lambda) \tag{3-2}$$

which has the property that $c_\varepsilon(o)=o$ and

$$c_\varepsilon(\lambda) \to c(\lambda), \quad \lambda \neq o.$$

With these conditions $W_\varepsilon(f)$ is self-adjoint and $\exp[iW_\varepsilon(f)]$ is unitary*. A useful identity reads

$$\exp[iW_\varepsilon(f)] = :\exp\{\int d\underset{\sim}{x}\int B_\varepsilon^\dagger(\underset{\sim}{x},\lambda)[e^{iwf(\underset{\sim}{x})}-1]B_\varepsilon(\underset{\sim}{x},\lambda)d\lambda\}:,$$

which leads immediately to the coherent state matrix elements

$$\langle\psi|e^{iW_\varepsilon(f)}|\psi'\rangle = e^{R_\varepsilon}\langle\psi|\psi'\rangle, \tag{3-3a}$$

where

$$R_\varepsilon \equiv \int d\underset{\sim}{x}\int[\psi^*(\underset{\sim}{x},\lambda) + c_\varepsilon^*(\lambda)][e^{iwf(\underset{\sim}{x})}-1] \times [\psi'(\underset{\sim}{x},\lambda) + c_\varepsilon(\lambda)]d\lambda. \tag{3-3b}$$

A necessary and sufficient condition for a sequence of uniformly bounded operators to converge weakly to an operator is for a total set of matrix elements to converge, which in the above application states that

$$\text{w-lim}[\exp iw_\varepsilon(f)] = V(f)$$

provided that R_ε converges for all ψ, $\psi'\varepsilon L^2$ or in a dense set of L^2. We may focus on the λ variable, and for R_ε to converge, we require that

$$[e^{iws}-1]c_\varepsilon(\lambda) \equiv \phi_\varepsilon(s,\lambda)$$

converge weakly for each s to a vector $\phi(s,\lambda)\varepsilon L^2(R)$ [s plays the role of $f(\underset{\sim}{x})$] and that

$$\int c_\varepsilon^*(\lambda)[e^{iws}-1]c_\varepsilon(\lambda)d\lambda \equiv J_\varepsilon(s)$$

converges for each s to $J(s)$. To ensure that $V(f)$ has the form $\exp[iW(f)]$, $W(f)$ self adjoint, we requires more, namely that $\phi(s,\lambda)$ and $J(s)$ are continuous in the variable s, and the group property, $V(f_1)V(f_2)=V(f_1+f_2)$, which is

---

*Vectors with finitely many "particles" and smooth wave functions (i.e., smoothe relative to w) form a dense set of analytic vectors for $W_\varepsilon(f)$.

fulfilled if and only if

$$\int \phi^*(-s,\lambda)\phi(t,\lambda)d\lambda = J(s+t)-J(s)-J(t).$$

It is not difficult to see that the group property implies, and is implied by, strong convergence of the vectors $\phi(s,\lambda)$. Thus we can replace the group law, weak convergence and continuity by strong convergence and continuity.

Since a sequence of unitary operators which converges weakly to a unitary operator also converges strongly, we have Theorem 3.1. Under the conditions (3-1) on $c_\varepsilon(\lambda)$,

$$\text{s-lim}\ \exp[iW_\varepsilon(f)] = \exp[iW(f)]$$

provided that

$$J(s) = \lim \int c_\varepsilon^*(\lambda)\{e^{iws}-1\}c_\varepsilon(\lambda)d\lambda \tag{3-4}$$

is a continuous function, and that

$$\text{s-lim}\{e^{iws}-1\}c_\varepsilon(\lambda) = \phi(s,\lambda) \tag{3-5}$$

is a continous (in s) vector-valued function.

Examples

Let us turn our attention to some examples to illustrate this theorem. Suppose first that $w=w(\lambda)$, multiplication by a real function of λ. To take care of the singularity [of the model function $c(\lambda)$] at the origin and possible domain questions [of $w(\lambda)$] at infinity, let us choose

$$c_\varepsilon(\lambda) = 0 \quad ; \ |\lambda|<\varepsilon,$$
$$c_\varepsilon(\lambda) = c(\lambda) \ ; \ \varepsilon\le|\lambda|\le\varepsilon^{-1},$$
$$c_\varepsilon(\lambda) = 0 \quad ; \ \varepsilon^{-1}<|\lambda|.$$

In this case

$$J(s) = \int\{e^{isw(\lambda)}-1\}c^2(\lambda)d\lambda$$

interpreted as a principal value integral ($⨍$) This integral exists and is continuous provided

$$⨍_{|\lambda|<1} w(\lambda)c^2(\lambda)d\lambda ,$$

$$\int_{|\lambda|<1} w^2(\lambda)c^2(\lambda)d\lambda$$

exist. In all cases here we assume that

$$\int \lambda^2/(1+\lambda^2)c^2(\lambda)d\lambda<\infty.$$

If we grant the conditions above, it follows that

$$\{e^{isw(\lambda)}-1\}c(\lambda) = \phi(s,\lambda)$$

is continuous in s and is square integrable as is required for the existence of the local operator defined by $w=w(\lambda)$.

Besides the obvious examples of the above situation there are some interesting "strange" ones. Suppose that $c(\lambda)=|\lambda|^{-\gamma}$, $\gamma>1/2$, and $w(\lambda)=|\lambda|^{\xi}$, $\xi>2\gamma-1>0$.
It follows that

$$\begin{aligned} J(s) &= \int (e^{is|\lambda|^{\xi}}-1)|\lambda|^{-2\gamma}d\lambda \\ &= |s|^{\alpha}J(1) \end{aligned}$$

where $\alpha=(2\gamma-1)/\xi$, and thus $0<\alpha<1$. Although $W(f)$ is a self-adjoint operator, neither the vacuum $|0\rangle$, nor any coherent state $|\psi\rangle$ with $\int|\psi(\underset{\sim}{x},\lambda)|^2|\lambda|^{\xi}d\lambda<\infty$ and $\int|\psi(\underset{\sim}{x},\lambda)||\lambda|^{\xi-\gamma}d\lambda<\infty$, lies in its domain. For if that were true

$$\langle\psi|W(f)|\psi\rangle = \lim_{t\to 0} \langle\psi|(it)^{-1}\{e^{itW(f)}-1\}|\psi\rangle$$

would be finite; but given the form of $J(s)$ the limit diverges. Similar examples arise for <u>any</u> $c(\lambda)$; it suffices to choose

$$\int_{|\lambda|\geq 1} w(\lambda)c^2(\lambda)d\lambda = \infty .$$

It is noteworthy in such cases that the <u>formal</u> expression

$$W(\underset{\sim}{x}) = \int A^{\dagger}(\underset{\sim}{x},\lambda)w(\lambda)A(\underset{\sim}{x},\lambda)d\lambda + \int w(\lambda)c(\lambda)A(\underset{\sim}{x},\lambda)d\lambda + \int A^{\dagger}(\underset{\sim}{x},\lambda)w(\lambda)c(\lambda)d\lambda + \int c(\lambda)w(\lambda)c(\lambda)d\lambda$$

is quite without meaning since the latter two terms are not, by themselves, operators. Nevertheless the formal sum of all four leads to a local self-adjoint operator as discussed above.

The preceding example demonstrates that the failure of $W(\underset{\sim}{x})$ to be a local operator cannot in general be predicted from the divergence of the c-number component, i.e., of the sequence $\langle o|W_{\varepsilon}(\underset{\sim}{x})|o\rangle$. However, this is a valid prediction for $w(\lambda)=1$ since convergence of (3-4) requires convergence of $\int|c_{\varepsilon}(\lambda)|^2 d\lambda$ which does not occur.

Two other operators are of special interest. The first example arises for $w=-i\partial/\partial\lambda$. In that case $W(\underset{\sim}{x})$ is not a local operator since the norm of

$$\{e^{iws}-1\}c_{\varepsilon}(\lambda) = c_{\varepsilon}(\lambda+s) - c_{\varepsilon}(\lambda)$$

fails to converge ($s\neq o$) for our choice of model functions,

$$\int|c_{\varepsilon}(\lambda+s) - c_{\varepsilon}(\lambda)|^2 d\lambda \geq \int_{|\lambda|\leq \frac{1}{2}|s|} |c_{\varepsilon}(\lambda) - c_{\varepsilon}(\lambda+s)|^2 d\lambda \to \infty,$$

no matter what choice for $c_{\varepsilon}(\lambda)$ is made. Thus the formal expression for the momentum,

$$\dot{\phi}(\underset{\sim}{x}) = \int B^{\dagger}(\underset{\sim}{x},\lambda)(-i\partial/\partial\lambda)B(\underset{\sim}{x},\lambda)d\lambda,$$

does <u>not</u> correspond to a local operator.

The second example arises for $w=-(i/2)[\lambda(\partial/\partial\lambda)+(\partial/\partial\lambda)\lambda]$. In that case

$$(e^{iws}-1)c_\varepsilon(\lambda) = e^{s/2}c_\varepsilon(e^s\lambda) - c_\varepsilon(\lambda)$$

converges strongly for the choice of $c_\varepsilon(\lambda)$ given in (3-2) provided $\gamma=\frac{1}{2}$ (but not if $\gamma>\frac{1}{2}$). In this particular case ($\gamma=\frac{1}{2}$)

$$J(s) = \lim\int c_\varepsilon(\lambda)[e^{s/2}c_\varepsilon(e^s\lambda) - c_\varepsilon(\lambda)]d\lambda$$

is continuous (which is a good exercise for the reader). We shall have occasion to make use of this example, and we introduce the notation

$$\kappa(\underset{\sim}{x}) = \int B^\dagger(\underset{\sim}{x},\lambda)\,\eta\,B(\underset{\sim}{x},\lambda)d\lambda, \tag{3-6a}$$

where

$$\eta = -\tfrac{1}{2}i(\lambda\frac{\partial}{\partial\lambda} + \frac{\partial}{\partial\lambda}\lambda). \tag{3-6b}$$

Extention of Test Function Space

If we assume that $W(\underset{\sim}{x})$ define a valid local operator according to Theorem 3.1, we can introduce another limiting operation to extend the space of test functions beyond C_0^∞. Let $f_n(\underset{\sim}{x})\varepsilon C_0^\infty$ be a sequence of test functions and consider

$$\langle\psi|\exp\{i\int f_n(\underset{\sim}{x})W(\underset{\sim}{x})d\underset{\sim}{x}\}|\psi'\rangle$$
$$= \exp R(f_n)\langle\psi|\psi'\rangle$$

where

$$R(f_n) \equiv \int d\underset{\sim}{x}\{J(f_n(\underset{\sim}{x})) + \int\psi^*(\underset{\sim}{x},\lambda)\phi(f_n(\underset{\sim}{x}),\lambda)d\lambda$$
$$+ \int\phi^*(f_n(\underset{\sim}{x}),\lambda)\psi'(\underset{\sim}{x},\lambda)d\lambda$$
$$+ \int\psi^*(\underset{\sim}{x},\lambda)\{e^{iwf_n(\underset{\sim}{x})}-1\}\psi'(\underset{\sim}{x},\lambda)d\lambda\}.$$

In order for $f_n(\underset{\sim}{x})\to f(\underset{\sim}{x})$ so that s-lim $\exp[iW(f_n)]\to\exp[iW(f)]$, it is necessary and sufficient that $R(f_n)\to R(f)$ and that $R(sf)$ be continuous in the real variable s. Since this is an abelian family of operators [for fixed $W(\underset{\sim}{x})$] the group law is automatically satisfied. Recall that ψ and ψ' may be chosen as smooth as desired which ensures that the last term is not the critical one. On the other hand, it is clear that converge of f_n in the metric

$$d(f_n;f_m) \equiv \int d\underset{\sim}{x}\,|J(f_n(\underset{\sim}{x})) - J(f_m(\underset{\sim}{x}))|$$
$$+ \{\int d\underset{\sim}{x} \int d\lambda\,|\phi(f_n(\underset{\sim}{x}),\lambda) - \phi(f_m(\underset{\sim}{x}),\lambda)|^2\}^{1/2}$$

fulfills the desired criteria, both as to existence of the limit $R(f)$ and the continuity of $R(sf)$ in the variable s. Thus for a given model function $c(\lambda)$ and operator w there is a permissable extension of the test function space to the space as completed in the metric d. Among the possible new test functions are characteristic functions which are admissable for any w and $c(\lambda)$ fulfilling the conditions of Theorem 3.1.

In the special case $w(\lambda)=\lambda$, corresponding to the field operator $W(\underset{\sim}{x})=\phi(\underset{\sim}{x})$, we have

$$d(f_n;f_m) = \int d\underset{\sim}{x}\,|\int(\cos\lambda f_n - \cos\lambda f_m)c^2(\lambda)d\lambda|$$
$$+ \{\int d\underset{\sim}{x} \int|e^{i\lambda f_n}-e^{i\lambda f_m}|^2 c^2(\lambda)d\lambda\}^{1/2} .$$

Convergence of d (to zero) is assured by convergence of

$$\int|f_n - f_m|^2 d\underset{\sim}{x}$$

provided $\int\lambda^2 c^2(\lambda)d\lambda<\infty$. If instead we only have

$$\int_{|\lambda|<1} \lambda^2 c^2(\lambda)d\lambda + \int_{|\lambda|\geq 1} |\lambda|^\tau c^2(\lambda)d\lambda<\infty$$

for $0<\tau<2$, then besides convergence in L^2 it is sufficient to have convergence of

$$\int |f_n - f_m|^\tau d\underset{\sim}{x} .$$

3.2 Properties of the Field Operator

Justification for Assumed Field Operator

We seek here to justify the form for the field operator which we have assumed. For this purpose, we suppose that the theory admits self-adjoint smeared field operators $\phi(f)$ for $f(\underset{\sim}{x})\epsilon C_0^\infty$ and a translationally invariant vector, $|o\rangle$, which is also the ground state of the presumed Hamiltonian. Then, the ultralocal nature of the models requires that

$$E(f) = \langle o|e^{i\phi(f)}|o\rangle = e^{-\int d\underset{\sim}{x}\, L'[f(\underset{\sim}{x})]} \tag{3-7}$$

since disjoint regions of space are dynamically independent of each other for all time. Thus, if $f(\underset{\sim}{x})=f_1(\underset{\sim}{x})+f_2(\underset{\sim}{x})$ with $f_1\cdot f_2=o$, it is necessary that $E(f)=E(f_1)E(f_2)$. An even potential, $V[-\phi_{cl}]=V[\phi_{cl}]$, leads to the symmetry $E(-f)=E(f)$ or $L'[-f]=L'[f]$.

From its definition, $E(f)$ must be a positive-definite functional since for any finite set of complex numbers $\{c_j\}$,

$$\Sigma\; c_j^* c_k E(f_k-f_j) = ||\Sigma c_k e^{i\phi(f_k)}|o\rangle||^2 \geq o\;.$$

The problem of positive-definite expectation functionals having the ultralocal form of (3-7) arises in classical stochastic variable theory in the context of generalized random processes with independent values at every point [4]. The most general form for L' consistent with our symmetry requirements is given by

$$L'[f] = bf^2 + \int_{|\lambda|>o} (1-\cos\lambda f)\,d\sigma(\lambda)$$

where $b\geq o$ and σ is a positive measure which satisfies

$$\int_{|\lambda|>o} \lambda^2/(1+\lambda^2)\,d\sigma(\lambda) < \infty \tag{3-8}$$

From the form of L'[f] we may divide the field into two statistically independent components, one associated with bf^2 and the second associated with the remainder. The former can be interpreted as belonging to a free, ultra-local Fock representation and as evolving in time independently from the latter component. This decomposition into uncoupled fields one of which is free suggests that we confine our attention to the nonfree component. Thus we set b=o and consider

$$L[f(\underset{\sim}{x})] = \int_{|\lambda|>o} \{1-\cos[\lambda f(\underset{\sim}{x})]\,d\sigma(\lambda)$$

$$= \int \{1-\cos[\lambda f(\underset{\sim}{x})]\}c^2(\lambda)\,d\lambda \ .$$

In the latter relation we have assumed that σ has only an absolutely continuous component so that

$$E(f) = \langle o|e^{i\phi(f)}|o\rangle = e^{-\int d\underset{\sim}{x}\int\{1-\cos[\lambda f(\underset{\sim}{x})]\}c^2(\lambda)d\lambda}$$

identical to the expression (2-4). From the point of view of the present models, discrete or singular continuous measures σ are not of interest as regards the field operator.

A basic reason for the divergence of $\int c^2(\lambda)d\lambda$ may be easily given*. On physical grounds, the field operator $\phi(f)$ is expected to have a continuous spectrum and not to have a discrete component. This requires, according to the Riemann-Lebesque Lemma that, for fixed $f(\underset{\sim}{x})\varepsilon C_o^\infty$,

$$\lim_{s\to\infty} E(sf) = 0,$$

---

*A more important reason is the nondegeneracy of the Hamiltonian ground state (see Sec. 3.3).

which implies

$$\int c^2(\lambda)d\lambda = \infty .$$

This relation coupled with (3-8) fixes our basic growth restrictions on $c(\lambda)$.

Cyclicity of Representation

The equality of the expectation functional with that found earlier means that we may adopt

$$\phi(f) = \int f(\underset{\sim}{x})d\underset{\sim}{x} \int B^{\dagger}(\underset{\sim}{x},\lambda)\lambda B(\underset{\sim}{x},\lambda)d\lambda . \qquad (3\text{-}9)$$

However, the real utility of this relation stems from the cyclicity of the field operator when acting on the vacuum. Specifically, if an overbar denotes the closed linear span, then

$$\overline{\exp[i\phi(f)]|o\rangle} = H = \overline{\exp\iint \psi(\underset{\sim}{x},\lambda)A^{\dagger}(\underset{\sim}{x},\lambda)dx\ d\lambda|o\rangle} .$$

We demonstrate this as a byproduct of our discussion of the operators affiliated with the von Neumann algebra generated by the field operator ϕ.

We have shown earlier that we can extend the space of test functions for (3-9) to include characteristic functions. Let

$$\begin{aligned} u(\underset{\sim}{x};\underset{\sim}{z},\delta) &= 1 ; \quad \text{all } |x_i - z_i| \le \tfrac{1}{2}\delta \\ &= 0 ; \quad \text{any } |x_i - z_i| > \tfrac{1}{2}\delta \end{aligned}$$

denote such a function where $x_i, 1 \le i \le s$, are the coordinates of $\underset{\sim}{x}$. The operator

$$Y_\delta \equiv \delta^{-s}\int h(\underset{\sim}{z})d\underset{\sim}{z}\{\exp[it\int u(\underset{\sim}{x};\underset{\sim}{z},\delta)\phi(\underset{\sim}{x})d\underset{\sim}{x}]-1\}$$

has coherent state matrix elements given by

$$\langle\psi|Y_\delta|\psi'\rangle = D_\delta\langle\psi|\psi'\rangle ,$$

where

$$D_\delta = \delta^{-s} \int h(z)dz\{e^{F_\delta}-1\}$$

and

$$F_\delta = \underset{\delta\text{-cube}}{\int\cdots\int}[\psi^*(x,\lambda) + c(\lambda)]\{e^{i\lambda t}-1\}[\psi'(x,\lambda) + c(\lambda)]d\lambda dx$$

where the integration for x_i extends from $z_i - \frac{1}{2}\delta$ to $z_i + \frac{1}{2}\delta$, for all i. As $\delta \to o$ we obtain

$$\lim_{\delta\to o} \langle\psi|Y_\delta|\psi'\rangle = D\langle\psi|\psi'\rangle$$

where

$$D = \int h(z)dz \int[\psi^*(z,\lambda) + c(\lambda)]$$

$$\times \{e^{i\lambda t}-1\}[\psi'(z,\lambda) + c(\lambda)]d\lambda.$$

Accordingly, we find that

$$Y = \text{w-lim } Y_\delta = \int h(z)dz\int B^\dagger(z,\lambda)\{e^{i\lambda t}-1\}B(z,\lambda)d\lambda.$$

This is the precise statement of the formal relation

$$Z^{-1}\{e^{itZ\phi(x)}-1\} = \int B^\dagger(x,\lambda)(e^{it\lambda}-1)B(x,\lambda)d\lambda$$

which is like that used in Sec. 2.

Let $y(\lambda)\varepsilon C_o^\infty$, $\tilde{y}(t)$ denote its Fourier transform, and form

$$\int\tilde{y}(t)Ydt = \int h(z)dz\int B^\dagger(z,\lambda)\{y(\lambda)-y(o)\}B(z,\lambda)d\lambda.$$

Such operators and their limits for different $y(\lambda)$ generate the local potential operators used in the models.

If we make a further superposition and take appropriate limits of operators of the above type for different $h(z)$ and $y(\lambda)$ we can generate

$$G = \int dz \int B^\dagger(z,\lambda)h(z,\lambda)B(z,\lambda)d\lambda$$

where $h(z,\lambda)\varepsilon C_o^\infty$ and $h(z,\lambda)\equiv o$ for λ in some neighborhood of zero. Now G, as well as exp G, are constructs of the field operator $\phi(\underset{\sim}{x})$. Moreover

$$e^G = !\, e^{\int d\underset{\sim}{z} \int B^\dagger(\underset{\sim}{z},\lambda)\{e^{h(\underset{\sim}{z},\lambda)}-1\}B(\underset{\sim}{z},\lambda)\,d\lambda}\,!\,,$$

and

$$e^G|o\rangle = e^g\, e^{\int d\underset{\sim}{z} \int A^\dagger(\underset{\sim}{z},\lambda)\{e^{h(\underset{\sim}{z},\lambda)}-1\}c(\lambda)\,d\lambda}|o\rangle$$

where

$$g \equiv \int d\underset{\sim}{z} \int \{e^{h(\underset{\sim}{z},\lambda)}-1\}c^2(\lambda)\,d\lambda .$$

Therefore the state $e^{G-g}|o\rangle$ is a coherent state (unnormed) with a smearing function

$$\{e^{h(\underset{\sim}{x},\lambda)}-1\}c(\lambda). \tag{3-10}$$

By choosing different h we can cover a dense set of elements in L^2 with functions such as (3-10). Let $\psi(\underset{\sim}{x},\lambda)/c(\lambda)\,\varepsilon C_o^\infty$, $\psi(\underset{\sim}{x},\lambda)=o$ for λ in some neighborhood of zero, and assume that for no $\underset{\sim}{x}$ and λ is $\psi(\underset{\sim}{x},\lambda)=-c(\lambda)$; such functions ψ are evidently dense in L^2. Then for each such element set

$$h(\underset{\sim}{x},\lambda) = \ln\{1 + \psi(\underset{\sim}{x},\lambda)/c(\lambda)\}.$$

In this way we have constructed a dense set of coherent states by applying a field functional on the vacuum; it follows that the closure of these states yields H, as was to be shown.

Global and Local Equivalence Conditions

For two field operators $\phi(\underset{\sim}{x})$ and $\tilde{\phi}(\underset{\sim}{x})$ to be unitarily equivalent there must be a unitary operator V such that

$$V\phi(\underset{\sim}{x})V^{-1} = \tilde{\phi}(\underset{\sim}{x}) \tag{3-11}$$

for all $\underset{\sim}{x}$, or more precisely for each $f\varepsilon C_o^\infty$ that

$$V\, e^{i\phi(f)}V^{-1} = e^{i\tilde{\phi}(f)}.$$

The sequence of operators $e^{i\phi(f_a)}$ where $f_a(\underset{\sim}{x})=(\underset{\sim}{x}+\underset{\sim}{a})$ approaches a limit as $|\underset{\sim}{a}|\to\infty$. It follows (imagine coherent state matrix elements with ψ, $\psi' \varepsilon C_o^\infty$) that

$$\text{w-lim}\; e^{i\phi(f_a)} = \langle o|e^{i\phi(f)}|o\rangle;$$

namely, a multiple of unity called a "tag". If two field operators are equivalent then so too are their tags. Hence, if two fields have any unequal tags they are inequivalent. For these particular tags to coincide for all $f\varepsilon C_o^\infty$ it is necessary that $J(s)=\tilde{J}(s)$ for all s which implies that $c(\lambda)=\tilde{c}(\lambda)$. Thus unequal fields are inequivalent fields as well.

Local equivalence asks that (3-11) hold for a compact region $0\varepsilon R^s$ rather than for all space. In that case, the results are different. Suppose, first, that $\Delta c\equiv\tilde{c}(\lambda)-c(\lambda)$ is square integrable. Then the fields $\phi(\underset{\sim}{x})$ and $\tilde{\phi}(\underset{\sim}{x})$ are locally equivalent and

$$V^{-1} = \exp\{\iint_O [A^\dagger(\underset{\sim}{x},\lambda)\Delta c(\lambda) - A(\underset{\sim}{x},\lambda)\Delta c(\lambda)]d\lambda d\underset{\sim}{x}\}.$$

To see this we observe that

$$VA(\underset{\sim}{x},\lambda)V^{-1} = A(\underset{\sim}{x},\lambda) + \Delta c(\lambda); \quad \underset{\sim}{x}\varepsilon 0,$$

or that

$$VB(\underset{\sim}{x},\lambda)V^{-1} = \tilde{B}(\underset{\sim}{x},\lambda) \equiv A(\underset{\sim}{x},\lambda) + \tilde{c}(\lambda); \quad \underset{\sim}{x}\varepsilon 0.$$

The stated equivalence is then immediate. The square integrability of Δc and finiteness of 0 are needed to ensure that V^{-1} is unitary.

When Δc is not square integrable the fields $\phi(\underset{\sim}{x})$ and $\tilde{\phi}(\underset{\sim}{x})$ are not locally equivalent. To show this we introduce a different class of tags defined locally rather than globally. Consider the set of unitary operators

$$T_a = \exp\{i \int_O d\underset{\sim}{x} \int_a^{2a} d\lambda \; B^\dagger(\underset{\sim}{x},\lambda) w_a(\lambda) B(\underset{\sim}{x},\lambda)\}$$

where a>o. We define the class of tags as a functional of $w_a(\lambda)$ according to

$$T[w] \equiv \underset{a\to o}{\text{w-lim}}\; T_a .$$

In evaluating this quantity it is convenient to employ a dense set of coherent states with arguments which vanish for λ in a neighbourhood of the origin. Then it is readily seen that T_a converges, if at all, to a multiple of unity given by

$$T[w] = \lim_{a\to o} \exp(-\bar{O} \int_a^{2a} \{e^{iw_a(\lambda)} - 1\} c^2(\lambda) d\lambda)$$

$$\equiv \exp\{-\bar{O}\; S[w]\}$$

where $\bar{O}$ is the volume of the compact region $O\varepsilon R^s$. The function

$$S[w] \equiv \lim_{a\to o} \int_a^{2a} \{e^{iw_a(\lambda)} - 1\} c^2(\lambda) d\lambda$$

are measures of the behavior of $c(\lambda)$ near the origin and may be zero, finite, infinite or undefined. In any case, they all are unitary invariants (since they are multiples of unity constructed in the same way from each field operator).

Assume that $w_a(\lambda)$ leads to a finite nonzero value for S[w]. For example,

$$w_a(\lambda) = w(\lambda) = |\lambda|^{-1}\; c^{-2}(\lambda)$$

leads to

$$S[w] = A^{-1}(e^{iA} - 1) \ln 2; \qquad \gamma = 1/2 ,$$

where

$$A \equiv \lim_{\lambda\to o} |\lambda|^{-1}\; c^{-2}(\lambda),$$

and to

$$S[w] = i \ln 2; \qquad \gamma > \tfrac{1}{2}.$$

Then for the field $\tilde{\phi}$ with model function $\tilde{c}(\lambda)$ we find

$$\tilde{S}[w] = \lim_{a \to o} \int_a^{2a} \{e^{iw(\lambda)} - 1\} \left(\frac{\tilde{c}(\lambda)}{c(\lambda)}\right)^2 c^2(\lambda)\, d\lambda.$$

Equivalence can only occur provided

$$\lim_{\lambda \to o} \frac{\tilde{c}(\lambda)}{c(\lambda)} = 1; \qquad (3\text{-}12)$$

violation of this condition leads to inequivalence and to divergence of the square integral of Δc. For the class of model functions of primary interest this actually completes the argument.

Suppose, however, that (3-12) is fulfilled and yet Δc has a divergent square integral, as is the case for

$$c(\lambda) \sim \frac{C}{|\lambda|^{\gamma}}; \qquad \tilde{c}(\lambda) \sim \frac{C}{|\lambda|^{\gamma}} + \frac{C'}{|\lambda|^{\gamma'}},$$

$\gamma > \gamma' \geq \tfrac{1}{2}$. In this case, choose

$$w_a(\lambda) = |\lambda|^{2\gamma'-1}; \qquad a < \lambda < f(a)$$

$$= -|\lambda|^{2\gamma'-1}; \qquad f(a) \leq \lambda < 2a.$$

Here $f(a)$ is chosen so that

$$\lim_{a \to o} \operatorname{Im} \int_a^{2a} \{e^{iw_a(\lambda)} - 1\} \frac{d\lambda}{|\lambda|^{2\gamma}} = 0,$$

while

$$\lim_{a \to o} \operatorname{Im} \int_a^{2a} \{e^{iw_a(\lambda)} - 1\} \frac{d\lambda}{|\lambda|^{2\gamma'}} \neq 0.$$

This procedure enables us to go beyond the leading singularity and seek the next leading singularity -- unequal behavior there, as is implicit above, leads to unequal tags and therefore inequivalence. If the first two leading singularities are identical, a more general $w_a(\lambda)$ (with more sign changes) lets the third leading term be probed, and so on. The procedure stops when $\gamma''''=\frac{1}{2}$ since any lesser power law is integrable and has zero tag contribution. Generalizations to non-power law singularities are evident, and we see that local equivalence is obtained if and only if $\Delta c=\tilde{c}(\lambda)-c(\lambda)$ is square integrable.

3.3 Hamiltonian and Dynamical Properties

Limitations on Form of Hamiltonian

The basic form of the Hamiltonian is dictated by the ultralocal nature of the dynamics which maintains the statistical independence of fields at distinct points. The coherent states $|\psi\rangle$ are the analog of product states in the formally similar problem of a countable number of degrees of freedom possessing independent dynamics. In the latter case, product states evolve in time into product states, and the parallel situation obtains here. Thats is, the ultralocal form of the dynamics leads to the matrix elements

$$\langle\psi|e^{-iHt}|\psi'\rangle = NN' \exp\{\iint\psi^*(\underset{\sim}{x},\lambda)\psi'(\underset{\sim}{x},\lambda,t)d\lambda d\underset{\sim}{x}\}.$$

The group property, and especially the implications for the one-particle sector, pmply the existence of a self-adjoint generator h mediating the transformation on the right side. Translation invariance and independence of distinct points implies that h acts on λ alone. This argument already gives

$$\langle\psi|e^{-iHt}|\psi'\rangle = NN'\exp\{\iint\psi^*(\underset{\sim}{x},\lambda)e^{-iht}\psi'(\underset{\sim}{x},\lambda)d\underset{\sim}{x}d\lambda\}$$

from which it directly follows that

$$H = \int d\underset{\sim}{x} \int A^{\dagger}(\underset{\sim}{x},\lambda) h A(\underset{\sim}{x},\lambda) d\lambda \ .$$

Positivity of H holds if and only if h is positive. Clearly $H|0\rangle = 0$, and uniqueness of the ground state of the Hamiltonian holds provided

$$\lim_{\beta\to\infty} \langle\psi|e^{-\beta H}|\psi'\rangle = \langle\psi|0\rangle\langle 0|\psi'\rangle = NN'$$

which holds, in turn, provided

$$\lim_{\beta\to\infty} \iint \psi^{*}(\underset{\sim}{x},\lambda) e^{-\beta h} \psi'(\underset{\sim}{x},\lambda) d\underset{\sim}{x} d\lambda = 0.$$

Thus h must be strictly positive, $h>0$, if $|0\rangle$ is to be a nondegenerate ground state of H. It is for this reason (also) that we must have $\int c^2(\lambda)d\lambda = \infty$. For if $c(\lambda) \varepsilon L^2$, it would follow that

$$h = -\tfrac{1}{2} c^{-1}(\lambda) \frac{\partial}{\partial\lambda} c^2(\lambda) \frac{\partial}{\partial\lambda} c^{-1}(\lambda)$$

would have $c(\lambda)$ as a <u>normalizable</u> eigenstate of zero energy. In that case, $e^{-\beta H} \to P_0$, a projection operator but <u>not</u> one which is one dimensional.

Two-Fold Level Degeneracy

We turn our attention to showing the two-fold degeneracy of the levels of h. Suppose we adopt $c(\lambda) = |\lambda|^{-\gamma} \exp[-y(\lambda)]$ which leads to

$$h = -\frac{1}{2} \frac{\partial^2}{\partial\lambda^2} + \frac{\gamma(\gamma+1)}{2\gamma^2} + e + v_0(\lambda).$$

If one considers $hu = \mu u$ as an eigenvalue equation in $L^2(R)$, then the singularity at the origin forces all (normalizable) solutions to have the form $(\lambda>0)$

$$u(\lambda) \sim \lambda^{1+\gamma};$$

the behavior at infinity, in turn, selects out the eigenvalue μ, which in general will depend on γ through its appearance in e and v_o. However, the required form near zero obeys the pair of boundary conditions

$$u(o) = o, \quad u'(o) = o.$$

These can be viewed as the initial data to another perfectly acceptable solution for $\lambda<o$, namely $u(\lambda)\equiv o$, $\lambda<o$. This shows the legitimate localization of eigenstates on either the positive or negative half line. The symmetry of v_o then implies the two-fold degeneracy of energy eigenstates. These may also be grouped into even and odd wave functions, the former having the behavior $u\sim|\lambda|^{1+\gamma}$ and the latter the behavior $u\sim\lambda|\lambda|^{\gamma}$ near the origin.

A satisfactory heuristic picture of this state of affairs is formed if one imagines replacing the repulsive term at the origin by a high smoothed-out barrier like $\gamma(\gamma+1)/2(\lambda^2+\delta^2)$, where δ is very small but positive. The eigenstates must penetrate through this barrier and are highly attenuated near the origin, the even states going roughly as

$$u(\lambda)\sim(\delta^2+\gamma\lambda^2)^{(1+\gamma)/2}$$

and the odd states as

$$u(\lambda) \sim \lambda[\delta^2 + \tfrac{1}{3}(\gamma+1)\lambda^2]^{\gamma/2}$$

near $\lambda=o$. The various energy levels are approximately two-fold degenerate. Heuristically, the formal limit $\delta\to o$ yields the quoted results.

The rigorous proof that h is self adjoint* on either

---

*Strictly speaking, it is the closure of h, $h^{\dagger\dagger}$, that is self adjoint.

half space, say $L^2(o,\infty)$, is based on the absence of square integrable, distribution solutions of both of the differential equations

$$h\psi(\lambda) = \pm i\psi(\lambda), \quad \lambda > o. \tag{3-13}$$

To demonstrate that there are no normalizable solutions we argue as follows. First we note that

$$\int_o^\infty \psi^* h\psi d\lambda = -\frac{1}{2}c\psi^* \frac{\partial}{\partial\lambda}c^{-1}\psi \Big|_o^\infty + \frac{1}{2}\int_o^\infty |c\frac{\partial}{\partial\lambda}c^{-1}\psi|^2 d\lambda$$

$$= \pm i \int_o^\infty |\psi|^2 d\lambda,$$

which implies that

$$\frac{1}{2}\mathrm{Im}\psi^*\psi'\Big|_o^\infty = \pm\int_o^\infty |\psi|^2 d\lambda$$

where $\psi' = \partial\psi/\partial\lambda$. Any solution ψ must behave as

$$\psi \sim a\lambda^{1+\gamma} + b\lambda^{-\gamma}$$

near $\lambda=o$. Since $\gamma \geq 1/2$, square integrability of ψ demands that $b=o$. In that case, $(\psi^*\psi')$ vanishes at $\lambda=o$ and

$$\frac{1}{2}\,\mathrm{Im}(\psi^*\psi')_{\lambda=\infty} = \pm\int_o^\infty |\psi|^2 d\lambda.$$

However if $v_o(\lambda) \sim \frac{1}{2}B\lambda^{2N}$ for large λ, any solution ψ has one of the asymptotic forms

$$\psi \sim \exp[\pm\sqrt{B}\,\lambda^{N+1}/(N+1)]$$

for large λ. The + sign is forbidden by square integrability, while the minus sign leads to the vanishing of $(\psi^*\psi')$ at $\lambda=\infty$. Hence the solutions of (3-13) either vanish identically or are not square integrable.

Dynamics in λ-Space

The localization of the eigenstates proved above means that

$$e^{-iht}\,P = Pe^{-iht}$$

where P is the λ-space projection of $L^2(-\infty,\infty)$ onto $L^2(0,\infty)$. Let us trace out the implications of this fact for the dynamical equation

$$\lambda(t) = e^{iht}\,\lambda e^{-iht}.$$

Define

$$\lambda = P\lambda P + Q\lambda Q \equiv \lambda_+ - \lambda_-$$

where Q=1-P. Here $\lambda_\pm$ both have non-negative spectrum, and

$$\lambda_+(t) \equiv e^{iht}\lambda_+ e^{-iht} = P\lambda(t)P$$

and similarly for λ_-. Evidently

$$\{P\lambda^2P\}^{1/2} = \lambda_+, \quad \{Q\lambda^2Q\}^{1/2} = \lambda_- .$$

Hence

$$\lambda(t) = \{P\lambda^2(t)P\}^{1/2} - \{Q\lambda^2(t)Q\}^{1/2}$$

which implies that λ(t) can be determined from a study of $\lambda^2(t)$. Clearly this argument extends to show that λ(t) can be recovered from any power.

In principle, there are advantages to studying equations for $\lambda^2(t)$ over those for λ(t). The equation of motion for λ(t),

$$\ddot{\lambda} = \gamma(\gamma+1)\lambda^{-3} - v_o'(\lambda)$$

is rather singular at the origin. However the equation for $\lambda^2 \equiv \rho$ is better behaved. It follows that

$$\dot{\rho} = i[h,\rho] = -i[\lambda(\partial/\partial\lambda) + (\partial/\partial\lambda)\lambda] \equiv 2\eta$$

and

$$\begin{aligned}\ddot{\rho} = 2i[h,\eta] &= 4h - 4[e+v_o(\lambda)] - 2\lambda v_o'(\lambda)\\ &\equiv 4h - w_o(\lambda)\\ &\equiv 4h - \bar{w}_o(\rho) .\end{aligned}$$

Here h is a constant of the motion and w_o is nonsingular. One equation of this type can be readily solved. Suppose $c(\lambda)=|\lambda|^{-\gamma}\exp(-\frac{1}{2}\mu^2\lambda^2)$, then $w_o(\lambda)=4\mu(\gamma-\frac{1}{2})+4\mu^2\lambda^2$ which leads to the linear equation of motion

$$\ddot{\rho} = 4h - 4\mu(\gamma-\tfrac{1}{2}) - 4\mu^2\rho \equiv 4h' - 4\mu^2\rho .$$

The solution

$$\rho(t) = \lambda^2(t) = (h'/\mu^2) + [\lambda^2 - (h'/\mu^2)]\cos 2\mu t + (\eta/\mu)\sin 2\mu t$$

has the oscillatory behavior of a free theory and the frequency (2μ) characteristic of a square.

A solution for $\lambda(t)$ can be used to determine the time dependence of the truncated vacuum expectation values according to (2-15) keeping in mind that $c(\lambda)$ is treated as a formal left and right eigenvector of e^{-iht}. For example, a term like $\lambda(t)c(\lambda)$ is to be interpreted as $e^{iht}\lambda c(\lambda)$. By the same argument $\lambda^2(t)c(\lambda)$ should be interpreted as $e^{iht}\lambda^2c(\lambda)$, etc. Since $\lambda(t)$ can be defined by its formal power series in t, the quantity $\lambda(t)c(\lambda)$ may well not be square integrable. On the other hand, $e^{iht}\lambda c(\lambda)$ is regarded as a vector in L^2, e.g.,

$$e^{iht}\lambda c(\lambda) = \sum_\ell e^{i\mu_\ell t} u_\ell(\lambda)\int u_\ell^* \lambda c d\lambda .$$

We have argued earlier that $h^n\lambda^\theta c(\lambda)\sim\lambda^{-2n}\lambda^\theta c(\lambda)$, which for large enough n fails to be square integrable, so this vector is not defined by a power series in t. However our estimate is incorrect for $\theta=1+2\gamma$ since the leading singularity drops out at each stage and one easily sees that $h^n\lambda^{1+2\gamma}c\varepsilon L^2$ for all an. Hence in this case

$$(d^n/dt^n)e^{iht}\lambda^{1+2\gamma}c=i^ne^{iht}h^n\lambda^{1+2\gamma}c$$

$$=i^ne^{iht}[h,[h,[h,\cdots[h,\lambda^{1+2\gamma}]\cdots]]c$$

which means that the power series for $\lambda^{1+2\gamma}(t)$ may be employed, and therefore $\lambda^{1+2\gamma}(t)c=e^{iht}\lambda^{1+2\gamma}c$. In the special "free" cases, where $c(\lambda)=|\lambda|^{-\gamma}\exp(-\frac{1}{2}\mu\lambda^2)$, $\lambda^{1+2\gamma}c$ is an eigenvector of h*, and we have

$$\lambda^{1+2\gamma}(t)c = e^{i(1+2\gamma)\mu t}\,\lambda^{1+2\gamma}c.$$

Hence, for the "appropriate vectors" $\lambda^{1+2\gamma}c$, there are less pathologies.

Consequences for Field Equations

In the case $\gamma=\frac{1}{2}$, $\lambda^2 c$ is the appropriate vector which makes the study of $\lambda^2(t)$ and its associated local field $\phi_r^2(\underset{\sim}{x},t)$ quite direct. Note first that λ^2 and $\frac{1}{2}d\lambda^2/dt\equiv\eta$ generate a Lie algebra,

$$[\lambda^2,\ \eta] = 2i\lambda^2\ .$$

As a consequence their associated local fields $\phi_r^2(\underset{\sim}{x})$ and $\kappa(\underset{\sim}{x})$ [cf. (3-6)] fulfill a local Lie algebra relation

$$[\phi_r^2(\underset{\sim}{x}),\ \kappa(\underset{\sim}{y})] = 2i\delta(\underset{\sim}{x}-\underset{\sim}{y})\phi_r^2(\underset{\sim}{x}), \tag{3-14}$$

while the Hamiltonian provides the connecting relation [6],

$$[H,\ \phi_r^2(\underset{\sim}{x})] = -2i\ \kappa(\underset{\sim}{x}). \tag{3-15}$$

A valid field equation can be found for $\phi_r^2(\underset{\sim}{x},t)$ namely

$$\ddot{\phi}_r^2(\underset{\sim}{x}) = 4H(\underset{\sim}{x}) - \Sigma\, w_{2n}\{\phi_r^{2n}(\underset{\sim}{x}) - \langle o|\phi_r^{2n}(\underset{\sim}{x})|o\rangle\}.$$

This equation is a direct consequence of the relation

---

*Eigenvectors and eigenvalues of h for the "free" case appear in Ref. 5, where it is shown that $\mu_{2\ell+1}=\mu_{2\ell+1}=\mu(2\ell+2\gamma+1)$. Discussions with J. McKenna regarding these differential equations are gratefully acknowledged.

$$\phi_r^2(\underset{\sim}{x},t) - \langle o|\phi_r^2(\underset{\sim}{x},t)|o\rangle$$
$$= \int A^\dagger(\underset{\sim}{x},\lambda)\lambda^2(t)A(\underset{\sim}{x},\lambda)\,d\lambda + \int A^\dagger(\underset{\sim}{x},\lambda)\lambda^2(t)c(\lambda)\,d\lambda$$
$$+ \int [\lambda^2(t)c(\lambda)]^* A(\underset{\sim}{x},\lambda)\,d\lambda$$

which holds for $\gamma=\frac{1}{2}$, and of the equation satisfied by $\lambda^2(t)$. Superficially this is an equation for $\phi_r^2(\underset{\sim}{x})$, but renormalized powers $\phi_r^{2n}(\underset{\sim}{x})$ enter the right-hand side so this is only formally an equation for $\phi_r^2(\underset{\sim}{x})$. Local field equations (superficially) for $\phi_r^{1+2\gamma}(\underset{\sim}{x})$ can be written for any γ, but none have the simplicity of that for $\gamma=\frac{1}{2}$.

Equations (3-14) and (3-15) in the case $\gamma=\frac{1}{2}$ can be approached from the postulate of an underlying current algebra. Ultralocal models have also been studied by C. Newman from this point of view who has given additional arguments for the association of $c(\lambda)=|\lambda|^{-\frac{1}{2}}\exp(-\frac{1}{2}\mu\lambda^2)$ with a free theory [7].

REFERENCES

1. See, e.g., J. Glimm and A. Jaffe in: Statistical Mechanics and Field Theory; C. DeWitt and R. Stora eds. (Gordon and Breach Science Publishers, Inc., New York and London, 1971).
2. J. R. Klauder, Commun. Math. Phys. 18, 307 (1970).
3. G. Wentzel, Helv. Phys. Acta 13, 269 (1940); L. I. Schiff, Phys. Rev. 92, 766 (1953); A. Araki, Princeton thesis, 1960, Chap. V, H. Ezawa, Commun. Math. Phys. 8, 261 (1968); 9, 38 (1968); R. F. Streater, Lectures at the 1968 Karpacz Winter School.

4. I. M. Gel'fand and N. Ya Vilenkin, Generalized Functions, Vol. 4: Applications of Harmonic Analysis, translated by A. Feinstein (Academic Press, New York, 1964), Chap. III.
5. G. Szego, Orthogonal Polynomials (American Mathematical Society, New York, 1959), pg. 99, 377.
6. J. R. Klauder, J. Math. Phys. 11, 6o9 (197o); Relativity, Carmeli et al. eds. (Penum, New York, 197o), pg. 1.
7. C. Newman, Princeton thesis (in preparation).

Acta Physica Austriaca, Suppl. VIII, 277–322 (1971)

NULL PLANE FIELD THEORY*

BY

F. ROHRLICH
Department of Physics, Syracuse University
Syracuse, New York

CONTENTS

1. MOTIVATION AND HISTORY

The purpose of these lectures is to present a coherent and logically consistent introduction into null plane field theory starting from first principles. We are dealing here with a well known subject to which have been

---

*Lecture given at X. Internationale Universitätswochen für Kernphysik, Schladming, March 1 - March 13, 1971.

added many new twists. The null planes open a variety of new aspects which are only beginning to be explored, the whole topic being just a few years old. Much of the material can be found in the published literature, especially where very recent publications are included; but there will also be a number of unpublished results making their first appearance.

Null surface field theory has been known to general relativists for years. Their favorite null surfaces are of course the null cones. The mathematical importance of these characteristic surfaces is obvious, their use as local specification of the geometry well known.

Null planes are especially useful in flat space, as was discovered in two quite different and independent contexts about four years ago.

(1) The first use of null plane quantum electrodynamics arose in connection with laser fields (1967). When treated as classical background fields to quantum electrodynamics they naturally define a null plane geometry: a coherent wave train of finite length appears in Minkowski space as a slab of electromagnetic radiation between two parallel null planes. The associated coordinatization simplifies the theory considerably and permits a wave packet description which is difficult or impossible in the usual choice of coordinates [1,2].

(2) The second context was that of current algebra [3]. The need for an "infinite momentum limit" soon brought about a series of papers in which field theory [4] and current algebra was done in the "infinite momentum frame". This frame is just the three-dimensional hyperplane tangent to a null-cone, i.e. a null plane.

In the present lectures we do not want to discuss null plane field theory from either of these two points of view. Rather, we wish to study this new formulation of quantum field theory as a theory in its own right, as a new- and hopefully better - approach to field theory in which old problems can be seen in a new light and can perhaps be brought closer to a solution.
As a long range goal there is also the important possibility of applications to physical processes where one can develop an approximation method in which the first approximation is the high energy limit. But we shall not be concerned with applications here.

In a crude way, null plane field theory can be characterized as the analogue of ordinary field theory, but in which the development of the system from one spacelike plane to the next is replaced by the development from one nullplane to the next.

At this point the intuition of the physicist will suggest the following three points.

(1) It seems very unphysical to replace the initial conditions at t=o (data on the hyperplane t=o) by initial conditions at t-z=o (data on a null hyperplane). There are two answers to this criticism. First, a function on a null plane does have a physical interpretation: it specifies the wave front of a light wave moving in the z-direction, for all times. Second, and more importantly, if one takes the S-matrix point of view that in last analysis only the asymptotic conditions matter, it is irrelevant which way one "slices" the Minkowski space-time, as long as one can find the mapping from the in-states to the out-states.

(2) If one makes a Lorentz transformation with velocity v of the initial plane t=o, and one then lets $v \to c$, one expects to obtain a null plane. But this limit does not exist for a Lorentz transformation. The answer is that this limit must be carried out in a more sophisticated way and does exist for a boost operation.

(3) Intuition tells one that in the infinite momentum limit masses should somehow be irrelevant, negligible in some sense. Can this notion be made precise? Here an exact answer will be given in the following lectures. This intuitive mass independence is in fact one of the most important and most interesting features of this formulation.

The first step in our study will be to establish a suitable mathematical tool for describing null planes invariantly.

2. INVARIANT NULL PLANE COORDINATIZATION

To begin with, we describe four-dimensional Minkowski space with respect to a special reference frame. In this frame the unit vectors in the x and y directions $e_{(1)}$ and $e_{(2)}$ remain the same as usual. But the unit vectors in the z and t directions, $e_{(3)}$ and $e_{(o)}$ are replaced by the null vectors

$$m \equiv \frac{1}{\sqrt{2}}(e_{(o)}-e_{(3)}), \quad n \equiv \frac{1}{\sqrt{2}}(e_{(o)}+e_{(3)}). \tag{2.1}$$

When we choose the usual Minkowski metric to have signature +2 then the tetrad m, $e_{(1)}$, $e_{(2)}$, n satisfies

$$m^2 = n^2 = o, \qquad e_{(1)}^2 = e_{(2)}^2 = 1 \tag{2.2}$$

$$e_{(i)}\cdot m = e_{(i)}\cdot n = 0 \quad (i = 1,2); \quad m\cdot n = -1. \tag{2.3}$$

The components of an arbitrary vector A with respect to this tetrad will be denoted by A_u, A_1, A_2, A_v, i.e.

$$\begin{aligned} A_u &= - m \cdot A \\ A_1 &= e_{(1)}\cdot A \\ A_2 &= e_{(2)}\cdot A \\ A_v &= - n\cdot A \ , \end{aligned} \tag{2.4}$$

so that A can be decomposed into four vectors:

$$A = mA_v + e_{(1)}A_1 + e_{(2)}A_2 + nA_u. \tag{2.5}$$

We shall also use the component representation

$$A = (A_v,\ A_1,\ A_2,\ A_u). \tag{2.6}$$

The customary procedure is to use this labelling to define the metric tensor

$$g^{\mu\nu} = \begin{pmatrix} 0 & 0 & 0 & -1 \\ 0 & -1 & 0 & 0 \\ 0 & 0 & 1 & 0 \\ -1 & 0 & 0 & 0 \end{pmatrix} \tag{2.7}$$

This would lead to a distinction between covariant and contravariant indices such as $A^u=-A_v$, $A^v=-A_u$, etc. We shall not proceed in this way, because it would single out a special reference frame.

If we make a homogenous Lorentz transformation and require that the four vectors of the tetrad are to transform as four-vectors, the orthogonality and normalization equations (2.2) and (2.3) will remain invariant. Thus, given any frame in which a tetrad satisfies these equations, there exists a Lorentz transformation from the old to the new tetrad.

The "components of the vector A", eq. (2.4), are then no longer the usual components, but invariants, scalar products of two four vectors. The null plane description can now be given completely in terms of these invariants for each vector. In this case there is no metric tensor. The subscripts refer to the projection on the tetrad and are not labels of covariant components, but of invariant components.

The scalar product of two vectors can be written in terms of these invariant components

$$A\cdot B = \underline{A}\cdot\underline{B} - A_u B_v - A_v B_u \qquad (2.8)$$

where we used the notation $\underline{A}=(A_1, A_2)$ and $\underline{A}\cdot\underline{B}=A_1B_1+A_2B_2$. For the coordinate vector it is convenient to define

$$u \equiv x_v \equiv -n\cdot x, \qquad v \equiv x_u \equiv -m\cdot x \qquad (2.9)$$

so that v is measured along n and u along m. With this notation we have

$$\partial_u \equiv -m\cdot\partial = -\frac{\partial}{\partial u}, \quad \partial_v \equiv -n\cdot\partial = -\frac{\partial}{\partial v}\,. \qquad (2.10)$$

The minus signs here are perhaps inconvenient but are required by consistency.

In the null plane formulation of field theory the physical system is developed from one null plane to the next in contradistinction to the usual formulation where the development proceeds from one spacelike plane to the next. If we choose the null plane family u = constant (analogous to t = constant) then the initial value problem will be posed on a three-dimensional null hyperplane $u=u_0$ and u will play the role of a time. In such a null plane we have vectors $\bar{x}$ characterized by the triple (x_1, x_2, v)

or $(\underline{x}, v)$, and volume elements

$$d^3\bar{x} = dx_1\, dx_2\, dv = d^2\underline{x}\, dv \,. \qquad (2.11)$$

The null planes u = constant are spanned by the triad of vectors $e_{(1)}$, $e_{(2)}$, n. The vector m "sticks out of the plane".

We conclude this largely notational discussion with a few useful relations. In the frame in which t, x, y, z are along $e_{(o)}$, $e_{(1)}$, $e_{(2)}$, $e_{(3)}$ we have

$$u = \frac{t-z}{\sqrt{2}}, \qquad v = \frac{t+z}{\sqrt{2}} \qquad (2.12)$$

$$p_u = \frac{p^o+p_z}{\sqrt{2}}, \qquad p_v = \frac{p^o-p_z}{\sqrt{2}} \,. \qquad (2.13)$$

In any frame the important scalar product

$$p\cdot x = \underline{p}\cdot\underline{x} - p_v u - p_v v \qquad (2.14)$$

will be needed for the Fourier transform. Eq. (2.8) implies the divergence of a vector to be

$$\partial\cdot A = \underline{\partial}\cdot\underline{A} - \partial_u A_v - \partial_v A_u . \qquad (2.15)$$

Finally, for a tensor $T_{\mu\nu}$ we shall use the notation $n^\mu T_{\mu\nu} = -T_{v\nu}$, $T_{\mu\nu}m^\nu = -T_{\mu u}$, etc.

3. THE CLASSICAL INITIAL VALUE PROBLEM

The Klein-Gordon equation in null coordinates is

$$2\partial_u\partial_v\phi(x) = (\underline{\partial}^2-m^2)\phi(x) \,. \qquad (3.1)$$

We shall here be interested not only in functions $\phi(x)$ which satisfy this equation but also in generalized functions $\phi(f)$ which are solutions of (3.1). Correspondingly, one distinguishes between classical and generalized solutions.

[A generalized function $\phi(f)$ is a continuous linear functional on a test space (fundamental space) F of which f is a vector. Since ϕ thus maps F to the complex numbers c, ϕ is an element of the space F' conjugate to F; $\phi(f)$ is an inner product and can be written $\phi(f) \equiv (\phi, f)$].

We assume that with f also $f_1 \equiv \partial_v f$ belongs to F. Then the operator ∂_v^{-1} is defined by

$$\phi(f) = \phi(\partial_v^{-1} f_1) = -\partial_v^{-1}\phi(f_1) \ . \tag{3.2}$$

Thus ∂_v^{-1} exists when acting on functionals over test functions in F. For a generalized solution, (3.1) can then be integrated over v without additional constants,

$$2\partial_u\partial_v\phi(f) = -2\partial_u\phi(\partial_v f) = -2\partial_u\phi(f_1) = (\underline{\partial}^2 - m^2)\phi(\partial_v^{-1} f_1)$$

or

$$2\partial_u\phi(f_1) = \partial_v^{-1}(\underline{\partial}^2 - m^2)\phi(f_1) \ . \tag{3.3}$$

[Mathematical remark: In general, test spaces do not have the property that $f\varepsilon F$ implies $Df\varepsilon F$ (D indicates the first derivative). But this property does hold for the well known countably normed spaces S and D of C^∞ functions which define the tempered and the Schwartz distributions, respectively. However, there also exist certain classes of Hilbert spaces with the property that every one of its elements is in the same L^2 space as its derivative (Sobolev Spaces)].

If ϕ is locally integrable, the operator ∂_v^{-1} has a representation as an integral operator,

$$\partial_v^{-1}\phi(f_1) = -\frac{1}{2}\int_{-\infty}^{\infty} \varepsilon(v-v')\phi(v')dv' f_1(v)dv \tag{3.4}$$

and one easily verifies $\partial_v \partial_v^{-1} \phi(f_1) = \phi(f_1)$.

The initial value problem on a spacelike plane $t = t_o$ (Cauchy problem) is well known. If the function $\phi(x)$ is to satisfy (3.1) and also $\phi(x) = \phi_o(x)$ for $t = t_o$ and $\partial_o \phi_o(x) = \phi_1(x)$, then $\phi(x)$ is uniquely determined for all t by

$$\phi(x) = \int_{t'=t_o} \Delta(x-x') \overleftrightarrow{\partial_o'} \phi(x') d^3x' \equiv (\Delta * \overleftrightarrow{\partial_o} \phi)_{t_o}(x) \qquad (3.5)$$

where $\Delta(x)$ is the well known "Green Function" which satisfies (3.1), $\Delta(o,\vec{x}) = o$ and $\partial_o \Delta(o,\vec{x}) = \delta_3(\vec{x})$.

But the initial value problem on a null plane is not so well known: to find a function $\phi(x)$ which satisfies (3.1) for $u > u_o$ and for which $\phi(x) = \phi_o(x)$ on $u = u_o$. Note that $\partial_v \phi(x)$ on $u = u_o$ is simply $\partial_v \phi_o(x)$ and is not a new piece of information. The solution to the problem is [5]

$$\phi(x) = - \int_{u'=u_o} \Delta(x-x') \overleftrightarrow{\partial_{v'}} \phi(x') d^3\bar{x}' = - (\Delta * \overleftrightarrow{\partial_v} \phi)_{u_o}(x) \qquad (3.6)$$

and requires the additional assumption

$$\lim_{v \to -\infty} \phi(x) = o \quad \forall \quad u \geq u_o. \qquad (3.7)$$

The proof can be sketched by means of Fig. 1.

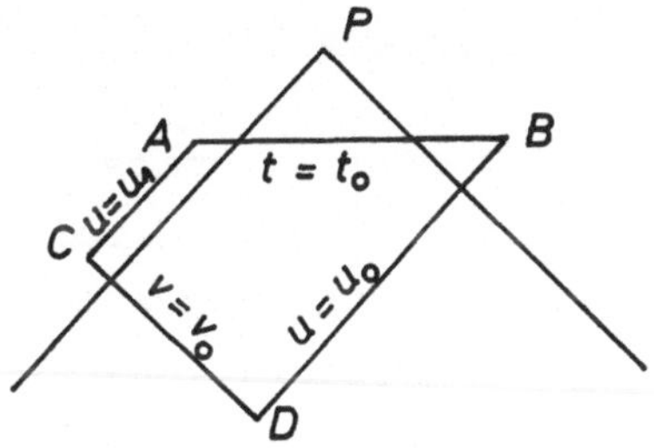

Fig. 1: Solution of the null plane initial value problem.

One uses Gauss' Theorem on the integral of the divergence-free quantity $\Delta(x-x')\overleftrightarrow{\partial}'_\mu\phi(x')$ over the four-dimensional volume whose cross section in the m-n-plane is A B D C and which is long enough at both ends to protrude out of the past light cone whose vertex is P,

$$\int_{V_4} \partial'^\mu[\Delta(x-x)\partial'_\mu\phi(x')]d^4x' = 0.$$

Since $(\Delta,f)=0$ for all $f\varepsilon F$ whose support is spacelike relative to the origin, only the three planes $t=t_0$, $v=v_0$, and $u=u_0$ contribute to the surface integral. The first of these yields $\phi(x)$ due to (3.6). Thus,

$$\phi(f) = -((\Delta * \overleftrightarrow{\partial}_u\phi)_{v_0},f) - ((\Delta * \overleftrightarrow{\partial}_v\phi)_{u_0},f). \tag{3.8}$$

If we choose v_0 in the distant past ($v_0 \to -\infty$) the first term will vanish whenever (3.7) holds and the result (3.6) obtains.

The asymptotic condition (3.7) can also be stated by saying that ϕ must be defined over test functions which for large negative v need not vanish faster than 1/v.

A trivial but important generalization of (3.1) to (3.3) is furnished by the differential equation

$$2\partial_u\partial_v\Phi(x) = L\Phi(x) + \Psi(x) \tag{3.9}$$

where L is a linear differential operator which does not involve ∂_u. Then (3.9) can be solved for $\partial_u\Phi$ exactly as in (3.3). We shall see that the Maxwell equations are not of the form (3.9); but they can be cast into this form by a special choice of gauge.

4. NULL PLANE CLASSICAL FIELD THEORY: SCALAR FIELD

In order to understand the main characteristic features of classical field theory formulated relative to null planes we shall investigate the simplest case, viz. the scalar field.

Formulated in Lagrangian form, classical field theory is based on the fundamental theorem by E. Noether [6]. For our purpose it is convenient to state this theorem in the following form:

Let $\phi_r(x)$ $(r=1,2,\dots,n)$ be the r th component of a classical n-component field defined over four-dimensional space-time; let G be a p-parameter Lie group of transformations and $T_i \varepsilon G$ an element of G associated with the i th parameter and connected to the identity; let t_i be the corresponding infinitesimal transformation so that under t_i

$$x^\mu \to x'^\mu = x^\mu + \delta_i x^\mu$$

$$\phi_r(x) \to \phi_r'(x') = \phi_r(x) + \delta_i \phi_r(x) \tag{4.1}$$

let $L[\phi, \partial_\mu \phi]$ be the Lagrangian of the system. Then, if under t_i the action integral is invariant up to a divergence, i.e.

$$\int_{V_4'} L'[\phi', \partial_\mu' \phi'] d^4x' = \int_{V_4} (L[\phi, \partial_\mu \phi] + \partial_\mu \delta_i \Omega^\mu(x)) d^4x, \tag{4.2}$$

then the field equations

$$\partial_\mu \frac{\delta L}{\delta\, \partial_\mu \phi_r} - \frac{\delta L}{\delta\, \phi_r} = 0 \quad (r = 1, \dots, n) \tag{4.3}$$

imply the p conservation laws

$$\partial_\mu \left(T^\mu{}_\nu \cdot \delta_i x^\nu - \sum_r \frac{\delta L}{\delta\, \partial_\mu \phi_r} \delta_i \phi_r - \delta_i \Omega^\mu\right) = 0 \qquad (i = 1, \ldots p) \tag{4.4}$$

where the energy tensor $T^\mu{}_\nu$ is defined by

$$T^\mu{}_\nu \equiv \sum_r \frac{\partial L}{\delta\, \partial_\mu \phi_r} \partial_\nu \phi_r - \delta^\mu_\nu L. \tag{4.5}$$

If ϕ_r is a complex field, ϕ_r and ϕ_r^* must be treated as independent and the summations $\sum_r$ in (4.4) and (4.5) must extend over all ϕ_r and all ϕ_r^*.

The differential conservation laws (4.4) can be written in the form

$$\partial_\mu G^\mu_i = 0 \qquad (i = 1, \ldots p) \tag{4.6}$$

and imply the existence of p generators of the infinitesimal transformations t_i,

$$G_i \equiv \int_\sigma d^3\sigma_\mu G^\mu_i(x) \qquad (i = 1, \ldots p) \tag{4.7}$$

where the three-dimensional hypersurface will be specified below. The G_i are the constants of motion associated with the invariance group G.

The usual representation of the generators G_i in (4.7) consists in an integral over a spacelike hyperplane σ. Consider a spacelike slab, i.e. a four-dimensional region in spacetime bounded by two parallel spacelike planes and extending to infinity in spacelike directions between them. Integration of (4.6) over such a region and use of Gauss' theorem yields [7] the conclusion that (4.7) with σ a spacelike hyperplane is independent of the choice of σ.

In null-plane field theory we consider null slabs (lightlike slabs) which are four-dimensional regions in spacetime bounded by parallel null planes and extending to infinity in null and spacelike directions between them.

When (4.6) is integrated over a null slab, Gauss' theorem tells us that (4.7) represents the generators G_i as integrals over null planes σ. These integrals are independent of the choice of these planes within the family of parallel null planes. It is this null plane representation of the generators which is an essential characteristic of null plane field theory.

[The existence of the generators G_i in (4.7) requires that the ϕ fall off fast enough for the integral to converge. In particular, on the u = constant plane both ϕ and $\partial_v \phi$ must be square integrable in v.]

In order to give specific examples to these general considerations, we shall apply the above to specific invariance groups (see (a) to (d) below). The ensuing results will be useful later.

(a) Translations: The simplest Lie group of physical importance is perhaps the abelian group of spacetime translations. The four infinitesimal parameters ε^μ yield the infinitesimal translations

$$\delta x^\mu = \varepsilon^\mu. \tag{4.8}$$

The parameter labelling i=1, ..., 4, can here be replaced by the labelling provided by the index $\mu=0,1,2,3$ when translations are carried out along the axes. A scalar field is by definition invariant under (4.8), so that $\delta\phi \equiv \phi'(x')-\phi(x)=0$. If we choose the usual free field Lagrangian

$$L = - \partial_\mu \phi^* \partial^\mu \phi - m^2 \phi^* \phi \tag{4.9}$$

we find that also $\delta\, \Omega^\mu=0$. Thus, the field equations

$$(\Box - m^2)\phi = 0 \tag{4.10}$$

yield according to (4.4), the well-known differential conservation law

$$\partial_{\mu} T^{\mu}{}_{\nu} = 0 \tag{4.11}$$

with

$$T^{\mu}{}_{\nu} = - \partial^{\mu}\phi^{*} \partial_{\nu}\phi - \partial_{\nu}\phi^{*} \partial^{\mu}\phi + \delta^{\mu}_{\nu} L \; . \tag{4.12}$$

The infinitesimal generators are the momenta

$$P_{\mu} = \int_{\sigma} d^{3}\sigma_{\alpha} T^{\alpha}{}_{\mu}. \tag{4.13}$$

Now we take σ to be a null plane u = constant, i.e.

$$d^{3}\sigma^{\mu} = n^{\mu} d^{3}\bar{x},$$

where $d^{3}\bar{x}$, (2.11), is an invariant, and n^{μ} are the components of the null vector defined in (2.1). Thus [5],

$$P_{\mu} = \int d^{3}\bar{x} \; n_{\alpha} T^{\alpha}{}_{\mu}, \tag{4.14}$$

or in terms of invariant components

$$\begin{aligned} P_{u} &= - \int d^{3}\bar{x} \cdot n \cdot T \cdot m = - \int d^{3}\bar{x} \, T_{vu} \\ P_{j} &= \int d^{3}\bar{x} \; n \cdot T \cdot e_{j} = - \int d^{3}\bar{x} \; T_{vj} \qquad (j = 1,\, 2) \\ P_{v} &= - \int d^{3}\bar{x} \; n \cdot T \cdot n = - \int d^{3}\bar{x} \; T_{vv} \quad . \end{aligned} \tag{4.15}$$

From (4.9) and (4.12) one finds

$$\begin{aligned} P_{u} &= \int (\underline{\partial} \; \phi^{*} \cdot \underline{\partial}\phi + m^{2} \phi^{*} \phi) d^{3}\bar{x} \\ P_{j} &= \int \partial_{j}\phi^{*} \partial_{r}\phi + \partial_{r}\phi^{*} \partial_{j}\phi) d^{3}\bar{x} \qquad (j = 1,\, 2) \\ P_{v} &= 2 \int \partial_{v}\phi^{*} \partial_{r}\phi d^{3}\bar{x} \; . \end{aligned} \tag{4.16}$$

(b) Homogeneous Lorentz transformations: These well-known transformations can be treated very briefly. The infinitesimal antisymmetric tensor $\omega_{\mu\nu}$ is determined by six parameters, the three (infinitesimal) Euler angles of space rotations and the three components of the (infinitesimal) velocity which determine the boosts. One has

$$\delta\, x^\mu = \omega^\mu{}_\nu\, x^\nu \tag{4.17}$$

and again $\delta\phi=0$ (by definition) and $\delta\,\Omega^\mu=0$ (by computation). Noether's theorem gives the differential conservation laws

$$\partial_\lambda J^{\lambda\mu\nu} = 0, \qquad J^{\lambda\mu\nu} \equiv T^{\lambda\mu}\, x^\nu - T^{\lambda\nu}\, x^\mu, \tag{4.18}$$

and the six generators

$$J^{\mu\nu} = \int d^3\bar{x}\; n_\lambda J^{\lambda\mu\nu} = -\, J^{\nu\mu} \tag{4.19}$$

as integrals over the null plane u = constant. The non-vanishing invariant components are

$$\begin{aligned}
J_{uv} &= \int d^3\bar{x}[(\underline{\partial}\phi^*\cdot\underline{\partial}\,\phi + m^2\phi^*\phi)u-2\partial_v\phi^*\partial_v\phi v] \\
J_{uj} &= \int d^3\bar{x}[(\underline{\partial}\phi^*\cdot\underline{\partial}\,\phi + m^2\phi^*\phi)x_j-(\partial_j\phi^*\partial_v\phi+\partial_v\phi^*\partial_j\phi)v] \\
J_{12} &= \int d^3\bar{x}[(\partial_1\phi^*\partial_v\phi+\partial_v\phi^*\partial_1\phi)x_2-(\partial_2\phi^*\partial_v\phi+\partial_v\phi^*\partial_2\phi)x_1] \\
J_{vj} &= \int d^3\bar{x}[2\partial_v\phi^*\partial_v\phi x_j-(\partial_j\phi^*\partial_v\phi+\partial_v\phi^*\partial_j\phi)u].
\end{aligned} \tag{4.20}$$

As was proven earlier, these integrals over the null plane u = constant are independent of the choice of u. Thus, one can put u=o in the above integrals; J_{uv} and J_{vj} are thereby considerably simplified.

The above two invariance groups together give the main part of the Poincaré group, $iL_+^\uparrow$. The full group would require the discrete transformation P and T which we shall not discuss here.

The desired solution of the field equation, (4.10), is conveniently expressed in terms of a complete set of fundamental solutions. This is an obvious generalization of the normal mode expansion of a periodic system. Thus, we write (with anticipation of later developments)

$$\phi(x) = \sum_\alpha (a_\alpha f_\alpha(x) + b^*_\alpha f^*_\alpha(x)). \tag{4.21}$$

The index α can be continuous ("plane wave expansion") or discrete ("wave packet expansion"); since the continuous case can be obtained from the discrete one by a limit, we shall adopt α as a discrete index.

The f_α together form a complete set of positive frequency solutions and also satisfy the Klein-Gordon equation. For this reason,

$$\sum_\alpha f_\alpha(x) f^*_\alpha(x') = - i \, \Delta_+(x-x'), \tag{4.22}$$

and the scalar product (not positive definite for general solutions, but positive definite for the f_α)

$$\langle f_\alpha, f_\beta \rangle \equiv i \int f^*_\alpha(x) \overleftrightarrow{\partial}_\mu f_\beta(x) \, d^3\sigma^\mu$$

is independent of σ. We note that only those functions $f_\alpha(x)$ are admitted for which this integral exists. Taking for σ the null surface u = constant we can choose f_α to be an orthonormal basis,

$$\langle f_\alpha, f_\beta \rangle = - i \int f^*_\alpha(x) \overleftrightarrow{\partial}_v f_\beta(x) \, d^3\bar{x} = \delta_{\alpha\beta}. \tag{4.23}$$

The numerical coefficients of the wave packet expansion (4.21) now become

$$a_\alpha = \langle f_\alpha, \phi \rangle, \qquad b_\alpha = - \langle f^*_\alpha, \phi \rangle \tag{4.24}$$

because

$$\langle f_\alpha, f^*_\beta \rangle = 0 \quad \forall \, \alpha, \beta. \tag{4.25}$$

The completeness relation of the positive frequency space, (4.22) gives a convenient way to find the transformation property of the field $\phi(x)$ induced by the coordinate transformations. Given the latter the transformation of the known function $\Delta_+(x-x')$ can be ascertained. Eq. (4.22) then tells us how $f_\alpha(x)$ transforms. But from the wave packet expansion (4.21) we learn that $\phi(x)$ transforms exactly as $f_\alpha(x)$ does. Thus the transformation of $\phi(x)$ is obtained.

It is easy to check that cases (a) and (b) above are consistent with this argument since Δ_+ is Poincaré invariant.

(c) Dilatation transformations [8]: These are scalings of the whole spacetime system,

$$x^{\mu\prime} = \ell x^\mu \tag{4.26}$$

so that the infinitesimal transformations yield with $\ell=1+\lambda+o(\lambda^2)$

$$\delta x^\mu = \lambda x^\mu . \tag{4.27}$$

This yields invariance of the action integral, (4.2), provided the scalar field has m=o and transforms according to

$$\phi'(x') = \frac{1}{\ell}\phi(x), \qquad \delta\,\phi(x) = -\,\lambda\,\phi(x) \tag{4.28}$$

and one has $\delta\,\Omega^\mu=o$. The same transformation follows with m=o from (4.21) and (4.22) since $D_+'(x')=D_+(x)/\ell^2$.

The conservation law follows from (4.4),

$$\partial_\mu D^\mu = o, \qquad D^\mu \equiv T^\mu{}_\alpha x^\alpha - \partial^\mu(\phi^*\phi), \tag{4.29}$$

i.e. it leads to a conserved "dilatation current". The integral conservation law is the constancy of the corresponding "dilatation charge"

$$D \equiv \int d^3\sigma_\mu D^\mu(x). \tag{4.30}$$

When expressed as a null plane integral,

$$D = -\int d^3\bar{x}D_v = \int d^3\bar{x}[\partial_v(\phi^*\phi)+\bar{x}\cdot\bar{\partial}\ \phi^*\partial_v\phi+\partial_v\phi^*\bar{x}\cdot\bar{\partial}\ \phi+T_{vu}u]. \tag{4.31}$$

Again, we can put u=o in this integral whenever (4.29) holds because then the integral is independent of u.

(d) Special Conformal Transformations [8]:

(Acceleration Transformations). These are defined by four parameters a^μ,

$$x'^\mu = \frac{x^\mu+a^\mu x^2}{\sigma(x,\ a)} \quad \sigma(x,a) \equiv 1+2a\cdot x+a^2x^2. \tag{4.32}$$

With a=1+α one obtains the infinitesimal transformations

$$x'^\mu = x^\mu+\alpha^\mu x^2-2\alpha\cdot x\ x^\mu$$

$$\partial x^\mu = \alpha^\nu(\partial^\mu_\nu x^2-2x_\nu x^\mu). \tag{4.33}$$

The volume element transforms as

$$d^4x' = \frac{d^4x}{[\sigma(x,a)]^4} = (1-8\alpha\cdot x)d^4x+o(\alpha^2) \tag{4.34}$$

and the derivative as

$$\partial'_\mu = \partial_\mu + 2(\alpha\cdot x\partial_\mu + \alpha_\mu x\cdot\partial - x_\mu \alpha\cdot\partial). \tag{4.35}$$

Finally, the transformation of the field follows from (4.21) and (4.22) by the argument given just preceding case (c). With m=o and for infinitesimal transformations,

$$\sum_\alpha f'_\alpha(x') f'^*_\alpha(o) = -\, iD'_+(x') = -\, i(1+2\alpha\cdot x)D_+(x).$$

Hence,

$$f'_\alpha(x') = (1+2\alpha\cdot x) f_\alpha(x), \quad \delta\phi = 2\alpha\cdot x\phi(x). \tag{4.36}$$

The action integral can now be investigated. Using (4.34) - (4.36) one finds the result (4.2) with

$$\delta\,\Omega^\mu = 2\alpha^\mu \phi^*(x)\phi(x) \tag{4.37}$$

provided m=o.

Noether's theorem therefore yields the differential conservation law

$$\partial_\mu C^{\mu\nu} = o, \quad C^{\mu\nu} \equiv T^\mu{}_\alpha(x^2 g^{\alpha\nu} - 2x^\alpha x^\nu) + 2\partial^\mu(\phi^*\phi)x^\nu - 2g^{\mu\nu}\phi^*\phi. \tag{4.38}$$

The conservation of this conformal tensor $C^{\mu\nu}$ corresponds to the constancy of a conformal vector

$$C^\mu = \int d^3\sigma_\alpha C^{\alpha\mu}(x) = -\int d^3\bar{x} C^{v\mu}(x), \tag{4.39}$$

when referred to a null plane. The invariant components of this vector are, using u=o:

$$C_u = -\int d^3\bar{x}[\underline{x}^2 T_{vu} - 2v(T_{vj}x_j - T_{vv}v) + 2(1+v\partial_v)(\phi^*\phi)]$$

$$C_j = -\int d^3\bar{x}[\underline{x}^2 T_{vj} - 2x_j(T_{vk}x_k - T_{vv}v) + 2x_j\partial_v(\phi^*\phi)]$$

$$C_v = -\int d^3\bar{x}\; \underline{x}^2 T_{vv}. \tag{4.40}$$

The invariant component of $T_{\mu\nu}$ are given in (4.15), (4.16).

If we have a zero mass field then all invariance properties (a) through (d) will hold and the theory will be invariant under the 15 parameter Liouville group (= general conformal group). If $m\neq o$ dilatation and the special conformal invariance will be broken so that the invariance group reduces to the 10 parameter Poincaré group.

Now it can happen that a conservation law associated with one symmetry implies a conservation law associated with a different symmetry of the same physical system. If this happens then the associated symmetry is also implied.

As an example, we consider the free scalar field again. Here momentum conservation, differentially expressed as $\partial_\mu T^{\mu\nu}=o$ implies $\partial_\lambda J^{\lambda\mu\nu}=o$, i.e. conservation of angular momentum and of center of mass. This follows from the form (4.18) of $J^{\lambda\mu\nu}$ and from the symmetry of $T^{\mu\nu}$. Thus, in this case invariance under homogeneous Lorentz transformations is actually implied by translation invariance! This will hold in any system in which translation invariance yields $\partial_\mu T^{\mu\nu}=o$ and $T^{\mu\nu}=T^{\nu\mu}$, the energy tensor being defined by (4.5).

A second example is furnished by the relation of dilatation and conformal invariance in the presence of translation invariance. In our system the conservation of the dilatation current is

$$o = \partial_\mu D^\mu = T^\mu{}_\mu - \Box(\phi^*\phi) \tag{4.41}$$

when momentum conservation holds. But the conservation of the conformal tensor,

$$o = \partial_\mu C^{\mu\nu} = - 2x^\nu(T^\mu_\mu - \Box(\phi^*\phi)) = - 2x^\nu \partial_\mu D^\mu, \tag{4.42}$$

is implied by (4.41). Thus, in this translation invariant theory, dilatation invariance implies conformal invariance. One can prove that this holds for a large class of relativistic theories [9].

5. SCALAR NULL PLANE QUANTUM FIELD THEORY

a) Fundamentals

Most fundamental classical theories are based on the action principle. This, in turn, is a special case of Noether's theorem. This theorem is therefore a most important basic tool in classical physics. Unfortunately, this tool is not available in quantum field theory. The reason is, of course, that the variational calculus does not exist for operators. This fact is very often ignored.

As a substitute, invariance properties in quantum field theory can be studied by means of Lie algebras. We shall lead up to this technique in this Section.

For the free quantized scalar field $\phi(x)$ we demand the same field equation (4.10) as for the classical field, so that we can take the same fundamental solutions. The expansion (4.21) is formally taken over,

$$\Phi(x) = \sum_{\alpha}(a_{\alpha}f_{\alpha}(x) + b_{\alpha}^{+}f_{\alpha}^{*}(x)) \tag{5.1}$$

but the a_{α} and b_{α}^{+} are now operators which are defined to satisfy the well-known relations

$$[a_{\alpha},a_{\beta}] = [b_{\alpha},b_{\beta}] = [a_{\alpha},b_{\beta}] = [a_{\alpha},b_{\beta}^{+}] = o;$$

$$[a_{\alpha},a_{\beta}^{+}] = [b_{\alpha},b_{\beta}^{+}] = \delta_{\alpha\beta}. \tag{5.2}$$

Thus, the scalar products (4.23) to (4.25) can be used,

$$a_{\alpha} = \langle f_{\alpha},\Phi\rangle; \qquad b_{\alpha} = - \langle f_{\alpha}^{*},\Phi\rangle. \tag{5.3}$$

They can be regarded as formal integrals expressing the field as an operator valued generalized function. The f_{α} then play the role of test functions. The test function space is defined over the three-dimensional null planes,

$$a_\alpha^+ = \langle f_\alpha, \Phi\rangle^+ = \langle \Phi, f_\alpha\rangle = -i\int \Phi^+(x)\overleftrightarrow{\partial}_v f_\alpha(x)\,d^3\bar{x} =$$

$$= -2i[\Phi(\partial_v f_\alpha)]^+. \tag{5.4}$$

Here we use the notation

$$\Phi(u,f) \equiv \int_u \Phi(x) f^*(x)\,d^3\bar{x} \tag{5.5}$$

and

$$\Phi(f) \equiv \Phi(o,f) . \tag{5.6}$$

The u-independence of a_α and b_α is ensured by the choice of the f_α as solutions of the Klein-Gordon equation.

The quantization (5.2) implies of course the well known commutation relations

$$[\Phi(x), \Phi^+(x')] = -i\Delta(x-x') \tag{5.7}$$

and v.v., (5.7) implies (5.2). However, if the theory is to be formulated as an initial value problem on a null plane, as in Section 2, then only the restriction of (5.7) to this null plane should be specified.

This restriction is easiest obtained from the Fourier representation,

$$\Delta(x) = \frac{i}{(2\pi)^3}\int e^{ip\cdot x}\varepsilon(p_v)\,\partial(p^2+m^2)\,d^2\underline{p}\;d\,p_v d\,p_u =$$

$$= \frac{1}{(2\pi)^3}\int e^{ip\cdot x}\sin(p_v v + \frac{\underline{p}^2+m^2}{2p_v}u)\,d\mu$$

where

$$d\mu \equiv \frac{dp_1 dp_2 dp_v}{p_v}\,\theta(p_v). \tag{5.8}$$

From this expression one finds for u=o

$$\Delta(o,\bar{x}) = \frac{1}{4}\,\varepsilon(v)\,\delta_2(\underline{x}). \tag{5.9}$$

The commutation relations on the null plane are therefore

$$[\Phi(u,\bar{x}),\Phi^+(u,\bar{x}')] = -\frac{i}{4}\varepsilon(v-v')\delta_2(\underline{x}-\underline{x}') \quad (5.10)$$

or
$$[\Phi(u,\bar{x}),\partial_v'\Phi^+(u,\bar{x}')] = -\frac{i}{2}\delta_3(\bar{x}-\bar{x}'). \quad (5.11)$$

By means of test functions this relation is written in a mathematically more meaningful way as

$$[\Phi(f),\Phi(\partial_v g)^+] = \frac{i}{2}(f,g). \quad (5.12)$$

The notation (5.6) is used here.

As a consistency check we verify easily that from (5.4) and (5.12) the operator algebra (5.2) is recovered if use is made of the orthonormality relation of the f_α, (4.23). For example,

$$[a_\alpha,a_\beta^+] = 4[\Phi(\partial_v f_\alpha),\Phi(\partial_v f_\beta)^+] = 2i(\partial_v f_\alpha,f_\beta) = \langle f_\alpha,f_\beta\rangle=\delta_{\alpha\beta}. \quad (5.13)$$

The fundamental commutation relation (5.12) for a free scalar field on a null plane is most remarkable:

(1) It is the canonical C. R. since the canonically conjugate to the field is $\pi(g)=-[(\partial_v\Phi)(g)]^+=\Phi(\partial_v g)^+$. The factor 1/2 on the right appears because on a null plane π and Φ are not dynamically independent; the usual canonical formalism must therefore be extended to include constraint relations. If this is done (5.12) can also be obtained by means of a canonical formulation of the classical field. (Of course one then obtains the corresponding classical relation, i.e. the Poisson bracket, from which (5.12) can then be deduced).

(2) Eq. (5.10) must be compared with the corresponding relation on a spacelike plane (t = constant) which yields are <u>abelian</u> algebra. The noncommutativity of the field along null directions is of course not surprising.

(3) From the geometrical fact that the normal of a null plane lies in that same null plane one sees that only the knowledge of Φ on the null plane enters the C. R. (5.12). In contradistinction to this, the usual C. R. $[\Phi(t,\vec{x}),\frac{\partial}{\partial t}\Phi^{+}(t,\vec{x}')]=i\delta_3(\vec{x}-\vec{x}')$ involves Φ as well as $\partial\phi/\partial t$. The latter cannot be deduced from the knowledge of ϕ on t = constant. An analogous situation was encountered in the classical initial value problem, Section 3.

Closely related to this fact is the observation that the knowledge of $\Phi(x)$ on the null plane suffices to determine both the a_α and the b_α, as is already implied in (5.4). In fact, in the notation (5.6),

$$a_\alpha = 2i\Phi(\partial_v f_\alpha), \qquad b_\alpha = -2i\Phi(\partial_v f_\alpha)^{+}. \tag{5.14}$$

In the usual case of spacelike planes both Φ and $\partial_o\Phi$ are necessary to determine the creation and annihilation operators, $\partial_o\Phi$ being independent of Φ.

These properties of the C. R. (5.12) can be most compactly summarized mathematically by saying that the fields Φ characterized by the algebra (5.12) are complete or irreducible. This and a related theorem were first proven by Klauder, Leutwyler,and Streit [10].

(A) The Fock space representations of (5.12) are irreducible.

(B) Two Fock space representations of (5.12) associated with different masses are unitarily equivalent.

The latter is the precise mathematical formulation of the intuitive feeling that in the infinite momentum frame mass should play no role.

For the proofs the reader is referred to the original paper. But the equivalence of (5.12) and (5.2) was already indicated in (5.13), suggesting theorem (A). The mass independence can be seen from the measure in momentum space , (5.8), and from the mass independence of the generators P_j and P_v. The point is as follows.

The mass shell test functions $f_\alpha(x)$ have a Fourier transform which is usually written with respect to a spacelike plane (t = constant, say)

$$f_\alpha(x) = \frac{1}{(2\pi)^{3/2}}\int\frac{d^3p}{2\omega}g_\alpha(\vec{p})e^{i\vec{p}\cdot x-i\omega t} \qquad (\omega \equiv +\sqrt{\vec{p}^2+m^2})$$

where the $g_\alpha(\vec{p})$ are any complete orthonormal set

$$(g_\alpha, g_\beta) \equiv \int\frac{d^3p}{2\omega}g_\alpha^*(\vec{p})g_\beta(\vec{p}) = \delta_{\alpha\beta}, \quad \sum_\alpha g_\alpha(\vec{p})g_\alpha^*(\vec{p}\,') = 2\omega\ \delta_3(\vec{p}-\vec{p}\,')$$

characterizing the wave packets.

But the Fourier transform can also be taken with respect to a null plane,

$$f_\alpha(x) = \frac{1}{(2\pi)^{3/2}}\int d\mu g_\alpha(\bar{p})e^{i(\bar{p}.\bar{x}-p_u u)} \qquad (p_u = \frac{p^2+m^2}{2p_v}) \qquad (5.15)$$

with $d\mu$ as in (5.8) in which case the $g_\alpha(\bar{p})$ are orthonormal with the scalar product

$$(g_\alpha, g_\beta) \equiv 2\int d\mu g_\alpha^*(\bar{p})g_\beta(\bar{p}) = \delta_{\alpha\beta} \qquad (5.16)$$

and complete,

$$\sum_\alpha g_\alpha(\bar{p})g_\alpha^*(\bar{p}\,') = \tfrac{1}{2}p_v\delta_3(\bar{p}-\bar{p}\,') \qquad (5.17)$$

as follows from the corresponding relations for f_α, (4.23) and (4.22).

Since p_v corresponds to the limit $p_3\to\infty$ it is physically reasonable to restrict oneself to g_α which vanish for p_v=o. (This restriction is in fact required if one wants g_α to be continuous, because g_α=o for p_v<o,

since the f_α are positive energy test functions). The Hilbert space $L^2(2d\mu)$ characterized in (5.16) is the "momentum test function space".

Comparing this space with the customary one with measure $d^3p/(2\omega)$ we see that in the latter the scalar product depends on the mass. But $d\mu$ is mass independent. This is the crucial feature which enters the proof of (B).

The Fourier transform (5.15) of the mass shell test function on the null plane provides the null plane wave packet expansion of the field. From (5.1),

$$\Phi(x) = \frac{1}{(2\pi)^{3/2}} \sum_\alpha \int d\mu \, (a_\alpha g_\alpha(\bar{p}) e^{ip\cdot x} + b_\alpha^+ g_\alpha^*(\bar{p}) e^{-ip\cdot x}). \tag{5.18}$$

We note that the popular plane wave limit of this expansion -- as in the spacelike plane case -- leads to ambiguities. One has

$$\sum_\alpha \to 2\int d\mu_\alpha, \; g_\alpha(\bar{p}) \to p_v \delta_3(\bar{p}-\bar{p}_\alpha), \; f_\alpha(x) \to \frac{1}{(2\pi)^{3/2}} \theta(p_v^\alpha) e^{ip_\alpha \cdot x} \tag{5.19}$$

and $\lim\limits_{|v|\to\infty} f_\alpha(x) = 0$ is not satisfied, but it is ambiguous. In any case, f_α can no longer be used as test functions, and the Hilbert space structure is destroyed.

b) Dynamics and Symmetries

It is well known that, if the free Hamiltonian density $\mathcal{H}(x)$ is translation invariant in four-dimensional Minkowski space, the vacuum will be annihilated by its four-dimensional integral,

$$\int \mathcal{H}(x) d^4x |o\rangle = o. \tag{5.20}$$

This is not so if one integrates only over a three-dimensional spacelike hyperplane. This becomes obvious when $\mathcal{H}(x)$ is written as a normal ordered product in p-space.

Translation invariance in (5.20) becomes equivalent to $\Sigma p^{\mu}=o$, so that both negative and positive mass shell momenta must be present, i.e. both creation and annihilation operators must be present; hence (5.20). But if one does not integrate over x^{o}, then $\Sigma p^{o}=o$ is not implied and

$$H \equiv \int \mathcal{H}(x)\, d^3x$$

does not annihilate the vacuum as a consequence of translation invariance.

For null planes this situation is different. The analogous operator of development of the system off the initial plane

$$P_u \equiv \int \mathcal{P}_u(x)\, d^3\bar{x} \tag{5.21}$$

does annihilate the vacuum,

$$P_u|o\rangle = o \tag{5.22}$$

as a consequence of translation invariance. The reason lies in the fact that (5.21) requires $\Sigma p_v=o$, so that both future and past null cone momenta must be present, and therefore also both, creation and annihilation operators. We note parenthetically that

$$\text{sgn}\, p^{o} = \text{sgn}\, p_v.$$

In our convention where the g_α have only $p_v>o$, the g_α^* have only $p_v<o$, and g_α and g_α^* are associated with annihilation and creation operators, respectively.

The property (5.22) of null plane field theory has a far reaching consequence. It implies that the analogue to the Dyson S-Matrix for null plane field theory leads to Feynman diagrams that vanish whenever they describe vacuum-vacuum transitions (as observed by Weinberg [4]). In the usual field theory these diagrams are known to be the most divergent ones.

The "null Hamiltonian" P_u is the generator of infinitesimal translations in the u-direction (the "time" - direction of null plane field theory). Finite translations are, of course, given by the unitary time development operator

$$\Phi(u,\bar{x}) = e^{iP_u u}\Phi(o,\bar{x})e^{-iP_u u}$$

which is, of course, a special case of the more general

$$\Phi(x) = e^{-iP\cdot x}\Phi(o)e^{iP\cdot x}. \tag{5.23}$$

The explicit form of P_u is easily inferred from the classical theory. From (4.16)

$$P_u = \int :(\underline{\partial}\,\Phi^+\cdot\underline{\partial}\,\Phi + m^2\Phi^+\Phi):d^3\bar{x}. \tag{5.24}$$

The knowledge of P_u, just as the knowledge of the Hamiltonia H in the usual theory, together with the initial information determinesthe dynamics: given the field algebra (5.12) on the initial null surface (u=o, say), and given P_u, then $\Phi(x)$ is determined by (5.23) for all x. This is equivalent to solving the "Heisenberg equations"

$$\partial_\mu\Phi(x) = i[\Phi(x),P_\mu] \tag{5.25}$$

along the unit vector m^μ with $\Phi(o,\bar{x})$ given. This is the quantum mechanical form of the initial value problem of Section 3.

The representation of P_u as a null plane integral can, of course, be extended to the other generators. Thus, one infers from the classical theory (4.14) the translation generators,

$$P_\mu = -\int d^3\bar{x}\; T_{v\mu}(x) \tag{5.26}$$

where

$$T_{V\mu}(x) = -:[\partial_V\Phi^+\partial_\mu\Phi+\partial_\mu\Phi^+\partial_V\Phi+g_{V\mu}(\partial_\alpha\Phi^+\partial^\alpha\Phi+m^2\Phi^+\Phi)]:, \qquad (5.27)$$

and the homogeneous Lorentz group generators from (4.19),

$$J_{\mu\nu} = -\int d^3\bar{x}J_{V\mu\nu},\ J_{V\mu\nu} = T_{V\mu}x_\nu - T_{V\nu}x_\mu. \qquad (5.28)$$

By means of the null plane field algebra (5.12) on the null plane one verifies that the Poincaré algebra is satisfied,

$$[P_\mu,P_\nu] = 0,\ [J_{\lambda\mu},P_\nu] = i(g_{\lambda\nu}P_\mu-g_{\mu\nu}P_\lambda)$$
$$[J_{\kappa\lambda},J_{\mu\nu}] = i(g_{\kappa\mu}J_{\lambda\nu}+g_{\lambda\nu}J_{\kappa\mu}-g_{\kappa\nu}J_{\lambda\mu}-g_{\lambda\mu}J_{\kappa\nu}). \qquad (5.29)$$

Conversely, and generally, the existence of generators P_μ, $J_{\mu\nu}$ which satisfy (5.29) as a consequence of the field algebra ensures the Poincaré invariance of the theory.

Similarly, one has the dilatation generator from (4.31) with u=o and $\partial_V(\Phi^+\Phi)$ integrated

$$D = \int d^3\bar{x}:[\bar{x}\cdot\bar{\partial}\ \Phi^+\partial_V\Phi+\partial_V\Phi^+\bar{x}\cdot\bar{\partial}\ \Phi]:, \qquad (5.30)$$

and the conformal vector from (4.39)

$$C_\mu = -\int d^3\bar{x}:[T_{V\alpha}(x^2\delta^\alpha_\mu-2x^\alpha x_\mu)+2x_\mu\partial_V(\Phi^+\Phi)-2g_{V\mu}\Phi^+\Phi]:, \qquad (5.31)$$

where $T_{V\alpha}$ is given by (5.27) without the mass term. If the field is massless, the full 15 parameter Liouville algebra will be satisfied. It consists of the Poincaré subalgebra and

$$[D,T_{\mu\nu}] = 0,\ [D,P_\mu] = iP_\mu,\ [D,C_\mu] = -iC_\mu,\ [C_\mu,C_\nu] = 0$$

$$[C_\mu,P_\nu] = -2i(g_{\mu\nu}D+J_{\mu\nu}),\ [J_{\lambda\mu},C_\nu] = i(g_{\lambda\nu}C_\mu-g_{\mu\nu}C_\lambda). \qquad (5.32)$$

Again quite generally, if generators P_μ, $J_{\mu\nu}$, D, C_μ can be found which satisfy the Liouville algebra (5.29),

(5.32) as a consequence of the field algebra, the theory is invariant under the Liouville group. We know, of course, that for this to happen the field must be massless. The explicite verification of the 105 equations of this algebra is left as an exercise.

But let us now assume that we have a field with mass, $m \neq 0$, which is Poincaré invariant. Since the dilatation generator (5.30) is mass independent and P_μ and $J_{\mu\nu}$ are mass independent in the null plane (i.e. as long as one does not take their products with m^μ!) the Poincaré + Dilatation algebra will be satisfied in the null plane. Thus, a null plane field theory is dilatation invariant in the null plane even if it has a mass. This fact is intuitively obvious from the mass independence of the field algebra (theorem (B) following (5.14)).

We have shown in the classical case that dilatation invariance implies conformal invariance at least in the scalar case. This expectation holds also in the quantum theory. In fact, as (4.40) shows, C_v and C_j are mass independent, since only the component T_{vu} of the energy tensor depends on m. Thus, the Liouville algebra holds on the null plane even for a finite mass field. Note that this is a ten parameter sub-algebra of the full 15 parameter Liouville algebra.

It is a property of an integrable Lie group that the Lie algebra implies the finite transformations connected to the identity. Thus, the unitary implementation of dilatation and conformal symmetry on the null plane is ensured by the above considerations.

When one propagates off the null plane with P_u of $m=o$ the Liouville invariance will be preserved. But if one propagates with P_u and $m \neq o$, the Liouville invariance will be broken and only Poincaré invariance will survive.

It is easy to check on the unitary implementation explicitly. For example, for dilatation invariance we have from (4.28)

$$f'_\alpha(x') = \frac{1}{\ell} f_\alpha(x) \tag{5.33}$$

and therefore for g_α from (5.15) with $p' = \frac{1}{\ell}p$

$$g'_\alpha(\bar{p}') = \ell g_\alpha(\bar{p}). \tag{5.34}$$

Thus,

$$\|g'_\alpha\|^2 = \int 2d\mu' |g'_\alpha(\bar{p}')|^2 = 2\int d\mu' \ell^2 |g_\alpha(\bar{p})|^2 =$$
$$= 2\int d\mu |g_\alpha(\bar{p})|^2 = \|g_\alpha\|^2. \tag{5.35}$$

The invariance of the norm ensures unitarity on the one particle subspace H_1 of Fock space. But since all of Fock space can be constructed from H_1,

$$H = \sum_{n=1}^{\infty} \oplus H_n$$

where H_n is the symmetrized outer product of n spaces H_1, it follows that unitarity is ensured on all of H. Thus, there exists a unitary operator $U(\ell)$ such that on the null plane u = constant

$$U(\ell)\Phi(x)U^+(\ell) = \frac{1}{\ell}\Phi\left(\frac{x}{\ell}\right). \tag{5.36}$$

Similarly, one can prove the unitary implementation of the whole Liouville group when restricted to the null plane, even though the field may have a mass [11].

Returning to the question how to describe the symmetry properties of a quantum field theory in view of the inavailability of Noether's theorem, we note the following. The existence of a symmetry under a Lie group requires the existence of the associated Lie algebra. Thus, one must construct the generators, G_i, in terms of the fields in such a way that they satisfy the desired Lie algebra (e.g. (5.29) and (5.32)) as a consequence of the given field algebra (e.g. (5.12)). Once the G_i are known, the transformation of the field under the infinitesimal transformation generated by them are also known: let t_i be an infinitesimal transformation, $x'=t_i x$; let $\delta_i\Phi$ be the transformation of Φ induced by t_i, $-\varepsilon_i\delta_i\Phi(x)\equiv\Phi'(x)-\Phi(x)$ where ε_i is the infinitesimal parameter of the transformation, then $\delta_i\Phi$ is given in terms of the associated G_i by

$$\delta_i\Phi = i[\Phi,G_i]. \tag{5.37}$$

As an example, we shall compute the transformation property of $\Phi(x)$ under dilatation:

$$\Phi'(x) = \Phi(x) - \lambda i[\Phi(x),D]. \tag{5.38}$$

From (5.11) and (5.30) we have on u=o

$$i[\Phi(x),D] = i\int d^3\bar{x}'[\Phi(x),:\{\partial_v'\Phi^+(x')\bar{x}'\cdot\bar{\partial}'\Phi(x')+\bar{x}'\cdot\bar{\partial}'\Phi^+(x') \times$$
$$\times\ \partial_{v'}\Phi(x')\}:]$$

since the term $\partial_v:\Phi^+\Phi:$ integrates to zero. The first term is easily seen to give $1/2\ \bar{x}\cdot\bar{\partial}\ \Phi$. In the second term we can shift $\partial_{v'}$ onto the other factors obtaining

$$[\Phi(x),-:\partial_{v'}(\bar{x}'\cdot\bar{\partial}'\Phi^+(x'))\Phi(x'):] =$$

$$-[\Phi(x),:\partial_{v'}\Phi^+(x')]\Phi(x'):-\bar{x}'\cdot\bar{\partial}'[\Phi(x),:\partial_{v'}\Phi^+(x')]\Phi(x'):$$

so that their integral gives

$$-\frac{1}{2}\Phi-\frac{1}{2}\int d^3\bar{x}'\bar{x}'\cdot\bar{\partial}'\delta_3(\bar{x}-\bar{x}')\Phi(x')$$

$$= -\frac{1}{2}\Phi+\frac{1}{2}\bar{\partial}\cdot(\bar{x}\Phi(x)) = -\frac{1}{2}\Phi+\frac{3}{2}\Phi+\frac{1}{2}\bar{x}\cdot\bar{\partial}\ \Phi.$$

Thus,

$$i[\Phi(x),D] = \Phi + \bar{x}\cdot\bar{\partial}\ \Phi$$

and

$$\Phi'(x) = \Phi(x)-\lambda(1+\bar{x}\cdot\bar{\partial})\Phi(x) \qquad (5.39)$$

on u=o.

This is in agreement with the classical result (4.28) which gives

$$\phi'(x) = \frac{1}{\ell}\phi(\frac{x}{\ell}) = (1-\lambda)\phi(x-\lambda x) = \phi(x)-\lambda(1+x\cdot\partial)\phi(x) \qquad (5.40)$$

and which reduces to the form (5.39) for u=o.

6. NULL PLANE QUANTUM ELECTRODYNAMICS

a) The Free Maxwell Field

The free Maxwell equations

$$\partial_\mu f^{\mu\nu} = o, \qquad f^{\mu\nu} = \partial^\mu a^\nu - \partial^\nu a^\mu \qquad (6.1)$$

can easily be cast into invariant null plane coordinate form. Then they read,

$$(\underline{\partial}^2-\partial_u\partial_v)a_u - \partial_u(\underline{\partial}\cdot\underline{a} - \partial_u a_v) = o \qquad (6.2)$$

$$(\underline{\partial}^2- 2\partial_u\partial_v)a_j - \partial_j(\underline{\partial}\cdot\underline{a} - \partial_v a_u - \partial_u a_v) = o \quad (j=1,2) \qquad (6.3)$$

$$(\underline{\partial}^2 - \partial_u\partial_v)a_v - \partial_v(\underline{\partial}\cdot\underline{a} - \partial_v a_u) = o\ . \qquad (6.4)$$

Recalling that u plays the role of time we see that the first of these is of second order in ∂_u, eq. (6.3) is of first order but not of the form (3.9), and only (6.4) is of first order in ∂_u and also of the form (3.9). But the gauge freedom which one obtains by the introduction of the potentials permits a considerable simplification of these equations. One can clearly choose a gauge in which one component of the potential vanishes. Our invariant coordinatization permits us to do this even in an invariant way.

We choose the gauge characterized by

$$a_v \equiv 0 . \tag{6.5}$$

This gauge will be called the null plane gauge because it restricts the vector a^μ to lie entirely in the null plane u = constant, it having no component in the direction of the unit vector m which sticks out of such a plane [2,12].

With this choice of gauge eqs. (6.3) and (6.4) reduce to

$$(\underline{\partial}^2-2\partial_u\partial_v)a_j-\partial_j(\underline{\partial}\cdot\underline{a}-\partial_v a_u) = 0 \quad (j = 1, 2) \tag{6.6}$$

and

$$\partial_v(\underline{\partial}\cdot\underline{a} - \partial_v a_u) = 0; \tag{6.7}$$

the first of these is of the form (3.9), the second contains no ∂_u operator at all. But eq. (6.2) is still not of the form (3.9).

However, the Maxwell equation (6.2) can easily be avoided altogether. Since the derivatives of (6.1) are related by an identity,

$$\partial_\nu\partial_\mu f^{\mu\nu} \equiv 0 ,$$

it follows that the ∂_v derivative of equation (6.2) is a consequence of the remaining equations, (6.3) and (6.4). Therefore, in the null plane gauge the equation

$$(\underline{\partial}^2-\partial_u\partial_v)\partial_v a_u - \partial_u\partial_v\ \underline{\partial}\cdot\underline{a} = 0$$

is a consequence of (6.6) and (6.7). But $\partial_v a^\mu(f)=-a^\mu(\partial_v f)$ is just as good as $a^\mu(f)$, since we assume both f and $\partial_v f$ to be in the test space of a^μ. Thus, this equation gives the same information as the undifferentiated equation. Therefore, (6.6) and (6.7) have all the content of the free

Maxwell field (in the null plane gauge and with our test space).

Eq. (6.7) is a constraint equation, since it is independent of ∂_u. In fact, in our gauge it is the same as the Lorentz gauge condition,

$$\partial_\mu a^\mu(\partial_v f) = -\partial_v(\underline{\partial}\cdot\underline{a} - \partial_v a_u)(f) = 0 . \qquad (6.8)$$

On the null plane this gauge and the gauge $a_v=0$ are therefore compatible.

The contraint equation (6.7) can be used to eliminate [12,2] a_u,

$$a_u(\partial_v f) = - \partial_v a_u(f) = - \underline{\partial}\cdot\underline{a}(f). \qquad (6.9)$$

This is another way of showing that the Maxwell equation (6.2) is not needed.

With the gauge $a_v=0$ and with a_u expressed in terms of $a_j (j=1,2)$ we find that <u>there are only two dynamically independent components in the Maxwell field</u>, viz. $a_j (j=1,2)$. Their equations are, in view of (6.8),

$$(\underline{\partial}^2-2\partial_u\partial_v)a_j = 0, \qquad (6.10)$$

i.e. the free wave equations.

It is satisfactory to see that the true nature of the Maxwell field, as represented by an irreducible representation of the Poincaré group of zero mass and spin 1 appears here explicitly in the number of independent components. This is not the case in the usual formulation.

The null plane commutation relations can be obtained exactly as in the scalar case. One finds [2,13]

$$[a_j(f), a_k(\partial_v g)] = \frac{i}{2}\delta_{jk}\ (f,g) \qquad (j,\ k = 1,\ 2) \qquad (6.11)$$

where the notation $a_j(f) \equiv a_j(o,f)$ is used as in (5.6). These commutation relations for the dynamically independent fields suffice, of course, since the C. R. for a_u follow from these by (6.9).

It is clear that these C. R. are consistent with the field equations, because these equations are obtained from the C. R. (6.11) together with the generator for translations in the "null-time" direction [2],

$$P_u = \frac{1}{2} \int d^3\bar{x}(f_{12}+(\underline{\partial}\cdot\underline{a})^2) \tag{6.12}$$

using (5.25),

$$\partial_u a_j = i[a_j, P_u]. \tag{6.13}$$

It is easy to verify that one indeed obtains (6.10).

The commutation relation (6.11) is also consistent with the gauge condition, since the latter can be regarded as the defining equation for the dependent component a_u. Thus, the difficulties associated with the supplementary condition encountered in the usual theory do not exist here.

b) The Free Dirac Field

The Dirac equation in null coordinates is [5]

$$(\underline{\gamma}\cdot\underline{\partial} - \gamma_u\partial_v - \gamma_v\partial_u + m)\psi = 0 \tag{6.14}$$

where

$$\gamma_u \equiv -m\cdot\gamma = (\gamma^0+\gamma^3)/\sqrt{2},\quad \gamma_v \equiv -n\cdot\gamma = (\gamma^0-\gamma^3)/\sqrt{2}$$

satisfy

$$\gamma_u^2 = 0,\ \gamma_v^2 = 0,\ \{\gamma_u,\gamma_v\} = -2,\ \{\gamma_u,\gamma_j\} = 0,\ \{\gamma_v,\gamma_j\} = 0 \quad (j = 1,\ 2). \tag{6.15}$$

Because of the null-potency of the matrices γ_u and γ_v the Dirac equation separates easily into two equations by multiplication with γ_u and γ_v.

If we define

$$\psi_u \equiv \gamma_u \psi, \; \psi_v \equiv \gamma_v \psi, \tag{6.16}$$

these equations are

$$\gamma_u \partial_u \psi_v + (\underline{\gamma} \cdot \underline{\partial} - m)\psi_u = 0 \tag{6.17}$$

or
$$\partial_u \psi_v = -\frac{1}{2}(\underline{\gamma} \cdot \underline{\partial} + m)\gamma_v \psi_u$$

and

$$\gamma_v \partial_v \psi_u + (\underline{\gamma} \cdot \underline{\partial} - m)\psi_v = 0 \tag{6.18}$$

or
$$\partial_v \psi_u = -\frac{1}{2}(\underline{\gamma} \cdot \underline{\partial} + m)\gamma_u \psi_v \; .$$

Only the first of these is a dynamical equation; the second one is independent of ∂_u and is therefore a constraint equation. It permits the elimination of ψ_u,

$$\psi_u = -\frac{1}{2}(\underline{\gamma} \cdot \underline{\partial} + m)\gamma_u \partial_v^{-1} \psi_v \tag{6.19}$$

the operation ∂_v^{-1} being defined in (3.2) to (3.4). The only dynamical components of ψ, viz. ψ_v then satisfy,

$$\begin{aligned} \partial_u \psi_v &= \frac{1}{4}(\underline{\gamma} \cdot \underline{\partial} + m)\gamma_v(\underline{x} \cdot \underline{\partial} + m)\gamma_u \partial_v^{-1} \psi_v \\ &= \frac{1}{2}(\underline{\partial}^2 - m^2)\partial_v^{-1} \; \psi_v . \end{aligned} \tag{6.20}$$

Thus, one obtains the Klein-Gordon equation for ψ_v, integrated over v.

The separation of ψ into ψ_u and ψ_v in (6.16) can also be written as

$$\psi = -\frac{1}{2}(\gamma_u \psi_v + \gamma_v \psi_u) \tag{6.21}$$

and is a separation of the 4-dimensional space spanned by the four solutions of the Dirac equation (6.14) into two-dimensional subspaces orthogonal to one another. This is easiest seen by means of the projection operators

$$\mathbb{P}_u \equiv -\frac{1}{2}\gamma_v\gamma_u, \quad \mathbb{P}_v \equiv -\frac{1}{2}\gamma_u\gamma_v \tag{6.22}$$

which satisfy

$$\mathbb{P}_u^2 = \mathbb{P}_u, \quad \mathbb{P}_v^2 = \mathbb{P}_v, \quad \mathbb{P}_u\,\mathbb{P}_v = \mathbb{P}_v\,\mathbb{P}_u = 0. \tag{6.23}$$

With their help (6.21) becomes

$$\psi = \mathbb{P}_u\psi + \mathbb{P}_v\psi \tag{6.21^1}$$

and (6.16) becomes

$$\psi_v = \gamma_v\,\mathbb{P}_v\psi, \quad \psi_u = \gamma_u\,\mathbb{P}_u\psi. \tag{6.24}$$

It follows that two of the four ψ are sufficient to determine all ψ. In a suitable representation these can be chosen to be the particle solutions. The antiparticle solutions then follow from these.

The commutation relations on the null plane can be obtained as in (5.12) and (6.11). One finds [5,13]

$$\{\psi_v(f),\ \psi_v(g)\} = 0, \quad \{\psi_v(f),\ \bar{\psi}_v(g)\} = i\gamma_v(f,g). \tag{6.25}$$

The dynamics is determined by the operator

$$P_u = \frac{1}{2}\int d^3\bar{x}{:}\bar{\psi}\,\gamma_u\overleftrightarrow{\partial}_v\psi \tag{6.26}$$

as can easiest be obtained from the Lagrangian formulation of the classical theory. If this operator is expressed in terms of ψ_v only, as is done by elimination of ψ_u via (6.19) one obtains

$$P_u = \frac{1}{4}\int d^3\bar{x}{:}\bar{\psi}_u\gamma_v\overleftrightarrow{\partial}_v\psi_u{:} = \frac{1}{8}\int d^3\bar{x}{:}\psi_v(\underline{\partial}^2-m^2)\gamma_u\overleftrightarrow{\partial}_v^{-1}\psi_v. \tag{6.27}$$

This operator and use of the commutation relations (6.25) gives the equation (6.20) from the dynamical equation

$$\partial_u\psi_v = i[\psi_v, P_u]. \tag{6.28}$$

c) <u>Interaction</u>

Consider a closed <u>classical</u> system of a Dirac field Ψ and an electromagnetic field with potential A^μ in interaction via minimal coupling. The well known Lagrangian, when written in null coordinates and in the null plane gauge $A_v=o$ leads to the Maxwell equations [13,12]

$$(\underline{\partial}^2-2\partial_u\partial_v)A_j-\partial_j(\underline{\partial}\cdot\underline{A}-\partial_v A_u)=-J_j\equiv-\frac{ie}{2}(\Psi_u\gamma_j\bar{\Psi}_v+\bar{\Psi}_v\gamma_j\Psi_u)$$

$$(j = 1,\ 2) \tag{6.29}$$

$$\partial_v(\underline{\partial}\cdot\underline{A}-\partial_v A_u) = J_v \equiv -\frac{ie}{2}\bar{\Psi}_v\gamma_u\Psi_v \tag{6.30}$$

and the Dirac equations for $\Psi_v\equiv\gamma_v\Psi$ and $\Psi_u\equiv\gamma_u\Psi$,

$$D_u\Psi_v = -\frac{1}{2}(\underline{\gamma}\cdot\underline{D}+m)\gamma_v\Psi_u \tag{6.31}$$

$$\partial_v\Psi_u = -\frac{1}{2}(\underline{\gamma}\cdot\underline{D}+m)\gamma_u\Psi_v. \tag{6.32}$$

Here $D^\mu=\partial^\mu+ie\,A^\mu$ expresses minimal coupling. Eqs. (6.30) and (6.32) are again constraint equations. They can be used to eliminate A_u and Ψ_u,

$$\partial_v A_u = \underline{\partial}\cdot\underline{A} - \partial_v^{-1}J_v \tag{6.33}$$

$$\Psi_u = -\frac{1}{2}\partial_v^{-1}(\underline{\gamma}\cdot\underline{D}+m)\gamma_u\Psi_v \tag{6.34}$$

so that the theory can be expressed entirely in terms of the dynamically independent quantities A_j (j=1,2) and Ψ_v.

Similarly, Noether's theorem leads to generators of infinitesimal transformations, expressed as integrals over a null plane, which can be written in terms of the A_j and Ψ_v.

In particular, we are interested in the "null Hamiltonian" P_u which determines the dynamics. While P_u

is easily written down in terms of Ψ and A^{μ}, it is not so easy to separate it into a free Hamiltonian $P_u^{(o)}$ and an interaction term $P_u^{(1)}$. The reason is that terms that seem to belong to $P_u^{(o)}$ such as $(\partial_v A_u)^2$ or $\bar{\Psi}\,\underline{\gamma}\cdot\overleftrightarrow{\underline{\partial}}\Psi$ actually contain coupling terms. This follows from (6.33) and (6.34) where it is evident that the dependent fields A_u and Ψ_u contain interaction terms. A lengthy calculation [13] finally leads to the simple result

$$P^{\omega} = -\frac{1}{2}\int d^3\bar{x}(J_{\mu}A^{\mu}+J^f_{\mu}A^{\mu}_f). \tag{6.35}$$

Here $J_{\mu}A^{\mu}=\underline{J}\cdot\underline{A}-J_vA_u$ since we are in the null plane gauge. $\underline{J}$ contains Ψ_u according to (6.29). Thus, one must use (6.33) and (6.34) to express the first term in terms of Ψ_v and A_j only. The second term is to be treated in the same way, except that the label f indicates that the dependent components are not A_u and Ψ_u of (6.33) and (6.34) but are related to the independent components in the same way as a_u and ψ_u of the free fields, (6.9) and (6.19).

The knowledge of this classical field theory permits now the usual formal calculations of the interaction picture (Dirac picture) of quantum electrodynamics. Thus, one starts with quantized free fields a^{μ} and ψ as in subsections a) and b) above and one impses on the state vectors the null-time dependence

$$-i\partial_u|u\rangle = P_u^{(1)}(u)|u\rangle, \tag{6.36}$$

where

$$P_u^{(1)}(u) = -\frac{1}{2}\int d^3\bar{x}:(j_{\mu}a^{\mu}+j^f_{\mu}a^{\mu}_f) \tag{6.37}$$

in the notation of (6.35). The elimination equations (6.33) and (6.34) must of course be written in interaction picture, but remain formally unchanged.

As a consistency check one can verify [13] that a transformation to the Heisenberg picture transforms the free field equations into the formal quantized analogue of (6.29) to (6.32).

The Dyson S-matrix for the null plane field theory,

$$S = U_{+}\exp(i\int P_u^{(1)}(u)du) \tag{6.38}$$

contains the null-time ordering and can be used as in the usual theory. Feynman receipes can be developed for the purpose of applications [14]. But this would be the beginning of another set of lectures.

d) Current Commutators

Another important application of null plane field theory (as demonstrated here for quantum electrodynamics) is the use of current commutators and current algebras for the study of high energy phenomena, as was indicated in the introduction. Since applications are not our concern in our lectures and Prof. Frishman has already emphasized this aspect of high energy physics in his lectures earlier this week, we shall only give the fundamental relations here.

The electromagnetic current

$$j_\mu(x) = -ie:\bar{\psi}(x)\gamma_\mu\psi(x): \tag{6.39}$$

has the commutation relation

$$[j_\mu(x),j_\nu(x')] = -ie^2:[\bar{\psi}(x)\gamma_\mu S(x-x')\gamma_\nu\psi(x')-\bar{\psi}(x')\gamma_\nu S(x'-x) \times \gamma_\mu\psi(x)]: \tag{6.40}$$

as has been known since the beginning of covariant field theory almost twenty five years ago.

The restriction of this commutator to the null plane u=u' requires the knowledge of S(x) on that plane. This was computed earlier [5] and is most conveniently written as

$$S(o,\bar{x}) = \frac{1}{2}[\gamma_u - (\underline{\partial}-m)\partial_v^{-1}+1/2\gamma_v(\underline{\partial}^2-m^2)\partial_v^{-2}]\delta_3(\bar{x}) \tag{6.41}$$

since $$\varepsilon(v) = -2\partial_v^{-1}\delta(v) . \tag{6.42}$$

Insertion of (6.41) into (6.40) yields the free current commutator on the null plane,

$$[j_\mu(x),j_\nu(x')]_{u=u'} = -\frac{ie^2}{2}: [\bar{\psi}(x)\gamma_\mu[\gamma_u-(\underline{\gamma}\cdot\underline{\partial}-m)\partial_v^{-1} +$$

$$+\frac{1}{2}\gamma_v(\underline{\partial}^2-m^2)\partial_v^{-2}]\delta_3(\bar{x}-\bar{x}')\gamma_\nu\psi(x') - \binom{x \leftrightarrows x'}{\mu \leftrightarrows \nu}] : . \tag{6.43}$$

The operators ∂_v^{-1} act only on the δ_3 function. It is important to note that the third term in $S(o,\bar{x})$ is nonlocal and makes this commutator nonlocal for those components in which it contributes. Consequently, if one attempts to write it as a sum of local operators [15], this sum must be infinite.

It is a trivial matter to deduce from (6.43) the commutator of specific current components. In particular, one finds

$$[j_v(x), j_v(x')]_{u=u'} = o \tag{6.44}$$

i.e. the null plane charge densities commute on the null plane. An immediate consequence of this fact is that, if Dirac type currents contain internal symmetries, the commutation relations on the null plane of the charge densities will be those of that symmetry. For example, if k denotes an SU(3) index, $j_\mu^k=-i\psi\ \gamma_\mu\lambda^k\psi$,

$$[j_v^k(x),j_v^\ell(x')]_{u=u'} = i\ c_m^{k\ell}\ j_v^m(x)\delta_3(\bar{x}-\bar{x}') \tag{6.45}$$

where $c_m^{k\ell}$ are the structure constants of SU(3).

Such current commutators can be used as the starting point for various applications, electroproduction in the Bjorken limit, sum rules, etc. This, too,would be a separate series of lectures.

7. CONCLUSIONS

In these lectures classical and quantum field theory were formulated as descriptions of physical systems that develop in a null direction from one null plane to another. While the fundamental commutation relations on a null plane, which are used here, are those for free fields, the theory developed is clearly applicable also to interacting fields, provided, either, a) the restriction of the field to a null plane makes sense and the C. R. on that plane are these of the free field, or b) a Dirac picture (interaction representation) exists, so that in this picture the fields can be treated as free.

If neither of these premises hold, we recall that null plane field theory could still be formulated with respect to null slices, i.e. four-dimensional spacetime regions sandwiched between two parallel null planes. It has been proven [10] that a Wightman field is complete in such a slice.

The main question which is posed here and whose answer has only barely been started is this: "Does the null plane formulation of quantum field theory lead to an improvement on the difficulties encountered in the more conventional formulations?"

The completeness and the mass independence of the free field algebra on null planes have a number of far-reaching implications. The most important of these discussed here are:

(a) The vacuum is an eigenstate of the null Hamiltonian with eigenvalue zero; this implies that all S-matrix elements connecting any state to the vacuum must vanish.

(b) The ten generator Poincaré algebra has a six generator subalgebra which is obtained by projecting the former onto a null plane. Similarly, the fifteen generator Liouville algebra (conformal algebra) has a ten generator null plane subalgebra. This latter subalgebra is unaffected by the mass terms of the theory which enter the null Hamiltonian. Dilatation and special conformal invariance continues to hold on the null planes and both are unitarily implementable there.

(c) The dynamically independent components of massive free fields are reduced from 2(2S+1) to 2S+1 on null planes. Massless fields can be reduced to only two components even in interaction, as is shown in the electromagnetic case. After elimination of the dependent components, the field equations are nonlocal, involving such operators as ∂_v^{-1}. This does not seem to pose a serious difficulty because of the distribution nature of the fields.

Finally, null plane field theory opens the way to an understanding of high energy limits from first principles. Expansions of scattering matrix elements in inverse powers of the energy beyond a leading term and similar high energy approximations are obvious applications of this theory. So are current commutators in the infinite momentum frame. A large and perhaps very fruitful area of research appears to be opening up.

In conclusion I want to thank Professor Urban for inviting me to Schladming and for giving me the opportunity to present this work. His efforts in making the Schladming Internationale Universitätswochen a success for ten consecutive years are invaluable.

REFERENCES

1. Robert A. Neville, Ph. D. thesis (Syracuse University, 1968).
2. R. A. Neville and F. Rohrlich, Phys. Rev., April 1971.
3. H. Bebie and H. Leutwyler, Phys. Rev. Lett. 19, 618 (1967);
   H. Leutwyler, Acta Phys. Austriaca, Suppl. V, 1968 and Springer Tracts in Modern Phys. 5o, 29 (1969).
4. L. Susskind, Phys. Rev. 165, 1535 (1968);
   L. Susskind and M. B. Halpern, Phys. Rev. 176, 1686 (1968);
   S. Chang and S. Ma, Phys. Rev. 180, 1506 (1969) and 188, 2385 (1969). The limit of the boost operator was mentioned by K. Bardacki and G. Segré, Phys. Rev. 159, 1263 (1967); the absence of vacuum diagrams in the limit was found by S. Weinberg, Phys. Rev. 150, 1313 (1966).
5. R. A. Neville and F. Rohrlich, Nuovo Cim.
6. For a simple case which however contains all the essential features this theorem is proven in F. Rohrlich, "Classical Charged Particles", Addison-Wesley, Reading, Mass. 1965, where references to other proofs are also given.
7. This is proven in Appendix 1 of reference 5.

8. Dilatation and conformal transformation were reviewed by T. Fulton, F. Rohrlich, and L. Witten, Rev. Mod. Phys. 34, 422 (1962); the representation theory can be found in H. A. Kastrup, Ann. Physik 9, 388 (1962), applications to quantum field theory in the excellent paper by J. Wess, Nuovo Cim. 18, 1086 (1960).
9. G. Mack and A. Salam, Ann. Physics (N.Y.) 53, 174 (1969);
D. J. Gross and J. Wess, Phys. Rev. D2, 753 (1970).
10. J. R. Klauder, H. Leutwyler and L. Streit, Nuovo Cim. 66A, 536 (1970).
11. F. Rohrlich and L. Streit (to be published).
12. J. B. Kogut and D. E. Soper, Phys. Rev. D1, 2901 (1970).
13. F. Rohrlich, Acta Phys. Austriaca 32, 87 (1970).
14. J. D. Bjorken, J. B. Kogut and D. E. Soper, Phys. Rev. D15, 1382 (1971).
15. We refer to Y. Frishman's lectures for references on Wilson's expansion.

Acta Physica Austriaca, Suppl. VIII, 323–357 (1971)

SOME EXTERNAL FIELD PROBLEMS IN QUANTUM ELECTRODYNAMICS*

BY

T. ERBER
Illinois Institute of Technology, Chicago

1. INTRODUCTION

A number of recent discoveries and experimental developments have provided renewed stimulus for the study of quantum electrodynamic processes which are essentially linked with intense electromagnetic fields. Most discussions in this direction have been concerned with possibilities raised by the improvement of lasers, flash X-ray xources, and various combinations of these devices with high energy particle beams. However recent estimates [1,2] have merely confirmed earlier conclusions [3,4] that decisive results in this vein are still several orders of magnitude beyond the present experimental capabilities. Much more fruitful lines of investigation are associated with the following developments:

(i) In the Summer of 1967 Miss Josephine Bell, who was working at that time at the Mullard Radio observatory, made the discovery of the first star that "ticks". Within a matter of month, after the publication of

*Lecture given at X. Internationale Universitätswochen für Kernphysik, Schladming, March 1 - March 13, 1971.

this result, astrophysicists decided that it was plausible to identify these objects as spinning neutron "lighthouses" whose illumination is furnished by magnetic Bremsstrahlung. For our present purposes the point of key interest is that these pulsars are supposed to be surrounded by enormous magnetic fields. i.e. $10^{11} \lesssim H \lesssim 10^{14}$G, with injected electron energies in the range $0.1 \lesssim E \lesssim 100$ MeV. Since the essential measure of the most non-linear electrodynamic effects is the critical field $m^2c^3/e\hbar \sim 4\times10^{13}$G, it is especially interesting that this value lies within the range of the hypothetical pulsar fields. Of course these pulsar models are only in the preliminary stages of development, and it is at present entirely unclear how these intense fields are sustained and how they were generated by flux compression [5]. A general review of magnetic stars and pulsars may be found in references [6] and [7].

(ii) In November 1970 the first series of trials combining megagauss generators with a high energy accelerator finished running at SLAC. In accordance with previous suggestions [8,9] this experimental work was mainly concerned with magnetic Bremsstrahlung. The principal parameters, and comparisons with previous work, are summarized in Table I. The instrumentation is described in reference [10].

Table I.

	GE '47	DESY '64	IIT-SLAC '70
Electron Energy (GeV)	0.08	6	19
Magnetic Field (MG)	0.009	0.006	1.1(2.2)
Radiation Rate (MeV/mm)	6.5×10^{-9}	1.6×10^{-5}	5.5
Photons/mm	5.6×10^{-3}	3.7×10^{-3}	0.68
Average Photon Energy (MeV)	1.18×10^{-6}	4.42×10^{-3}	8.12

Both of these developments have made it progressively more respectable to reexamine a variety of external field effects that used to be considered science fiction [11, 12, 13, 14, 15, 16]. However even in the generous time provided for these lectures, one cannot meaningfully review all of these possibilities so it will be necessary to restrict the discussion to the following topics:

(a) First of all, since the megagauss-accelerator techniques are so new, it will be useful to describe the overall physical set up of an experiment. This will make it clear what parameters are actually accessible to measurement.

(b) Next we present a summary of the most useful formulas for describing the spectral and angular characteristics of magnetic Bremsstrahlung. Where available, both the relativistic classical and relativistic quantum mechanical results are given.

(c) Then we come to the heart of this entire lecture series: the special relevance of magnetic Bremsstrahlung to the question of the existence of radiation reaction effects and the associated problem of detecting a possible non-point electron form factor. Specifically in Section 4 we make the following points:

(i) It is now technically possible - by combining high energy accelerators and megagauss targets - to enhance the classical $(2e^2/3c^3)\dddot{x}$ radiation reaction effects to the point where one can verify experimentally whether or not they exist.

(ii) If - as we believe - measurements will show that these shifts exist in some form, then we must face the question - which has lain dormant since 1962 [17]: What is the quantum theory which yields these radiation reaction effects in the classical limit? It is particularly appropriate to recall at these Schladming Lectures that Fritz Rohrlich has stressed for years that we do not really understand the correspondence limit of RQED. The lack of a proper description of radiation reaction is a concrete illustration of this problem. Specifically we know the correct classical relativistic form of the magnetic Bremsstrahlung spectrum [see (3.11)], and we also know the corresponding quantum mechanical result [see (3.17)]. It is obvious that the quantum corrections depend explicitly on $\hbar$ [see (2.23a) of [8]]. However it is dimensionally trivial that the classical radiation reaction modifications of the spectrum are independent of $\hbar$ [18]. The reconciliation of these calculations is an open question.

(iii) We propose that a fruitful solution may be found in a reexamination of the Kramers-van Kampen Hamiltonian [19]. This treats the electromagnetic field in dipole approximation but is otherwise exact in the coupling constant e. It is therefore the only quantum mechanical Hamiltonian which leads to radiation reaction terms which are analogous to those in the classical Lorentz-Dirac equation. It is well known that these terms are not analytic in e and can therefore not be recovered from the usual perturbation expansions. The immediate goal of generalizing the Kramers-van Kampen approach would be to develop a theory that would give a unified description of radiation damping and quantum mechanical effects. However the long term goal is to understand the experimental consequences of the form-factor information which is built into the Kramers-van Kampen Hamiltonian. Since the Bremsstrahlung spectrum effectively provides an electromagnetic Fourier resolution of the electron-photon vertex - and it is experimentally feasible by this means to study lengths down to $2\times 10^{-3}\ e^2/(mc^2)$ - we expect that the mega-gauss-accelerator program will eventually proceed with the electron form-factor search that was originally envisaged for the storage rings.

(d) In Section 5 we turn our attention to a completely different kind of non-linear external field effect: Vacuum polarization Cerenkov radiation. This is expected to occur when charged particles cross intense-field regions at velocities exceeding the phase velocity of light. Processes of this kind are of great conceptual interest since they would be manifestations of collective radiative effects of the vacuum polarization [20]. We supplement an earlier discussion [21] with an explicit formula for the radiation rate. As an illustration of the orders of magnitude, we present calculations for one example which indicates that an experimental check may be feasible.

2. MEGAGAUSS BREMSSTRAHLUNG (EXPERIMENTS)

A schematic of the set-up is shown in Fig. 1.

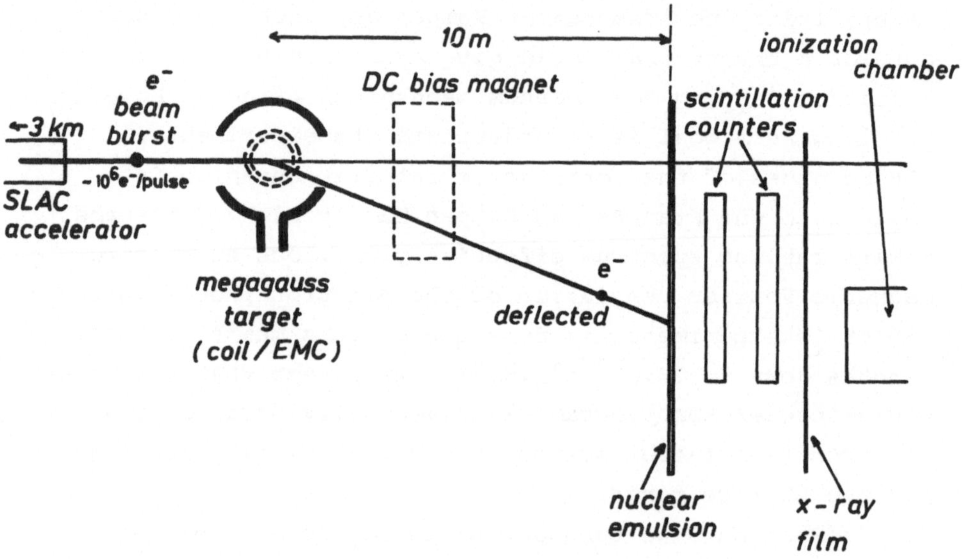

Fig. 1: Schematic of bremsstrahlung experiment.

Looking downstream along the beam-line the polarity of the megagauss targets was arranged to deflect the electron pulses to the right; the auxiliary DC bias magnet deflected the exiting electrons down. The essential trick in this arrangement is to insure adequate space-time synchronization of the accelerator beam burst and the functioning of the megagauss targets. In virtue of careful collimation the diameter of the electron beam was reduced to 1.3 mm at the target and contained in a cone of full angle 3×10^{-4} radian. The megagauss generators were of two kinds:

(a) single turn strap coils; typically 4.6 mm ID (at peak field) × 10 mm height, and 2.1 mm thickness, made out of copper;

(b) electromagnetic (Cnare) imploders with the following geometry

Coil (Copper)		Foil (Aluminum)	
inner diameter	41.27 mm	outer diameter	38.10 mm
height	25.4 mm	height	19.05 mm
wall thickness	2.1 mm	wall thickness	0.457mm

In both cases the copper coils had beam holes of diameter $\gtrsim$2 mm drilled through to permit undisturbed passage of the electron beam. The targets and beam were aligned to an overall precision of ±0.1 mm before each shot.

Since the energy density of a megagauss field is $\sim10^3$ cal/cm^3, or roughly comparable to high explosives, it is intuitively clear that megagauss generators must be pulsed, single-shot devices. In particular the coils reach peak field in about 2.5 μsec, and dwell for approximately 0.5 μsec. On the other hand the implosion devices exhibit a gradual field rise over 10 μsec, and then have a sharp ($\sim$0.1 μsec) peak corresponding to the maximum of flux compression. To achieve time synchronization, our beam line was converted to single shot operation, and the beam burst sharpened to 0.1 - 0.3 μsec. The jitter of the overall synchronization of the accelerator burst and the operation of the megagauss targets was $\lesssim$50 n sec.

The megagauss targets operated in the range 1.1 - 2.2 MG. The accelerator energy was at 19 GeV for practically all runs. The systems were not evacuated since the beam divergence already exceeded the multiple scattering effects. In the implosion experiments the electron beam intercepted the compressed -- and thickened -- Aluminum foil. It is shown in the Appendix that this has a negligible effect on the spectral distribution in the range of interest.

Due to the fact that this arrangement leads to tremendous "counting rates", i.e. 10^6 e^-/beam burst $\Longrightarrow$ 5×10^6 Bremsstrahlung photons/0.1 μsec, nuclear emulsions were used as the prime detector. A fraction ($\sim4\times10^{-3}$) of these γ's converts into e^+-e^- pairs in the emulsions, and the original Bremsstrahlung spectrum must then be tediously reconstructed from the pair energy distribution. In future work we expect to retain emulsions because of their excellent spatial resolution [see (4.11)], but introduce pair spectrometers for the spectral analyses.

The instrumentation for this experiment is described in [10]. General background information on megagauss generators may be found in [22] and [5].

3. MEGAGAUSS BREMSSTRAHLUNG (THEORY)

The relevant geometry and notation is shown on Fig. 2. There should be no difficulty in mentally superposing this on the experimental schematic indicated on Fig. 1. We begin by recalling that the total magnetic Bremsstrahlung energy radiated by an electron in traversing a path length Δ normal to a magnetic field H is given by

$$I(H,E,\Delta) = \frac{2}{3}\,\alpha mc^2\,\frac{\Delta}{\lambda\!\!\!^{-}_c}\,Y^2, \qquad (3.1a)$$

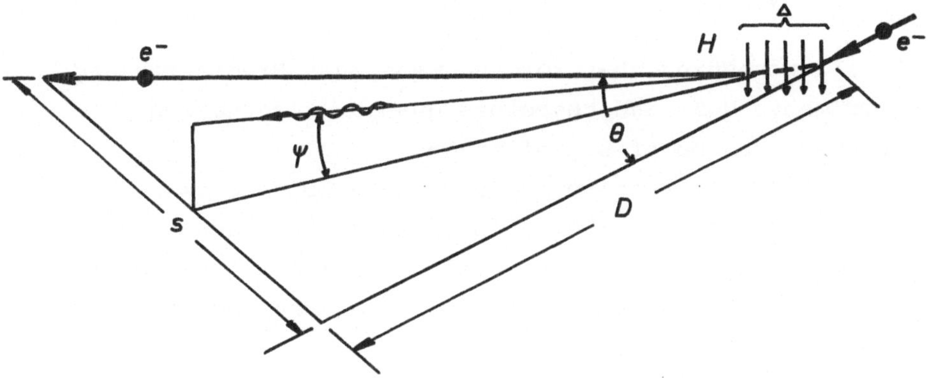

Fig. 2: Kinematics of magnetic bremsstrahlung.

where

$$Y = \frac{E}{mc^2}\cdot\frac{H}{H_{cr}}, \quad H_{cr} \equiv \frac{m^2c^3}{e\,\hbar} \cong 4.414\times10^{13}\ \mathrm{G};$$

and E is the energy of the incident electron [23,24]. All other quantities have their usual significance, and α^{-1}="137". {Note:$H_{cr}=\alpha\times m^2c^4/e^3=\alpha\times1.81\times10^{18}$ V/cm, the field at the "surface" of the classical electron.}. In practical units

$$I_{MeV} = 1.263\times10^{-2}\Delta_{mm}[H_{MG}E_{GeV}]^2. \tag{3.1b}$$

Although these formulas are only valid in the relativistic classical regime, one can easily derive the corresponding expectation value for the total number of emitted photons:

$$N(H,E,\Delta) = \frac{5}{2\sqrt{3}}\,\alpha\,\frac{\Delta}{\lambda\!\!\!^{-}_c}\,\frac{H}{H_{cr}}\ . \tag{3.2a}$$

In practical units this is

$$N(H,E,\Delta) = 0.618\ \Delta_{mm}\ H_{MG}. \tag{3.2b}$$

It is interesting that for the megagauss Bremsstrahlung experiments $N \sim 5$, and therefore quantum fluctuation phenomena can be of significance.

It is instructive to compare these results with the quantum mechanical calculations. When the exact Dirac wave functions for an electron in a uniform field of infinite extent are used, one finds that the radiation rate is given by [25,8,26]

$$\left(\frac{dE(Y)}{dx}\right)_{\text{GeV/mm}} = 6.438\times10^{3} g(Y), \tag{3.3}$$

where

$$g(Y) = \begin{cases} Y^2(1-5.953\quad +\ldots), & Y \ll 1; \\ 0.5563\quad ^{2/3}, & Y \gg 1. \end{cases}$$

In other words, quantum effects in the overall radiation rate are negligible as long as

$$6\ Y \ll 1\ . \tag{3.4}$$

From (3.3) we also infer that under hypothetical pulsar conditions, say $H \sim H_{cr}$, $E \sim 5$ MeV, the radiation rate is 1.2×10^{4} GeV/mm. This implies that the principal energy losses occur over a distance of 10^{-6} mm!

By combining (3.1a) and (3.2a), we obtain an average photon energy:

$$\langle\hbar\omega\rangle = \frac{4}{5\sqrt{3}}\ E\ Y \cong 0.461\ E\ Y, \tag{3.5a}$$

or in practical units

$$\langle\hbar\omega\rangle_{MeV} = 2.05\times10^{-2}\ H_{MG}\ E_{GeV}^{2}\ . \tag{3.5b}$$

This average is about 10% higher than the most probable photon energy,

$$\hbar\omega_p \cong 0.420\, E\, Y \,, \tag{3.6}$$

which is associated with the peak of the spectral distribution [compare (3.11)].

A quantity of basic physical interest is the photon distribution - doubly differential in energy and angle. From Fig. 2 it is clear that the relevant angle is ψ. We find that the corresponding radiation rate is given by

$$\frac{d^2N}{d(\hbar\omega)\,d\psi} = \frac{\alpha}{\pi^2}\frac{\Delta}{\lambda\!\!\!^{-}_c}\left[1+\left(\frac{E}{mc^2}\psi\right)^2\right]^{1/2} \times$$

$$\times \frac{\xi}{E}\left[K^2_{2/3}(\xi) + \frac{\left(\dfrac{E}{mc^2}\psi\right)^2}{1+\left(\dfrac{E}{mc^2}\psi\right)^2} K^2_{1/3}(\xi)\right], \tag{3.7}$$

where

$$\xi = \frac{\hbar\omega}{3\,Y\,E}\left[1+\left(\frac{E}{mc^2}\psi\right)^2\right]^{3/2} .$$

The K's are Bessel functions of imaginary argument. Due to the asymmetric dependence on E, H, and ψ, there is a complicated interplay between the kinematics and angular distribution. Some representative curves which illustrate the significance of these expressions are sketched on Fig. 3. Just as in the corresponding work of Bedo et al. [27] on the energy distribution, it appears that the softer photons ($\hbar\omega \sim 0.1$ MeV) are spread out more than an order of magnitude beyond the characteristic angle $\theta \sim mc^2/E$.

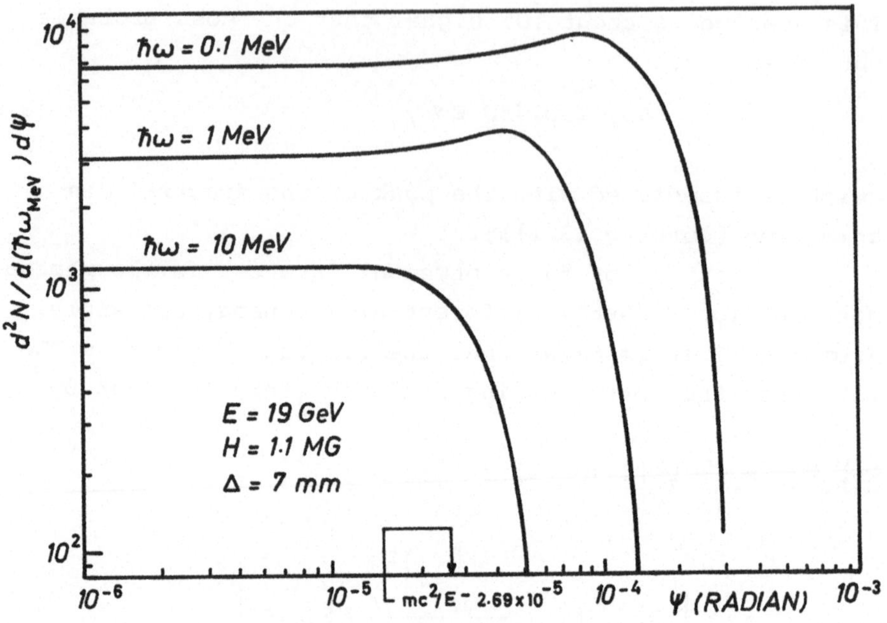

Fig. 3: Photon distribution as function of photon energy ($\hbar\omega$) and emission angle (ψ) [see (3.7)].

The analysis of the angular variation is somewhat simplified if we integrate over the energy spectrum and consider the singly differential photon distribution:

$$\frac{dN}{d\psi} = \int_0^{\hbar\omega_{max}} d(\hbar\omega)\, \frac{d^2N}{d(\hbar\omega)\,d\psi} =$$

$$= \frac{\alpha}{\pi\sqrt{3}}\,\frac{\Delta}{\lambda_c}\,Y\,\frac{2+3\left(\frac{E}{mc^2}\psi\right)^2}{\left[1+\left(\frac{E}{mc^2}\psi\right)^2\right]^2}\,. \tag{3.8}$$

Further, if we take into account the symmetry about $\psi = o$, then the total number of photons emitted into the band between $-\psi$ and $+\psi$ is given by

$$N_\gamma(E,H,\Delta,\psi) = \frac{\alpha}{\pi\sqrt{3}} \frac{\Delta}{\lambda\!\!\!^{-}_c} \frac{H}{H_{cr}} \{5\tan^{-1}[\frac{E}{mc^2}\psi] - \\ - \sin(2\tan^{-1}[\frac{E}{mc^2}\psi])\} . \quad (3.9)$$

The limiting case $\psi \gg mc^2/E$ of course leads back again to (3.2a). For the sake of completeness, we note that the corresponding expression for the energy radiated into the band between $-\psi$ and $+\psi$ is

$$\mathcal{E}(E,H,\Delta,\psi) = \alpha mc^2 \frac{\Delta}{\lambda\!\!\!^{-}_c} \Upsilon^2 \{\sin(\tan^{-1}[\frac{E}{mc^2}\psi]) - \\ - \frac{1}{3}\sin^3(\tan^{-1}[\frac{E}{mc^2}\psi]) - \\ - \frac{1}{8}\sin(\tan^{-1}[\frac{E}{mc^2}\psi])\cos^4(\tan^{-1}[\frac{E}{mc^2}\psi])\}. \quad (3.10)$$

When the differential distribution (3.7) is integrated over all angles, we obtain the photon spectrum

$$\frac{dN}{d(\hbar\omega)} = \frac{\sqrt{3}}{2\pi} \frac{\alpha}{\hbar\omega} \frac{\Delta}{\lambda\!\!\!^{-}_c} \frac{H}{H_{cr}} \kappa(\frac{2}{3\Upsilon} \frac{\hbar\omega}{E}), \quad (3.11)$$

where κ is the Bremsstrahlung function

$$\kappa(z) \equiv \int_z^\infty dx\, K_{5/3}(x) \simeq \begin{cases} 2.149\, z^{1/3}, & z \ll 1; \\ 1.253\, z^{1/2} e^{-z}, & z \gg 1. \end{cases} \quad (3.12)$$

In practical units

$$\frac{dN}{d(\hbar\omega)} = \frac{0.12}{\hbar\omega} H_{MG}\Delta_{mm}\kappa\left(\frac{15\ \hbar\omega_{MeV}}{H_{MG}E^2_{GeV}}\right). \tag{3.13}$$

The peak of the spectrum is given by (3.6). According to (3.12) the spectrum decreases exponentially for energies in the range

$$\hbar\omega_{MeV} >> H_{MG}\ E^2_{GeV}\ /\ 15\ . \tag{3.14}$$

The actual number of photons radiated into the spectral region between $\hbar\omega_1$ and $\hbar\omega_2$ is

$$N_\gamma(E,H,\Delta,\hbar\omega_1,\hbar\omega_2) = 0.12\ H_{MG}\Delta_{mm}\int_{\hbar\omega_1}^{\hbar\omega_2}\frac{d\omega}{\omega}\ \kappa\left(\frac{15\ \hbar\omega_{MeV}}{H_{MG}E^2_{GeV}}\right)\ . \tag{3.15}$$

The low end of the spectrum can be determined directly from the asymptotic expansion (3.12). For $\hbar\omega<<1.5\ E\ Y$, we have

$$N_\gamma(E,H,\Delta,\hbar\omega_1,\hbar\omega_2) \cong 0.68\ H_{MG}\Delta_{mm}(E\ Y)^{-1/3}\{(\hbar\omega_2)^{1/3}-(\hbar\omega_1)^{1/3}\}. \tag{3.16}$$

The exact relativistic quantum mechanical counterpart of (3.11) has also been computed. In case $\hbar\omega<<E$, this can be expressed in terms of the simple replacement

$$\kappa\left(\frac{2}{3\ Y}\frac{\hbar\omega}{E}\right) \to \left(1-\frac{\hbar\omega}{E}\right)\kappa\left(\frac{2}{3\ Y}\frac{\hbar\omega/E}{1-\hbar\omega/E}\right), \tag{3.17}$$

where Y may be much greater than unity on the RHS [25,8]. Since κ has a very sensitive exponential fall-off above its maximum, (3.17) implies the existence of a shift in the spectrum relative to the classical relativistic result. This shift can be appreciable even when the inequality (3.4) is fulfilled. Furthermore it is easy to see that all spectral ranges corresponding to the inequality

$$\frac{\hbar\omega}{E} >> \frac{0.42\ Y}{1+0.42\ Y}\ , \tag{3.18}$$

are much more strongly damped in the RQED case.

Although the RQED results may be applied to megagauss accelerator experiments, considerable caution is necessary in extracting information on hypothetical pulsar models. Again supposing $H \sim H_{cr}$, and $E \sim 5$ MeV, then (3.17) indicates that the peak of the spectrum occurs in the vicinity of $\hbar\omega \sim E/3$; however this result is entirely untrustworthy since it violates the assumption $E-\hbar\omega >> mc^2$ which has been built into the calculations to avoid analytical problems near the Bremsstrahlung tip.

4. RADIATION REACTION

This is a most controversial subject since a variety of basic equations of motions have been proposed [28,29], and it has been seriously argued that all are superfluous [if not actually erroneous] since the only correct way of describing radiation damping effects is to deduce them from quantum mechanics. In this section we show that the advent of megagauss Bremsstrahlung measurements will finally permit experimental decisions among a number of choices. We also critically discuss the problems which will have to be surmounted in deriving strong radiative corrections from quantum mechanics.

We shall base our discussion on the classical Lorentz-Dirac equation

$$F_k + \frac{2}{3}\frac{e^2}{c^3}(\ddot{u}_k - u_k\dot{u}_i^2) = m\dot{u}_k, \tag{4.1}$$

where F_k is the Minkowski force, u_k the 4-velocity, and the dots denote proper time derivatives. It is well known that this equation is afflicted with runaway solutions and other ambiguities; nevertheless it at least has the virtue of familiarity. In fact we can accent this feature by restricting ourselves (momentarily) to one dimensional motion and substituting

$$v = c \sinh (w/c). \tag{4.2}$$

In this case (4.1) can be formally reduced to the simpler non-relativistic equation

$$\vec{F}_{ext} + \vec{F}_{RR} = m\dot{\vec{w}}; \; F_{RR} = \frac{2}{3}\frac{e^2}{c^3}\ddot{\vec{w}}; \tag{4.3}$$

where the dots still indicate proper time derivatives. A simple gauge of the relative importance of radiation reaction effects is the ratio

$$\frac{|\vec{F}_{RR}|}{|\vec{F}_{ext}|} = \delta . \tag{4.4}$$

In the magnetic Bremsstrahlung case it is easy to show on the basis of model independent momentum transfer considerations that [8]

$$\delta \simeq \frac{2}{3} \alpha \frac{\dot{H}}{H_{cr}}\left(\frac{E}{mc^2}\right)^2 = 4.16\times10^{-4} E^2_{GeV} H_{MG} . \tag{4.5}$$

One can readily check that for the parameters reported in [10], e.g. $H\sim 2MG$, $E\sim 20$ GeV, this ratio is

$$\delta \sim 0.34 << 1; \tag{4.6a}$$

and so for the first time we have created (classical) conditions where the problems of the Lorentz-Dirac theory cannot be trivially by-passed by perturbation methods. In this connection it is instructive to recall that the corresponding ratio for the Lamb shift, according to a rudimentary quantized version of the theory [30], is negligibly small;

$$\delta_L \sim 4\times10^{-7}. \qquad (4.6b)$$

There is an interesting method for circumventing (4.6a). Since (4.4) is merely the ratio of 3-vectors, it does not have any invariant significance. In particular if we recompute this quantity in the instantaneous particle rest system we find the more conservative value

$$\left(\frac{|\vec{F}_{RR}|}{|\vec{F}_{ext}|}\right)_{rest} \sim \frac{2}{3}\,\alpha\,\frac{H}{H_{cr}}\,\frac{E}{mc^2}\ , \qquad (4.7)$$

where E and H retain their meaning as energy and field strength measured with respect to the laboratory system. Of course if this "laboratory" happens to be a pulsar, then even this ratio may exceed unity, and we are beyond the bounds of self-consistency of classical electrodynamics. In the case represented by (4.6a), it is now plausible to solve the problem by perturbation theory in the instantaneous rest system and then to transform back into the laboratory. This involves Lorentz transforms between enormously accelerated systems: in particular if we introduce as a natural measure the characteristic acceleration of classical electrodynamics, mc^4/e^2, we find that the relative accelerations are $\alpha H/H_{cr}$. A calculation of this kind appropriate for the magnetic Bremsstrahlung experiments has been carried out by Shen [31] and Latal [32]. The principal result for the kinematics is that _in addition_ to the normal deflection of an electron in a magnetic field (see Fig. 2)

$$s = D\,\frac{mc^2}{E}\,\frac{H}{H_{cr}}\,\frac{\Delta}{\lambda\!\!\!^{-}_c}\left[1 + \frac{1}{2}\left(\frac{mc^2}{E}\right)^2\right] , \tag{4.8}$$

there is a radiation reaction induced shift

$$\delta s = D\,\frac{\alpha}{3}\left(\frac{H}{H_{cr}}\right)^3\left(\frac{\Delta}{\lambda\!\!\!^{-}_c}\right)^2 ;$$

$$= \frac{D}{3}\left(\frac{H}{H^c_{cr}}\right)^3\left(\frac{\Delta}{r_o}\right)^2 . \tag{4.9}$$

The second form, containing the classical critical field

$$H^c_{cr} = \frac{m^2c^4}{e^3} \simeq 6.05\times10^{15}\mathrm{G}, \tag{4.10}$$

and the classical electron radius, has been inserted as a reminder that all results derived from the Lorentz-Dirac theory must be independent of $\hbar$. In practical units we have

$$\delta s_{mm} = 1.06\times10^{-4}\; D_m\; H^3_{MG}\; \Delta^2_{mm} ; \tag{4.11}$$

and this finally makes it evident that magnetic Bremsstrahlung can enhance the radiative shifts to practically measurable values. For the design of experiments it is also useful to consider the ratio

$$\frac{\delta s}{s} = \frac{\alpha}{3}\,\frac{\Delta}{\lambda\!\!\!^{-}_c}\,\frac{E}{mc^2}\left(\frac{H}{H_{cr}}\right)^2 ,$$

$$\cong 6.33\times10^{-6}\; E_{GeV}\; H^2_{MG}\; \Delta_{mm} . \tag{4.12}$$

This indicates that <u>strong</u> radiation reaction conditions, corresponding to $\delta s \gtrsim s$, can be reached at

$$E_{GeV}\; H^2_{MG}\; \Delta_{mm} \sim 1.6\times10^5 . \tag{4.13a}$$

This is marginally compatible with the suppression of quantum effects (3.4), i.e

$$E_{GeV}\, H_{MG} << 4\times 10^{3}. \qquad (4.13b)$$

Skipping all the obvious possibilities, it is amusing to consider (4.13a) for values of $E \sim 500$ GeV; we then find the field requirement

$$H_{MG}^{2}\, \Delta_{mm} \sim 300. \qquad (4.13c)$$

Even this is technically within reach of current megagauss equipment. Corresponding calculations are now under way to determine alterations in the magnetic Bremsstrahlung spectrum [18].

Since it is clear that experimental information bearing on these questions will become available during the next decade, it is useful to anticipate a variety of possible results and to consider their significance:

1.) Suppose the measurements disagree with the Lorentz-Dirac predictions even though (4.13b) is satisfied. We can then discard 60 years of classical radiation reaction theory.
2.) Suppose however that shifts do exist and show up either in the scattering (4.8) and/or the spectrum (3.11). We then have to check whether the quantum corrections in the spectrum (3.17) or the scattering (not yet computed!) agree. If there are discrepancies, we should also compute the first order radiative quantum corrections, i.e. the diagram

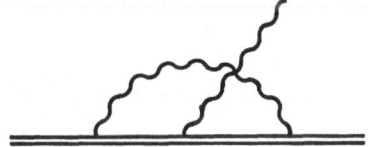

where the double lines represent the exact relativistic wave functions in the magnetic field.

3.) Now it is in fact possible and plausible - just as in the investigation of analogous questions with synchrotrons [24] - that the experiments will exhibit a mixture of "quantum" and "radiation damping" effects. Quantitatively the distinction arises from the fact, already stressed in the Introduction, that the quantum mechanical calculations are scaled by $\hbar$ [except for fortuitous cancellations (see however (4.23))], whereas the classical damping terms do not contain this factor. We therefore ought to find a version of quantum electrodynamics that yields both of these effects as consequences of a unified description.

We now propose that the only quantum mechanical Hamiltonian which is at least an adequate starting point is the Kramers-Van Kampen Hamiltonian [19]. In standard notation this is

$$H = \frac{1}{2m_o}(\vec{p}-e\vec{\tilde{A}})^2 + V(\vec{R}) + \frac{1}{8\pi}\int\{\vec{E}^2+(\text{curl }\vec{A})^2\}d\vec{r}, \tag{4.14}$$

where m_o is the bare mass, V is a position dependent potential, and

$$e\,\vec{\tilde{A}} = \int \vec{A}(\vec{r})\rho(\vec{r}-\vec{R})d\vec{r}\ . \tag{4.15}$$

If we interpret ρ as an electron form factor, then this term in principle takes into account the particle-field interaction. (The rigorous classical result is given in [33].) Van Kampen makes two simplifying assumptions:

(i) self-field contributions to $\vec{A}$ which have an $\vec{R}$-dependence are discarded;

(ii) one considers only electron motions which are confined to a region around the origin that is small compared to all relevant wavelengths (dipole approximation). In this case (4.15) simplifies to

$$e\,\vec{\tilde{A}} = \int \vec{A}(\vec{r})\,\rho(\vec{r})\,d\vec{r}. \tag{4.16}$$

To isolate the actual dipole contributions, the field is expanded as follows (sphere of radius L):

$$\vec{A}(\vec{r}) = \sum_{n=1}^{\infty} \sqrt{3/L}\,\vec{q}_n \frac{\sin(\nu_n r)}{r}, \quad \nu_n = \frac{n\pi}{L}. \tag{4.17}$$

Then assuming $\rho(\vec{r})=\rho(|\vec{r}|)$ we get

$$e\,\vec{\tilde{A}} = \sqrt{4/3L}\sum_{n=1}^{\infty} \vec{q}_n \int 4\pi r\rho(r)\sin(\nu_n r)\,dr \equiv \sum \varepsilon_n \vec{q}_n. \tag{4.18}$$

The appearance of the ε_n structure constants is one of the essential feature of this theory. Van Kampen represents these in the form

$$\varepsilon_n = \delta_n \nu_n \frac{2e}{\sqrt{3L}};\quad \delta_n \begin{cases} \to 1, & \text{point electron;} \\ \to \lim\limits_{n\to\infty} \delta_n \to 0, & \text{extended electron.} \end{cases} \tag{4.19}$$

If one now introduces the observable electron mass m by means of

$$m = m_0 + \sum \left(\frac{\varepsilon_n}{\nu_n}\right)^2; \tag{4.20}$$

then it is possible to carry out a sequence of canonical transformations which leads to the Hamiltonian

$$H = \frac{p'^2}{2m} + V(R' + \frac{e}{m}\sum \frac{2}{\sqrt{3L_n}} \frac{\cos\eta_n}{k_n} p_n'') + $$

$$+ \frac{1}{2}\sum (p_n''^2 + k_n^2\, q_n''^2). \tag{4.21}$$

The electron structure constants ε_n are implicit in the objects k_n, η_n and L_n. Just as in classical electrodynamics, if one now makes the transition to a point electron, the sum $\Sigma(\varepsilon_n/\nu_n)^2$ becomes very large and ultimately exceeds the observed mass m - in the point limit it of course diverges. Precisely at this threshold m_o must become negative. The quadratic forms in the field variables are then no longer positive definite and anomalous (runaway) solutions appear. In particular in the point electron limit Van Kampen's canonical transformations yield the equation

$$\tan(Lk)=\frac{2}{3}\, r_o k, \qquad (4.22)$$

for the k_n's. There are precisely two imaginary roots

$$ik_o = \pm \frac{2}{3}\, r_o, \qquad (4.23)$$

which lead to self-accelerations. From this point on the development of the formalism is straightforward, and it is perhaps best to summarize the correspondence between the classical and quantum theories in tabular form.

Table II.

Physics	Classical Radiation Reaction [28]	Kramers-van Kampen Hamiltonian [19]
The desired physical damping terms appear	yes	yes ("Lamb" shift)
Unwanted runaways scaled by $2e^2/3mc^3$ appear	yes	yes (Wildermuth-Baumann [34]) note-no factor α!
Suppression of runaways	adjust initial conditions ↓ pre-acceleration (acausality)	adjust commutators ↓ pre-acceleration (acausality) [36]
	non-point form factors ↓ runaways become oscillators ↓ isobar electrodynamics [35]	counter terms in Hamiltonian ↓ runaways become oscillators ↓ isobar quantum electrodynamics (Coleman and Norton [17])
Electromagnetic fields associated with runaways (oscillators)	localized within e^2/mc^2	localized within e^2/mc^2 no factor α!

On this basis it seems highly plausible that if a confrontation between theory and experiment leads to case 3 (above), then a relativistic extension of the Kramers-van Kampen Hamiltonian beyond dipole approximation is the appropriate framework for further calculations. Since the scattering matrix of this theory will explicitly depend on structural assumptions (the ε_n's), we expect that this finally will lead to a determination of the form factor of the electron.

Another attractive prospect in such a program is that it should settle the question as to whether nature allows for an isobar electrodynamics. As indicated in Table II, a formal RQED model of this kind has already been constructed [17]. Here we briefly indicate the essentials of the classical version.

For simplicity we consider the case of one dimensional motion of a spherically symmetric rigid charge structure; then the total self-force is

$$\vec{F}_{self} = -\,32\pi^2 e^2 \int_0^\infty d\tau\,\dot{\vec{\xi}}(t-\tau) \int_0^\infty dk\,\frac{|\hat{f}(k)|^2}{s^2}\,\times$$

$$\times\,\cos(ck\tau)\left[\frac{\sin(ks)}{ks} - \cos(ks)\right], \qquad (4.24)$$

where

$$s = |\vec{\xi}(t) - \vec{\xi}(t-\tau)|,$$

and $\hat{f}(k)$ is the form factor

$$\hat{f}(k) = k^{-1} \int_0^\infty dr\, r\rho(r)\sin(kr). \qquad (4.25)$$

Isobars correspond to non-trivial solutions of $\vec{F}_{self}=0$.

It can be shown that sufficient conditions for their existence involve the form factor as well as Hilbert transforms of the form factor [35]. Specifically:

$$\hat{f}([m+1]k_o) = 0; \qquad m = 0, 1, \ldots 2N;$$

$$n = 0, 1, \ldots N;$$

$$\int_{-\infty}^{+\infty} dk \, \frac{k^{2(n+1)} |\hat{f}(k)|^2}{k-(m+1)k_o} = 0; \tag{4.26}$$

where $k_o = \omega/c$, the wavenumber corresponding to the oscillation frequency.

It has been proposed several times - most seriously and in most detail by Bohm and Weinstein - that these isobars should be related to particle physics [37]. (In the case of non-rigid, macroscopic current distributions they correspond to beam self-oscillations.) The hope is that form factor information and electrodynamic constraints might be sufficient to determine excited mass states - somewhat similar to the fashion that the buckling of beams can be determined without any knowledge of elastic properties. This hope lives on because of the marvellous numerical coincidence that a heuristic quantization using $(2e^2/3mc^3)^{-1}$ as the characteristic frequency yields Nambu's empirical formula [38]:

$$m_\mu c^2 = m_e c^2 [1 + \frac{3}{2\alpha}] = 105.55 \text{ MeV}$$
$$(105.65 \text{ MeV-exp}). \tag{4.27}$$

5. VACUUM POLARIZATION CERENKOV RADIATION

Suppose we consider the propagation of light through a strong external magnetic field in a direction normal to the lines of flux. Because of interactions of the type

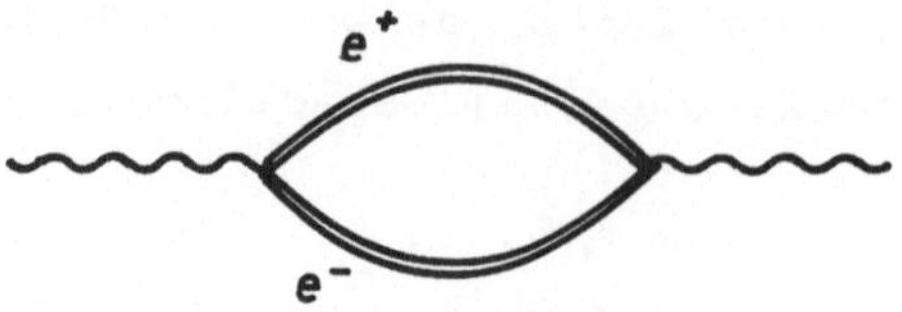

the phase velocity will be reduced below its free space value: $c \to c/[1+\Delta n(\nu)]$ where Δn represents the dispersive effects due to vacuum polarization. If a high energy charged particle crosses the field such that its velocity exceeds the phase velocity of light we expect a vacuum polarization Cerenkov effect. It can be shown that for a magnetic field the index of refraction averaged over both polarization states is [39,4,8,16]

$$\Delta n(\nu) = \frac{\alpha}{4\pi}\left(\frac{H}{H_{cr}}\right)^2 J\left(\frac{1}{2}\,\frac{h\nu}{mc^2}\,\frac{H}{H_{cr}}\right), \tag{5.1}$$

where,

$$J(x) \to \begin{cases} 0.22+0.30x^2, & x \ll 1; \\ -0.56x^{-4/3}, & x \gg 1. \end{cases} \tag{5.2}$$

These modifications may play a subsidiary role in pulsar physics. Here we would like to draw attention to the possibility that this kind of effect might have detectable consequences in an accelerator experiment. The essential point is that strong vacuum polarization fields are certainly available in the vicinity of atomic nuclei. We want to assign these an index of refraction and compute the characteristics of the expected vacuum polarization Cerenkov radiation. In order to carry out this program we first would have to evaluate the imaginary part of the scattering amplitude, i.e. pair production in the Coulomb field. Next we would have to determine the associated dispersive part - the forward coherent Delbrück amplitude. Fortunately all of this hard work - at least in unscreened Born approximation - was done years ago by Rohrlich and Gluckstern [40], [41]. On the basis of their results we infer that we can assign an effective index of refraction of the order of

$$n(\omega) = 1 + (\alpha Z)^2 R(\frac{\hbar\omega}{mc^2}) \tag{5.3}$$

to an atom of atomic number Z. In principle we would of course have to take into account the spatial inhomogeneity of the vacuum polarization effects. However it can be argued that in first approximation an index of refraction of this magnitude prevails in a sphere of radius $\lambda\!\!\!^{-}_c$ (Compton length) surrounding each nucleus. If we consider $\hbar\omega >> 1$ MeV, then the Cerenkov radiation can be visualized as occurring in a kind of dilute fog suspension whose individual "droplets" are Compton spheres with this index of refraction [21].

The function R(x) has the limiting forms

$$R(x) \rightarrow \begin{cases} \frac{\alpha}{\pi^3} \frac{73}{3\times 2^{12}}, & x<<1, \\ \frac{\alpha}{\pi^3} \frac{7}{3\times 2^5} \frac{1}{x}, & x>>1. \end{cases} \tag{5.4}$$

Some appropriate values are given in Table III.

Table III.

$\hbar\omega$ (MeV)	0	0.41	1.33	1.69	2.61	5.11	25.5	51.1	511	5110
$R\times10^6$	1.38	1.39	1.54	1.56	1.53	1.20	0.32	0.16	0.013	0.0017

In case $|\Delta n|<<1$, and $E>>mc^2$, the threshold energy for Cerenkov emission - including relativistic and quantum mechanical recoil corrections - is given by

$$\frac{E_{min}(\hbar\omega)}{mc^2} = \frac{1}{2}\{\frac{\hbar\omega}{mc^2} + [(\frac{\hbar\omega}{mc^2})^2 + \frac{2}{\Delta n(\omega)}]^{1/2}\}. \tag{5.5}$$

From this we can construct Table IV which lists the values appropriate for a Uranium radiator.

Table IV.

(table entries in MeV)

$\hbar\omega$	5.11	25.5	51.1	511	5110	511×10^2	255×10^3	511×10^3
E_{min} (Uranium)	493	963	1370	4360	15770	73.6×10^3	285×10^3	540×10^3

The corresponding opening angle of the Cerenkov cone is given by

$$\theta = [2\Delta n(\omega)(1-\frac{\hbar\omega}{E}) - (\frac{mc^2}{E})^2]^{1/2}. \tag{5.6}$$

From an experimental point of view the largest angles are of principal interest. These are covered by the following table:

Table V.

$\hbar\omega_{MeV}$	5.11	25.5	51.1	511	5110
θ_{mrad}(E=6 GeV)	1.04	0.53	0.37	0.06	-
θ_{mrad}(E=20 GeV)	1.04	0.54	0.38	0.10	0.02
θ_{mrad}(E=300 GeV)	1.04	0.54	0.38	0.11	0.04

Suppose we consider an electron with energy E traversing a thickness ℓ of material characterized by atomic number Z, atomic weight A, and density ρ. Let $N(\hbar\omega_1, \hbar\omega_2)$ denote the corresponding number of Cerenkov photons emitted into the energy band between $\hbar\omega_1$ and $\hbar\omega_2$. We then find

$$N(\hbar\omega_1,\hbar\omega_2) = 4\pi^2\alpha\lambda_c^2\ell N_o\rho A^{-1}[1-(E_{min}/E)^2]\int_{x_1}^{x_2}\Delta n(x)dx, \qquad (5.7)$$

where $x=\hbar\omega/mc^2$. We illustrate these formulas with a case of practical interest: Suppose we consider a Uranium radiator and let $E\gtrsim300$ GeV. Then in the interval $10\lesssim\hbar\omega\lesssim40$ MeV, the Cerenkov opening angle is $\theta_c\gtrsim0.4$ mrad. Note that the Coulomb Bremsstrahlung will be concentrated into a cone of half-angle $\theta_B\sim mc^2/E\sim1.7\mu$rad. Assuming a thin radiator 0.1 mm ($\gtrsim1/30$ radiation length), we can easily check that N (10 MeV, 40 MeV)$\gtrsim2.4\times10^{-4}$ photons per electron. This is a practical "counting" rate. It remains to be seen whether there are any other mechanisms that could give rise to such an extraordinarily wide γ-halo at these high energies.

APPENDIX

COULOMB BREMSSTRAHLUNG BACKGROUND IN MEGAGAUSS IMPLODERS

Fig. 4 shows the generation of magnetic Brems-strahlung and primary and secondary background radiation in the walls of a typical implosion configuration.

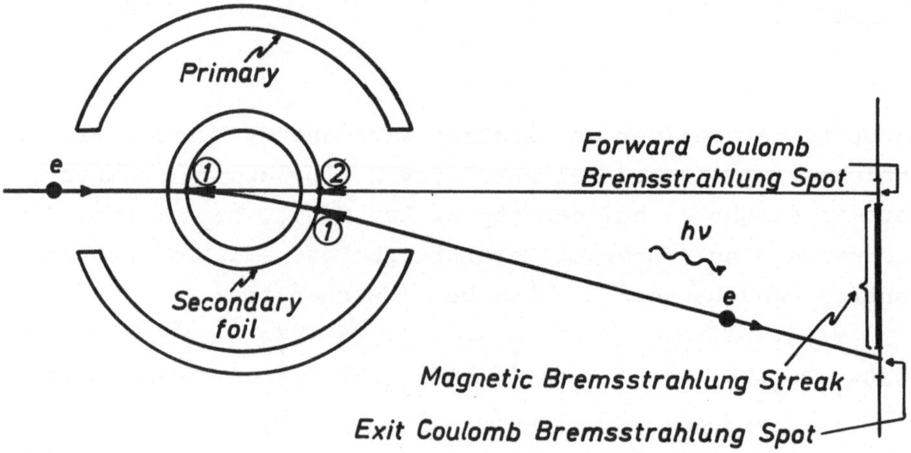

Fig. 4: Background radiation from primary ① and secondary ② sources in magnetic bremsstrahlung experiments with EMC targets.

Due to the presence of the DC bias sweep magnet downstream from the target [compare Fig. 1] we need to concern ourselves mainly with the production of photons due to processes other than magnetic Bremsstrahlung. The principal background source at the points marked ① on the Figure is Coulomb Bremsstrahlung. This will mainly be emitted in a cone of half angle $\Theta_c \sim mc^2/E$. The magnetic Bremsstrahlung angle Θ [Fig. 2] is given by

$$\Theta = \frac{\Delta}{\bar{\lambda}_c} \frac{H}{H_{cr}} \frac{mc^2}{E} ; \tag{A-1}$$

and so we immediately find the ratio

$$\frac{\Theta}{\Theta_c} \cong \frac{\Delta}{\bar{\lambda}_c} \frac{H}{H_{cr}} = 58.8 \; \Delta_{mm} \; H_{MG} >> 1 \tag{A-2}$$

It is clear therefore that these kinematics afford generous protection.

The actual number of Coulomb Bremsstrahlung photons per incident primary electron in the ultra-relativistic case is given by

$$n_\gamma(\hbar\omega_1, \hbar\omega_2) \approx 0.58 \; \bar{\Delta}_{mm} \; h(Z) \ln(\omega_2/\omega_1) , \tag{A-3}$$

where $\bar{\Delta}_{mm}$ is the material thickness in millimeters, and h(Z) denotes the function displayed in Table VI.

Table VI.

	Beryllium	Air	High Explosive	Aluminum	Copper
Z	4	~ 7	8	13	29
h(Z)	5×10^{-3}	6×10^{-6}	10^{-2}	0.02	0.2

In case the detector [emulsions] has an energy discrimination, we are principally concerned with the ratio [compare (3.15)]

$$\frac{N_\gamma}{n_\gamma} \simeq \frac{0.2\, H_{MG}\, \Delta_{mm}\, \int_{\hbar\omega_1}^{\hbar\omega_2} \frac{d\omega}{\omega}\, \kappa \left(\frac{15\, \hbar\omega_{MeV}}{H_{MG}\, E^2_{GeV}}\right)}{h(Z)\, \bar{\Delta}_{mm}\, \ln(\omega_2/\omega_1)} \; . \tag{A-4}$$

This can be simplified still further if we consider the spectral range

$$0.15\, Y \gtrsim \hbar\omega/E \lesssim Y\,, \tag{A-5}$$

since κ is then slowly varying and roughly equal to 0.8. So for this special case

$$\left.\frac{N_\gamma}{n_\gamma}\right|_{(A-5)} \sim 0.17\, \frac{H_{MG}}{h(Z)}\, \frac{\Delta_{mm}}{\bar{\Delta}_{mm}} \; . \tag{A-6}$$

In the implosion experiments we have the approximate values: $H_{MG} \sim 1.7$, $\Delta_{mm} \sim 3$, $\bar{\Delta}_{mm} \sim 3$, $h(Z) = 0.02$ (A1). We therefore find

$$N_\gamma \sim 14.5\, n_\gamma >> n_\gamma \; . \tag{A-7}$$

Some care has to be taken in estimating the additional background emanating from the second wall at point ② . Very high energy primary background can still be ignored since the kinematic factor mc^2/E will confine most of the associated secondaries essentially to the forward direction. But somewhere this argument must fail, and if we arbitrarily choose the upper limit [compare (A-2)]

$$\theta_c < \theta/10 \;, \tag{A-8}$$

it becomes clear that all primary photons with energy $\gtrsim 1$ GeV impinging on ② must be taken into account. This number follows easily from (A-3):

$$n_\gamma^{(1)} \simeq 4h(Z)\bar{\Delta}_{mm} \,. \qquad \text{(A-9)}$$

A brutal overestimate of all secondaries $[\gamma, e^+, e^-]$ issuing from ② is then given by

$$n^{(2)} \simeq n_\gamma^{(1)} \tau \, \bar{\Delta}_{mm} \qquad \text{(A-10)}$$

where τ is the linear attenuation coefficient corresponding to photoeffect + pair production + Compton scattering. For Aluminum the upper limit $\tau \lesssim 0.01\ mm^{-1}$ applies over the entire region of interest, i.e. $1 \lesssim \hbar\omega \lesssim 10^3$ MeV. We therefore find $n^{(2)} \lesssim 0.007$. The corresponding total number of magnetic Bremsstrahlung photons is ~ 3.5. A megagauss field is therefore much more "opaque" than Aluminum for the emission of "low" energy photons. The separation of Coulomb Bremsstrahlung and magnetic Bremsstrahlung has also been confirmed by preliminary analysis of the implosion experiments.

REFERENCES

1. E. Brezin and C. Itzykson, Phys. Rev. 2D, 1191 (1970).
2. E. Brezin and C. Itzykson, Phys. Rev. 3D, 618 (1971).
3. P. P. Kane and G. Basavaraju, Rev. Mod. Phys. 39, 52 (1967).
4. T. Erber, Nature 190, 25 (1961).
5. T. Erber and H. G. Latal, Repts. on Progr. in Physics 33, (in press).
6. R. C. Cameron, ed. "The Magnetic and Related Stars", Mono Book Corp. Baltimore, 1967.
7. F. Pacini, "Neutron Stars, Pulsar Radiation, and Supernova Remnants", Gordon and Breach, New York, 1971.

8. T. Erber, Rev. Mod. Phys. 38, 626 (1966).
9. T. Erber, Zeit. Angew. Phys. 24, 188 (1968).
10. F. Herlach, R. McBroom, T. Erber, J. J. Murray and R. Gearhart, Proc. of Particle Acceler. Conference Chicago, March 1971 (in press).
11. R. G. Newton, Phys. Rev. 3D, 626 (1971).
12. B. Jancovici, Phys. Rev. 187, 2275 (1969).
13. V. I. Ritus, Zh. Eksperim. i Teor. Fiz. 57, 2176 (1969) [JETP 30, 1181 (1970)].
14. I. M. Ternov, V. G. Bagrov, V. A. Bordovitzin and O. F. Dorofeev, Zh. Eksperim. i. Teor. Fiz. 55, 2273 (1968) [JETP 28, 1206 (1969)].
15. Z. Bialynicki - Birula and I. Bialynicki - Birula, Phys. Rev. 10D, 2341 (1970).
16. S. L. Adler, Annals of Phys. (in press); S. L. Adler, J. N. Bahcall, C. G. Callan and M. N. Rosenbluth, Phys. Rev. Letters 25, 1061 (1970).
17. S. Coleman and R. E. Norton, Phys. Rev. 125, 1422 (1962).
18. C. S. Shen and H. G. Latal, private communication.
19. N. G. van Kampen, Dan. Mat. Fys. Medd. 26, no. 15 (1951).
20. J. M. Jauch and K. M. Watson, Phys. Rev. 74, 950; 1485 (1948).
21. T. Erber and H. C. Shih, Acta Phys. Austriaca 19, 17 (1964).
22. F. Herlach, Rept. on Progr. in Physics 31, 341 (1968).
23. J. Schwinger, Phys. Rev. 75, 1912 (1949).
24. A. A. Sokolov and I. M. Ternov, "Synchrotron Radiation", Akademie Verlag, Berlin, 1968.
25. N. P. Klepikov, Zh. Eksperim. i Teor. Fiz. 26, 19 (1954).
26. J. J. Klein, Rev. Mod. Phys. 40, 523 (1968).

27. D. E. Bedo, D. H. Tomboulian and J. A. Rigert, Jr. Appl. Phys. 31, 2289 (1960).
28. T. Erber, Fortschr. der Physik 9, 343 (1961).
29. F. Rohrlich, "Classical Charged Particles", Addison-Wesley, Reading, 1965).
30. E. Fermi, Accad. Lincei, Atti 5, 795 (1927).
31. C. S. Shen, Phys. Rev. Letters 24, 410 (1970).
32. H. G. Latal, Acta Phys. Austriaca (in press).
33. H. C. Shih and T. Erber, Acta Phys. Austriaca 28, 325 (1968).
34. K. Wildermuth and K. Baumann, Nucl. Phys. 3, 612 (1957).
35. T. Erber and S. M. Prastein, Acta Phys. Austriaca 32, 224 (1971).
36. R. E. Norton and W. K. R. Watson, Phys. Rev. 116, 1957 (1959).
37. D. Bohm and M. Weinstein, Phys. Rev. 74, 1789 (1948).
38. Y. Nambu, Progr. Theor. Phys. 7, 595 (1952).
39. R. Baier and P. Breitenlohner, Acta Phys. Austriaca 25, 212 (1967); Nuovo Cimento 47, 117 (1967).
40. F. Rohrlich and R. L. Gluckstern, Phys. Rev. 86, 1 (1952).
41. F. Rohrlich, Phys. Rev. 108, 169 (1957).

Acta Physica Austriaca, Suppl. VIII, 358 (1971)

PERTURBATION THEORY FOR A CLASS OF NONPOLYNOMIAL LAGRANGIAN FIELD THEORIES * **

BY

H. LEHMANN
University of Hamburg, Germany

Recent work [1] on nonpolynomial Lagrangians is discussed with emphasis on the ambiguity problem in higher-order perturbation theory. Some results are stated for the case of a scalar field with exponential self-interaction. A less academic model is the derivative coupling of a spinor field with a neutral pseudoscalar field transformed to exponential form which can be treated by the same methods.

REFERENCE

1. H. Lehmann and K. Pohlmeyer, Comm. Math. Phys. 20, 101 (1970).

---

* Lecture given at X. Internationale Universitätswochen für Kernphysik, Schladming, March 1 - March 13, 1971.

** A short version of this lecture series was presented at the Coral Gables Conference on Fundamental Interactions (January 1971). Since it is published in the proceedings of that conference only the above abstract is included here.

Acta Physica Austriaca, Suppl. VIII, 359 (1971)

ELECTROMAGNETIC INTERACTIONS AN EXPERIMENTAL SURVEY *

BY

G. BUSCHHORN
DESY, Hamburg

Since the editors did not receive the manuscript of the lecture in time for a speedy publication of these proceedings, this contribution will be published in one of the forthcoming regular issues of Acta Physica Austriaca.

---

* Lecture given at X. Internationale Universitätswochen für Kernphysik, Schladming, March 1 - March 13, 1971.

Acta Physica Austriaca, Suppl. VIII, 360 (1971)

LIGHT CONE EXPANSIONS AND APPLICATIONS [x]

BY

Y. FRISHMAN
Department of Physics
Weizmann Institute, Rehovot, Israel

Since the editors did not receive the manuscript of the lecture in time for a speedy publication of these proceedings, this contribution will be published in one of the forthcoming regular issues of Acta Physica Austriaca.

[x] Lecture given at X. Internationale Universitätswochen für Kernphysik, Schladming, March 1 - March 13, 1971.

Acta Physica Austriaca, Suppl. VIII, 361–366 (1971)

SUMMARY - FIRST WEEK*

BY

H. PIETSCHMANN
Institut für Theoretische Physik
Universität Wien

When it was decided, that I was going to summarize the first week of this years Schladming Winter School, I was in a way very pleased. Little disturbes me more deeply than poor lecturing but all of the talks during this years first week were of very high standard or at least not below average. For this fine performance I would like to thank all lecturers collectively and very sincerely.

It has been talked over repeatedly that this years first week in Schladming had put quite an emphasize on mathematical problems of high energy physics. I agree with this statement but I think it is not the consequence of a definite selection of speakers but rather reflects the general trend - or shall we say fashion - in current high energy physics. Let me add a personal comment right away: I think, mathematical rigor is necessary but not sufficient for physics to make progress.

---

*Summary given at X. Internationale Universitätswochen für Kernphysik, Schladming, March 1 - March 13, 1971.

From a very general consideration of all lectures one got the impression that extreme efforts are made not only towards mathematical rigor but in all directions. Models are carried to extremes where calculations are so cumbersome that they hardly compare with the trust put into the model at the beginning. What is the reason?

To analyze this situation let us start with a triviality: Physics is unthinkable without experiment. (Otherwise you end up with something like Eddingtons Fundamental Theory!). However right now we are in a time where not many experiments uncover essentially new things. As a matter of fact, besides quantitative refinement we have not had any qualitatively new discovery for quite some time. It seems to me, that the reason for this is the simple fact, that the "old" generation of accelerators is now more then ten years in operation and the "new" generation is not yet functioning. Therefore, we find ourselves in a sort of "intermission" between two periods of intense experimental discovery. It is only natural, that this is a period of "creative pause" in which existing models or theories are polished and carried all the way to their extreme consequences. It should also be a period of careful preparation of new ideas in such a way, that their predictions are clearly listed so that they can be subject to experimental verification or falsification.

I would like to state very clearly that I am sometimes depressed about the way, models are cranked out without any care whether or not their predictions are even noticed by experimentalists. Also this situation becomes more crucial in our "intermission" time and will hopefully be swept away by a new wave of exciting empirical findings.

Let me now try to view the lectures of this years first week in Schladming from this particular angle.

A) <u>Mathematical Physics</u>

In a series of lectures we have heard about current struggles for either more mathematical rigor or more mathematical ingenuity to improve or reformulate what is (as far as physics is concerned) already contained in present theories.

a) <u>Nilsson</u>:

Instead of the normal Minkowski space, Jan Nilsson uses larger spaces in which the Poincaré group is represented. He gets a set of additional integration variables but also a new quantum number describing the non-locality of a particle of new type. It was very nice that he went as straight as possible to an actual computation in which he could try out his mathematical model. Non-local theories are one of the many hopes to improve over the persistent troubles of infinities in present theories.

b) <u>Rohrlich</u>:

In his null-plane quantum field theory, Fritz Rohrlich essentially investigates whether progress is facilitated by choosing a more suitable set of coordinates and variables. It turns out indeed in a surprising way that he gets conformal invariance in a very direct manner. The undertaking is very interesting but still on its way and we are eager to see what its final outcome will turn out to be.

c) <u>Klauder</u>:

Klauder succeeded in presenting his rather mathematical lectures in such a lively and inspiring way that one was fascinated from beginning to the very end.

Most of the troubles in QFT arise from the use of perturbation theory in which one separates the potential term from the Hamiltonian, solves the free problem and then adds the potential term in a perturbative fashion. Instead, Klauder separates the derivative term $(\nabla\phi)^2$ from the total Hamiltonian. He solves the (ultra-local) remainder and hopes that by putting on the derivative term he shall encounter less serious troubles than in usual perturbation theory. He gave a wonderful insight into the mechanism of what is going on when you perturbe energy levels. We can only wish him the best of luck for the rest of his challenging journey towards a consistent QFT.

Before I enter into the other topics, let me recall a platitude which is so often repeated in connection with mathematics: people say that the task of a mathematician is easy because he can make all the assumptions, necessary for his proof. While this is, of course, true, we should not forget that most physicists facilitate their own task in a similar - but unacceptible - way by not hesitating to assume any number of elementary particles (mostly infinitely many) without even suggesting that they might be searched for! This remark joins nicely to our next subject:

B) Strong Interactions

A phantastic amount of material has been accumulated over the years but nothing really qualitatively new has emerged recently. Therefore, one is trying to refine, reshape and prolong models until they become "ammonites"* and eventually die out.

---

*Ammonites are the shells of a special kind of octopussy which developed at some early stage. Their shells took such bizarr forms that they could not possibly evolute and were bound to die out.

a) Zachariasen:

Most of us remember the happy pioneerspirit which governed the field of Regge poles before 1963. It was then realized that also in the l-plane cuts had to be taken into account which shied away a vast number of theoreticians who went to investigate SU_3 and subsequently higher symmetries. It was extremely comforting to learn from Zachariasen's clear lectures that cuts can be very simply introduced into models by means of pairs of complex conjugate poles. Let us hope that the hangover after the feast in 1963 is not repeated too soon.

b) Drummond:

Drummond gave a very humoristic and nice presentation of the multiveneziano model which has perhaps become very much like an "ammonite". A vast number of particles is predicted and fancyness is not missing in any place. To use a technical term of Dialectics, it seems that in this field "quantity has not yet turned into quality". But this should not understate the immense importance of this kind of investigation, in particular since it is - or could be - so close to experiments.

C) Weak Interactions

a) Kummer:

Wolfgang Kummer presented an alternative model to the V-A-theory which is able to produce identical results at present energies. This is typically something which has to be brought up in times of "intermission" because alternative models in general have to be supported or ruled out by experiments at different energies, say. It may be that they are only good to be ruled out as soon as higher energies become available but they may very well turn out to be closer to truth. Therefore these investigations are very worth while.

D) Electromagnetic Interactions

a) Leader:

Let me first state that Leaders lecture was wonderful because he organized them very clearly and pedagogical. Problems were brought out very clearly. From the (unblurred) spinless case he proceeded to the physical case. In general, one got the impression that electromagnetic interactions are proceeding at a nice pace. This fits well into my picture because electron machines are still fully experimenting in a widely open area and there is not yet any "intermission".

In order to soften some of my critical remarks let me end with the joke of the Count who tells his adventures as a lion hunter. When he is asked by one of his breathless listeners, how many lions he has killed so far, he answers: "None". His reply to the further question: "Isn't this a bit few?" is: "Not for lions!"

That the tenth Schladming Winter School was again such a wonderful success is entirely due to Professor Urban and his crew to whom the sincere thanks of all of us are now dedicated.

Acta Physica Austriaca, Suppl. VIII, 367–374 (1971)

SUMMARY - SECOND WEEK*

BY

F. ROHRLICH
Department of Physics, Syracuse University
Syracuse, New York

It is certainly an honor to be asked to summarize an international meeting like the Schladming Universitätswochen, especially when it is the tenth anniversary. A special aspect of the Schladming meetings - as has become abundantly clear after ten consecutive events - is the didactic value of the lectures as our director and organizer, Professor Urban, emphasizes at all times. This emphasis results in a rather broad sweep of topics which range from an experimental talk all the way to the very mathematical subject of current algebraic quantum field theory. It also obliges the lecturers to be clear and informative, a matter which may force the speaker to play down his own contribution if it is too specialized and narrow.

The wide range of subject matter discussed in the second week may appear somewhat overwhelming to the novice theorist. He feels suddenly confronted by a

---

*Summary given at X. Internationale Universitätswochen für Kernphysik, Schladming, March 1 - March 13, 1971.

"Theoriesalat" which seems indigestable. This summary is an effort to help him and to prevent indigestion, an endeavor clearly within the didactic spirit of this meeting. At the same time I cannot be too long, being scheduled between a very beautiful (and tiresome) skiing afternoon and - supper time. Thus, I shall not be able to do justice to many of the lectures.

As a guiding light let us keep in mind the interrelation between fundamental theory (the ideal toward which we all aim) and phenomenological theory (the practical, effective, easier-at-hand ordering scheme of experimental data). Both are necessary, but the latter always comes first and often helps evolve the fundamental theory. But very often we find that we cannot push the fundamental theory far enough and we build a phenomenological one on top of it. These are, not suprisingly, the most effective phenomenologies.

The work on quark field theory by the DESY group which was presented by Miss Krammer is a fine example: the Bethe-Salpeter equation for two quarks (at first assumed spinless for simplicity) is difficult to handle except in the ladder approximation. Not being able to push beyond that, a potential with several adjustable parameters is assumed and inserted in that approximation. A suitable choice of parameters compensates at least partially for the lack of an approximation better than "ladder" and actually yields very good results. Thus emerges a phenomenological approach based on more fundamental theory which will undoubtedly survive and be extremely useful. We shall encounter other, similar, examples later on.

The experimental survey by Buschhorn was impressive by its large amount of compiled data. At the same time he put us theorists to shame: how far behind the experimentalists we are, not to be able to put much rhyme or reason into this work! The vector dominance model, the Regge model, SU(3), parton theory and scaling, all these are pieces of a jig-saw puzzle which we cannot put together. All of these are only partially confirmed by experiment and we know no clear validity limits of any of these. And so we ask our experimental friends to go to higher energies and to higher accuracy in the hope of making more sense of our poor theorizing.

One can also attempt to go to fields of higher intensity. This is being attempted by Erber who probes our "best" theory, quantum electrodynamics, at 1-2 Mgauss magnetic fields. The hope is twofold, (1) to check the validity of QED under these conditions, and (2) to explore the mysterious and much neglected penumbra between classical and quantum electrodynamics that has been with us for so long. Despite various attempts, a satisfactory classical limit of QED has not yet been given. Consequently, we do not understand the classical Schott term ($\ddot{v}$ term of Lorentz) of so-called radiation reaction in the Lorentz-Dirac equation. The latter is an inescapable consequence of Maxwell's equations and point particle relativity. The Lorentz-Dirac equation

$$m\dot{v}^{\mu} = F^{\mu}_{ext} + \frac{2}{3}e^2\ddot{v}^{\mu} - Rv^{\mu} \qquad R \equiv \frac{2}{3}e^2\dot{v}^{\alpha}\dot{v}_{\alpha}$$

can be combined with the asymptotic condition (free particles in the distant future) to yield the integro-differential equation

$$m\dot{v}^{\mu}(\tau) = \int_0^{\infty} e^{-\alpha}(F^{\mu}_{ext} - Rv^{\mu})_{\tau+\alpha\tau_0} d\alpha, \qquad \tau_0 \equiv \frac{2}{3}\frac{e^2}{m} .$$

This eliminates the Schott term (as well as all run-away solutions) and provides a true equation of motion because it is of first order (2nd order when written for x^μ). But it is non-local in time (describing preacceleration). Is this nonlocality observable in a classical scattering experiment? Its observation is equivalent to the observation of the Schott term. But can one actually "see" the preacceleration it causes? If not, what is the corresponding QED effect? We have no answers.

Erber's present experiment of magnetic bremsstrahlung will hopefully be improved some years hence and we are all looking forward to at least partial answers to some of these questions.

In contrast to this work which is important but rather off the main path of current endeavors there is the very active field of high energy limits, infinite momentum frames and scaling. Frishman reported on this matter.

Of course, large momenta correspond to short distances so that the recent work of K. Wilson and many others on the product of two operator valued distributions, $A(x)B(y)$ as $x-y \to o$ is relevant here. The proposal to expand or at least approximate such products by finite sums of local operators has been extended by Frishman to the case when the distance $x-y$ approaches the light cone. There follows the application to current commutators, to electroproduction, and to the scaling laws. We observe that quantum field theory cannot be pushed far enough to permit detailed predictions. But one obtains the structure of transition amplitudes (in terms of certain functions W_1 and W_2, for example) which can form a useful basis for a parametrization of the experimental results, leading to a phenomenological theory based on more fundamental considerations.

Another example of this situation emerges from the nonlocal approach to weak interactions, reported by Pietschmann. As a nonrenormalizable field theory the standard theory of weak interactions is in trouble. High energy predictions cannot be relied upon and we are barred from a "deeper" understanding (i.e. short distance behavior of weak interactions). The well-known way out is to generalize the interaction Lagrangian by putting a $K^{\mu\nu}(x-y)$ instead of a $g^{\mu\nu}\delta(x-y)$ between the two interacting currents. But the meanings of this formal Ansatz can be manifold, the two most interesting ones being the not so recent intermediate vector boson (IVB) and a quite recent nonpolynomial Lagrangian (Fivel and Mitter). To test these models various weak processes were computed explicitly in the hope of being soon able to measure them. But these models all involve parameters which can be used to fit at least some of the data. Thus, while neither model may be eventually confirmed, the associated parametrization may indeed be very useful for a phenomenological description of future experiments.

We come now to the land of superpropagators and super-Feynman diagrams. Since this field of nonpolynomial Lagrangians is governed much more by mathematical than physical considerations it takes superfaith to believe in its eventual usefulness, and so the brave theorists working on it should all be known as super-men. Lehmann reported on various aspects of this topic.

If a field is defined as the Wick product of a non-polynomial function of free fields, its two-point function is a non-polynomial function of Δ_+, $F(\Delta_+)$, and its propagator is $F(\Delta_c)$, where Δ_c is the causal function (Feynman function). The immediate problem is to give mathematical meaning to such functions in view of the fact that not even integral powers of Δ_c are defined. Formally,

$$[\Delta_c(x)]^n = [\theta(x^o)\Delta_+(x)+\theta(-x^o)\Delta_+(-x)]^n =$$

$$= \theta(x^o)\Delta_+^n(x) + \theta(-x^o)\Delta_+^n(-x).$$

Now Δ_+^n exists rigorously,

$$\Delta_+^n(x) \sim \int\rho_n(\kappa^2)\Delta_+(x,\kappa^2)d\kappa^2$$

the $\rho_n(\kappa^2)$ being known functions. But taking the θ-functions inside the integral leads to the undefined result

$$\Delta_c^n(x) \sim \int\rho_n(\kappa^2)\Delta_c(x,\kappa^2)d\kappa^2.$$

The momentum representation of Δ_c shows immediately that Δ_c^n diverges like $\int t^{n-3}dt$. The integral thus requires n-1 subtractions, i.e. it is determined up to a polynomial of order n-2 (in p-space), corresponding to

$$\sum_{m=o}^{n-2} c_m \Box^m \delta \ ,$$

and involving n-1 arbitrary parameters. (The usual self-energy decomposition involves two subtractions, a mass renormalization and a "wave function" renormalization, and then gives finite results for $n \leq 3$ internal lines, corresponding to a $\lambda\phi^{n+1}$ interaction which is then renormalizable for these n only).

Nonpolynomial functions would then be expected to have an infinite nunber of subtractions (non-renormalizable theory). This non-uniqueness is the basic problem of the superpropagator. It was resolved for exponential self-interaction and for the exponential form of spinor-pseudoscalar derivative coupling by Lehmann and Pohlmeyer by requiring "minimum singularity" which (in the first case) amounts to the absence of $\square^n\delta$ type terms. It is most surprising that such a minimality requirement works for exponential functions but not for polynomials.

While there is of course no good reason to assume that all physical interactions are polynomial (it is manifestly false for gravitation) there is presently much too much freedom in the possible non-polynomial interactions available. The restriction due to causality (Jaffe) to entire function of order less than two is clearly not enough. Additional criteria will have to be found.

It is unfortunate that Salam could not keep his plans to speak here. He would presumably have told us more about the physics of this interesting new venture.

The most sophisticated lectures were undoubtedly those by Mayer on algebraic field theory. As is well known the exponentiated Heisenberg commutation relations (the Weyl relations) have, for the case of a finite number of degrees of freedom, only one unitary equivalence class of representations. But in the infinite dimensional case (field theory) a non-denumberable infinity of unitary equivalence classes exists, of which the Fock representation is only one. This together with the need for a deeper understanding of partial and broken symmetries makes an algebraic approach very attractive. Until recently, this approach was restricted to the C^*-algebras of observables typically represented by bounded operators on Hilbert space. But our theories do contain non-observable elements, potentials, spinor fields, etc. And so recent attempts (e.g. the papers by Doplicher, Haag and Roberts) asked to what extent the observables (i.e. their algebra) determine the non-observables (i.e. the larger algebra of fields). The bridge is furnished by the gauge group. Mayer's lectures laid the foundations for a weakening of this bridge: duality theorems relating a compact group (as the gauge group is assumed to be) with "W*-bigebras", a generalization of a W*-algebra, may permit one to work with the latter instead of the former. This would be a step closer to observations because one observes certain unitary, finite dimensional representations (e.g. in SU(3) symmetry) and not the whole group. The latter is only inferred.

Whether this attempt will be successful, in which case we would all have to learn about W*-bigebras, or Krein algebras, remains to be seen.

This concludes my very cursory survey of a great deal of information and learning. I think that everyone here profited in some measure from these lectures and will return home with new ideas and long memories of Schladming that exceed even the fabulous winter landscape this little town has to offer. I am sure that I am speaking in the name of everyone if I express the sincere thanks of all the visitors from about twenty countries to our generous hosts.

Special thanks are of course due to Professor Urban whose choice of lecturers proved as successful as it has for the preceding nine years, and who set the format and informality of the Universitätswochen. But we cannot forget the many people who did all the leg work and took care of the many details and unexpected changes, especially Drs. Latal, Pesec and Widder, who collaborated so efficiently. All of us are indebted to them for the success of this meeting. Thank you.

Acta Physica Austriaca, Suppl. VIII, 376–424 (1971)

SEMINARS

C. ALABISO: Bound States, Pade Approximants and Variational Principles.

N. ARTEAGA-ROMERO: Photon-Photon Collisions: A New Field in High Energy Physics.

L. CASTELL: Conformal-Invariant Mass Zero Scattering.

G. CICUTA: Validity of the Regge-Eikonal Model in $\lambda\phi^3$ Theory.

J. CLEYMANS: Veneziano-Type Representations for the Pion Form Factor.

B. ENFLO: Covariant Spin Analysis of Multiparticle Amplitudes.

P. HABERLER: Chiral Invariance, Dilatation Invariance and the Bootstrap Condition in Quantum Field Theory.

Z. HORVATH: Exactly Soluble Models for Interacting Vector and Scalar Fields.

W. HURLEY: Progress in Relativistic Wave Equations of Higher Spin.

A. JUREWICZ: Multiperipheral Approach to the Analysis of High-Energy High-Multiplicity Pion-Nucleon Interactions.

J. KLAUDER: Tachyon Quantization.

M. KRAMMER: Some Properties of the Mesons from a Relativistic Scalar Quark Model with Strong Binding.

T. PALEV: On Some Algebraic Properties of the Para-Fermi Operators and their Representations.

H. TILGNER: A Mathematical Note on the Conformal Group of a Pseudo-Orthogonal Vector Space.

M. WELLNER: Compensation Theory.

N. ZOVKO: Dipole Type Formulas for Electromagnetic Form Factors.

BOUND STATES, PADE APPROXIMANTS AND VARIATIONAL PRINCIPLES*

BY

C.ALABISO*, P.BUTERA, G.M.PROSPERI
Istituto di Scienze Fisiche dell'Università, Milano
*also Istituto di Fisica dell'Università, Parma

We consider the problem of the calculation of bound states in potential theory. This is a well suited test ground to understand and possibly overcome difficulties which can arise also in more complicated theories.

At fixed energy z (real and negative) the Schrödinger equation

$$(H_o+gV)|\psi> = z|\psi> \tag{1}$$

can be easily translated in either of the variational principles over the coupling constant g:

$$\frac{1}{g} = - \frac{<\psi|V\ R_o(z)V\ R_o(z)V|\psi>}{<\psi|V\ R_o(z)V|\psi>} \tag{2}$$

$$\frac{1}{g} = - \frac{<\psi|V\ R_o(z)V|\psi>}{<\psi|V|\psi>} \tag{2'}$$

where $R_o(z)=(H_o-z)^{-1}$. The operators involved in (2) and

---

*Abstract of Seminar presented by C. Alabiso at X. Internationale Universitätswochen für Kernphysik, Schladming, March 1 - March 13, 1971.

(2') are of course $R_o^{1/2} V R_o^{1/2}$ and $V^{1/2} R_o V^{1/2}$ respectively. For z real and negative and under very broad hypothesis on the potential the former operator is compact and self-adjoint and the latter is compact and self-adjoint if the potential has a definite sign. We can apply to these forms well known variational theorems and look for stationary and extremum points of (2) and (2') using as trial functions:

$$|\psi\rangle = \sum_{n=o}^{N-1} a_n (-R_o(z)V)^{n+j} |k,\ell\rangle \tag{3}$$

where a_n are numerical coefficients and $|k,\ell\rangle$ is a free state of definite angular momentum ℓ and magnitude of momentum k. If we substitute (3) in (2) and (2') and make these stationary in the a_n's, a consistency condition follows from which we obtain the coupling constant

$$g = g(z,k) \tag{4}$$

corresponding to a bound state at energy z. These consistency conditions are precisely the conditions for the vanishing of the denominators of the Padé Approximants [N,N+2j+1] and [N,N+2j] to the scattering matrix $T(z) = gV - gR_o(z)V\,T(z)$ in the angular momentum representation.

Due to the above mentioned theorems, we can use the parameter k of (4) as a further vatiational parameter and look for stationary and extremum points. Convergence theorems of the method can be given.

In the usual approximation procedure with Padé Approximants, the parameter k in (4) was choosen "on shell", that is $k=\sqrt{z}$; as a consequence in each term of the perturbative expansion the well known left-hand singularities in the z complex plane affected the research of the zeroes of the Padé Approximants' denominators. In our approach, the variational principles prescribe the best value of k independently of z, and this fact avoids all difficulties related to the left-hand singularities.

Numerical calculations with various potentials show the power of the method.

REFERENCES

C. Alabiso, P. Butera and G. M. Prosperi: "Variational principles and Padé Approximants. Bound states in potential theory".
Preprint Milano, February 1971. Submitted to "Nuclear Physics", and references quoted there.

PHOTON-PHOTON COLLISIONS: A NEW FIELD IN HIGH ENERGY PHYSICS[x][xx]

BY

N. ARTEAGA-ROMERO[xxx]
Laboratoire de Physique Atomique
Collège de France, Paris

This is a work which has been performed by P. Kessler, A. Jaccarini, J. Parisi and myself. We have considered, to lowest order in electrodynamics, all Feynman diagram contributions to the reactions of the type:

$$e^-e^+ \to e^-e^+ A^-A^+ \quad \text{and} \quad e^-e^+ \to e^-e^+ X\ .$$

Where $A=e,\mu,\pi,K,p$ and $X=\pi^o,\eta,\eta'$. Detecting both the outgoing electron and positron at small angles with respect to the beam direction the most important contribution to these processes comes from the Feynman diagram where the two photons exchanged are space-like. Consequently for calculating the total cross sections we have taken into account only this diagram's contribution. Considering the value $\theta_{max}=4.10^{-3}$ radian as a cut-off angle for the electron and positron scattered, and varying the energy

---

[x] Abstract of Seminar given at X. Internationale Universitätswochen für Kernphysik, Schladming, March 1 - March 13, 1971.

[xx] To be published in Physical Review.

[xxx] On leave of absence from the Central University of Caracas, Venezuela.

E_o of the incident beam between 1 and 10 GeV, we have calculated the total cross sections for all these processes as a function of E_o; they increase with rising beam energy in sharp contradistinction with the behaviour of the cross sections for annihilation processes. For $A=e,\mu,\pi$ the cross sections are high enough to ensure that reasonably high counting rates will be obtained with electron-positron colliding beams of beam energy of 2 or 3 GeV and luminosities of the order of 10^{32} $cm^{-2}sec^{-1}$. If M is the invariant mass of the pair of particles A^-A^+ produced, and q, q' the four momenta of the photons exchanged, we have always, under the conditions defined above:

$$|q|^2, \ |q'|^2 << M^2. \tag{1}$$

For example, for E_o=2 GeV, $\theta_{max}=4.10^{-3}$ radian, $|q|^2_{max}$ = = $|q'|^2_{max}$=64 MeV^2, whereas the values of M^2 considered were at least of order 10^4 MeV^2. We have made an exact calculation of the differential cross sections taking into account the relation (1). We have found a formula of the "Weizsäcker-Williams type" which is more precise than the old Weizsäcker-Williams formula for the equivalent photon spectra of the electron and positron vertices. This formula allows us to study the reactions considered above, under these conditions, as practically equivalent to $\gamma\gamma\to A^-A^+$ and $\gamma\gamma\to X$, where both γ's are treated as free photons and to analyse the experimental results in terms of photon-photon collision cross sections. We must insist upon the fact that the Weizsäcker-Williams approximation is a small transfer approximation and that it is illegal to extend it to the calculations of the total cross sections without angular cut-off for the electron and positron [2] and for studying the asymptotic behaviour [3].

We have also calculated the angular distributions of the particles produced. The curves are the more peaked (forward and backward) the smaller the mass of A is. We are studying now a possible contamination of Frascati's experiments on multiparticle production for these type of processes. We have considered some applications; for the near future: $\gamma\gamma\to e^- e^+$ and $\gamma\gamma\to\mu^-\mu^+$ which can serve as control experiments or for detecting a possible anomaly of the muon; $\gamma\gamma\to\pi^-\pi^+$, in this case it will be interesting to look for the 0^{++} resonance (called σ or ε, with mass about 600-700 MeV) where the ρ is completely absent and without any spectator hadron; $\gamma\gamma\to X$ with $X=\pi^0,\eta,\eta'$, in this case the values of the decay rates $\Gamma(\pi^0\to 2\gamma)$ and $\Gamma(\eta\to 2\gamma)$ may be improved and $\Gamma(\eta'\to 2\gamma)$ determined; for a future generation of more powerful electron-positron storage rings: $\gamma\gamma\to p\bar{p}$, we have just finished the analytical part of a calculation taking into account the anomalous magnetic moment; $\gamma\gamma\to K^-K^+$, we can study here (as in the pion case) all states with parity +1, charge conjugation quantum number +1 and even angular momenta; $\gamma\gamma\to$ any hadrons, we have the possibility of studying all nonstrange mesonic resonances with charge conjugation +1 without any spectator hadron; $\gamma\gamma\to\gamma\gamma$ which is a field of study in itself; we can make some speculations: $\gamma\gamma\to W\bar{W}$ where W is the intermediate vector boson of weak interactions; $\gamma\gamma\to\ell^*\ell^*$ where ℓ^* is a heavy lepton; $\gamma\gamma\to$ quark antiquark etc. We can conclude saying that in annihilation processes only states 1^{--} are produced (at least as long as we suppose only one photon exchanged) whereas here all angular momenta and parity states can be created (the only restriction is C=+1). This type of processes provides a powerful stimulation for building new electron-positron storage rings.

REFERENCES

1. See also at the same authors: Comptes Rendus á l'Academie des Sciences, Paris, 269B, 153 and 1129 (1969). Internal Report PAM 70-02 (Collège de France, April 1970, Paris);
Lettere to Il Nuovo Cimento 4, 933 (1970);
A. Iaccarini and I. Parisi, Thesis (Paris 1970, 1969, unpublished).
2. In connection with this type of extrapolation see: Balakin, Budnev and Ginzburg, JETP letters (English traduction) 11, 388 (1970);
Brodsky, Inoshita and Terazawa, Phys. Rev. Letters 25, 972 (1970);
Serbo, JETP letters (English traduction) 12, 39 (1970).
3. See Brodsky et al., ref. (1).

CONFORMAL-INVARIANT MASS ZERO SCATTERING*

BY

L. CASTELL
Max-Planck-Institut für Physik und Astrophysik
München

1) The unitary irreducible representations of the universal covering group of the identity component of the Poincare group characterized by mass zero m=o, positive energy $p^4>o$, and a fixed value of the helicity $\lambda=o$, $\pm 1/2$, ± 1, $\pm 3/2$,... form also unitary irreducible representations [1] of the spin-covering group $SU_o(2.2)$ of the conformal group $SO_o(4,2)/C_2$. Therefore we can reduce the initial state of two incoming mass zero particles with helicity λ and λ' with respect to irreducible representations of $SU_o(2,2)$. If the S-matrix is approximately conformal-invariant at asymptotic high energies, the reduced S-matrix elements for a certain final state depend only on the values of the Casimir operators of $SU_o(2,2)$ for the intermediate states in the limit $s\to\infty$, and possibly on a discrete parameter of degeneracy. For the initial state of two mass zero particles we obtain for elastic scattering

*Abstract of Seminar given at X. Internationale Universitätswochen für Kernphysik, Schladming, March 1 - March 13, 1971.

$$e^{2i\delta_\ell(s)} \to c_\ell = \text{const.}, \quad \ell = |\lambda-\lambda'|+k, \; k = o, 1, 2, \ldots .$$

2) From the assumed conformal invariance of the S-matrix at asymptotic energies, it follows that the different final states of the scattering process can be simply parametrized by irreducible representations of $SU_o(2,2)$. That means the "particles" of the final state consists of mass m=o particles, or have to be described by "physical" massive unitary irreducible representations $o<m^2<\infty$, $p^4>o$ of $SU_o(2,2)$. This follows already from dilatational invariance [2].

3) It has been conjectured [3] that all unitary irreducible massive representations, which are physical, i.e., for which $o<m^2<\infty$, $p^4>o$, if restricted to the Poincare group, are contained in the direct product of a finite number ≥2 of mass m=o, $p^4>o$ representations of $SU_o(2,2)$. Thus the investigation of all inelastic channels is reduced to the investigation of all inelastic mass m=o channels. This problem has been solved for the given initial state [3].

4) The simplest way to solve the above problem is to label all states of each irreducible representation by the <u>discrete</u> quantum numbers of the maximal compact subgroup $SU(2)\times SU(2)\times U(1)$ of $SU_o(2,2)$, and carry out the reduction of the direct products in this basis. We obtain as the main result for γ-γ scattering:

a) $\lambda = -\lambda' = \pm 1$.

The possible final states consist only of two-mass m=o particles with opposite helicity $\lambda_1=-\lambda_2, |\lambda_1|=o, 1/2, 1, 3/2, \ldots, \ell/2$.

b) $\lambda = \lambda' = \pm 1$.

Here we have two possibilities of getting real inelastic channels, in addition to two-mass m=o final states. In the ℓ=o channels the following final states are allowed: $\lambda_1=\lambda_2 = = \lambda_3=\lambda_4=0$; $\lambda_1=\lambda_2=1/2$, $\lambda_3=0$; $\lambda_1=\lambda_2=-1/2$, $\lambda_3=0$. In the ℓ=1 channel we obtain correspondingly $\lambda_1=\lambda_2=\pm 1/2$, $\lambda_3=0$. However, this means that we could never create two massive physical particles with spin 1/2. Also π-π production would be forbidden [4].

If we neglect weak and gravitational interactions, i.e., two neutrino or graviton final states, we expect from the above result that γ-γ scattering will become purely elastic at very high energies. It follows in addition that the matrix element for pair annihilation into 2γ should vanish at high energies.

These results cannot be obtained group-theoretically from Poincaré and dilatational invariance alone. The strong restrictions on the inelastic channels of certain relativistic scattering processes are a very specific feature of full conformal symmetry.

The author would like to thank Professor H.D.Doebner and Dr.K.Ringhofer for discussions in Schladming.

REFERENCES

1. G. Mack and J. Todorov, J. Math. Phys. 10, 2078 (1969); L. Castell, Nuclear Phys. B13, 231 (1969); J. Michelson and J. Niederle, to be published.
2. U. Ottoson, Ark. Phys. 33, 523 (1967).
3. L. Castell, "Restrictions on Inelastic Channels from Conformal Invariance", Symposium on De Sitter and Conformal Groups, University of Colorado (1970).
4. L. Castell, Phys. Rev. D2, 1161 (1970).

VALIDITY OF THE REGGE-EIKONAL MODEL IN $\lambda\phi^3$ THEORY*

BY

G.CICUTA
Istituto di Scienze Fisiche dell'Università
Milano

A number of field theoretical models have recently been proposed to describe the interactions of hadrons at high energy. In such models, eikonal ideas, familiar in potential theory, are extended in the relativistic domain. A class of Feynmann diagrams has been studied which represent two fast hadrons which "exchange" any number of virtual particles (ladders and crossed ladders). An eikonal representation is obtained for the scattering amplitude. If the interaction Lagrangian is $L_{int}=\lambda\phi^3$ it reads:

$$T(s,t) \sim (-su)^{1/2} \int d^2b \; e^{-i\vec{q}\cdot\vec{b}}[e^{i\chi(\vec{b})}-1]$$

where $t=q^2$ is fixed, and

$$\chi(\vec{b}) \sim (-su)^{-1/2} \int d^2k \; \frac{e^{i\vec{k}\cdot\vec{b}}}{k^2+\mu^2} .$$

---

*Abstract of Seminar given at X. Internationale Universitätswochen für Kernphysik, Schladming, March 1 - March 13, 1971.

This representation fails to provide a satisfactory description of the scattering of two fast hadrons for a number of reasons:

1) $T(s,t)$ reduces to the Born term in the high energy limit.
2) The "eikonal approximation" necessary to derive the former representation fails to provide the high energy limit of the Feynman graphs beyond the 6th order in λ.
3) The scattering amplitude satisfies the elastic unitarity equation.

While the reason no. 1 and 2 are pertinent to a scalar theory and are cured by considering the "exchange" of virtual vector particles, the third point holds in any field theory model. One is led to a more systematic analysis of all Feynman diagrams in $\lambda\phi^3$ theory and to the definition of a criteria to choose the most relevant diagrams first [1].

Having chosen the criteria of summing, in the high energy limit, first the graphs with the highest power of log s, for a given order in λ, one finds the class of ladder graphs which sums in a well known way to a Regge pole $\sim\beta(t)s^{\alpha(t)}$.

Next one is led to study all diagrams in which to ladders are "exchanged" and the leading graphs are then found to be those pointed out by Mandelstam to give rise to a Regge cut. Such graphs are computed in the high energy limit and they result in a contribution consistent with the Regge-eikonal model by Arnold [2], and different from the result quoted in the literature for Mandelstam-type graphs.

The analysis is extended to the "n ladder exchange" which is found again consistent with the Regge-eikonal model. All the contributions may be summed into a closed integral form, which reads:

$$T(s,t) \sim (-su)^{1/2} \int d^2b \; e^{-i\vec{q}\cdot\vec{b}} [e^{iR(\vec{b})} - 1]$$

where

$$R(\vec{b}) \sim (-su)^{1/2} \int d^2k \; e^{i\vec{k}\cdot\vec{b}} [(-s)^{\alpha(k)} + (-u)^{\alpha(k)}] \beta(k).$$

REFERENCES

1. G. Cicuta and R. L. Sugar, Phys. Rev. D, (Feb. 15, 1971).
2. Arnold, Phys. Rev. 153, 1523 (1967).

VENEZIANO-TYPE REPRESENTATIONS FOR THE PION FORM FACTOR*

BY

J. CLEYMANS**
III. Physikalisches Institut
RWTH - Aachen

The Veneziano model [1,2,3] predicts large couplings of heavy vector mesons to the $\pi\pi$ system (and also suggests large total widths). It seems therefore surprising why such small couplings of these mesons to photons have been observed [4]. We want to show that this has a possible explanation [5] in the context of Veneziano-type representations [6-14] for the pion form factor. The approximate validity of ρ-meson universality does not allow, namely, a strong coupling of the photon to heavy vector mesons.

In the presence of an infinite series of heavy vector mesons one has for the pion form factor:

$$F_\pi(t) \simeq \sum_{i=1}^{\infty} \frac{C_i}{m_{\rho i}^2 - t} \qquad (1)$$

Since all these particles have an appreciable coupling to the $\pi\pi$-system in the Veneziano model, there is a priori no reason why they should decouple from the photon.

---

* Abstract of Seminar given at X. Internationale Universitätswochen für Kernphysik, Schladming, March 1 - March 13, 1971.

** Alexander von Humboldt Foundation fellow.

Therefore we assume <u>all</u> C_i's in (1) to be different from zero.

Exploiting our knowledge of the mass spectrum we rewrite equation (1) as:

$$F_\pi(t) \sim \sum_{i=1}^{\infty} \frac{\alpha' C_i}{\alpha(m_{\rho i}^2) - \alpha(t)} \tag{2}$$

where α' is taken to be the slope of the ρ-Regge trajectory and $\alpha(t)=\alpha_0+\alpha' t$. Since $m_{\rho i}^2=(2i-1)m_\rho^2$ we have $\alpha(m_{\rho i}^2)=i$. Equation (2) can be written in an integral form:

$$F_\pi(t) = \int_0^1 dx\; x^{-\alpha(t)}\; f(x) \tag{3}$$

where

$$f(x) = \sum_{i=1}^{\infty} x^{i-1}\; \alpha' C_i \tag{4}$$

Veneziano-type representations can be obtained by making a simple ansatz for f(x):

$$f(x) \sim (1-x)^m \tag{5}$$

where m is at first arbitrary. However, negative integer values of m are excluded since they give rise to unwanted poles in $F_\pi(t)$. Positive integer values are also excluded since the series (5) would then have only a finite number of terms which is inconsistent with our starting assumption.

We thus can write:

$$F_\pi(t) = \frac{\Gamma(\frac{n-1}{2})}{\sqrt{\pi}} \frac{\Gamma(1-\alpha(t))}{\Gamma(\frac{n}{2}-\alpha(t))} \tag{6}$$

Expression (6) has been normalised such that $F_\pi(0)=1$ and

$n \neq$ even integer.

This expression for the form factor can be tested by studying the following points:

1) Residue at the ρ-pole: $g_{\rho\pi\pi}/g_\rho$
2) Residue at the ρ'-pole: $(g_\rho/g_{\rho'})^2$
3) Asymptotic behaviour: $F_\pi(t) \sim t^x$.

Since (6) contains only one parameter, we can get some non-trivial information. We obtain:

n	3	17/4	5	13/2	7	Expt. [4-15]
$\frac{g_{\rho\pi\pi}}{g_\rho}$	0.64	1.07	1.26	1.6	1.7	1.07
$\left(\frac{g_\rho}{g_{\rho'}}\right)^2$	0.055	0.0034	0.055	0.3	0.5	<0.05
x	-0.5	-9/8	-1.5	-9/4	-2.5	

Comparison of these numbers shows that it is possible to choose n in such a way as to fit both the $g_{\rho\pi\pi}/g_\rho$ and the $(g_\rho/g_{\rho'})^2$ ratios in a satisfactory way.

Although the value $x=-9/8$ is not inconsistent with the experimental data on the asymptotic behaviour of the form factor, (this is mainly due to the large errors on the data points), it gives a very bad fit to nucleon form factors. Therefore this value is not appealing. We do not necessarily expect however to obtain the correct asymptotic behaviour because the translation of the Harari-Freund

ansatz to form factors would suggest that background could give the most important contribution there.
The following two conclusions follow from the analysis:

1) Veneziano type form factors look like the simplest generalization of the vector dominance model. By this we mean that we can have $g_{\rho\pi\pi} \sim g_{\rho}$ and at the same time very small values for the couplings of photons to other heavy vector mesons.

2) They do not solve the problem of the asymptotic behaviour of form factors.

REFERENCES

1. G. Veneziano NC. 57A, 190 (1968).
2. C. Lovelace Pl. 28B, 264 (1968).
3. J. Shapiro PR. 179, 1345 (1969).
4. D. R. Earles et al. PRL. 25, 1312 (1970) and references contained therein.
5. J. Cleymans and R. Rodenberg, Aachen preprint.
6. P. Di Vecchia and F. Drago LNC, 1, 917 (1969).
7. P. Frampton PR. D1, 3141 (1970).
8. Y. Oyanagi, Progr. Theor. Phys. 42, 898 (1969).
9. H. Suura PRL. 23, 551 (1969).
10. H. Schnitzer PRL. 22, 1154 (1969).
11. R. Arnowitt et al. PRL. 22, 1158 (1969).
12. D. Geffen PRL. 23, 897 (1969).
13. T. C. Chia, M. Hama and D. Kiang, PR. D1, 2126 (1970).
14. Amatya and Taha, PR. D1, 2147 (1970).
15. J. J. Sakurai, Proceedings 4th Int. Symp. on Electron and Photon Interactions at high energies, Daresbury 1969.

COVARIANT SPIN ANALYSIS OF MULTIPARTICLE AMPLITUDES*

BY

B.O. ENFLO AND B.E. LAURENT
Institute of Theoretical Physics
University of Stockholm

A general scheme for treating the spin covariantly in arbitrary processes is proposed. As fundamental quantities in building up the amplitude we use, for each particle with spin s, a $2\times(2s+1)$ - component spinor function $W(e,m,\chi)$ where $(e)\equiv(\overset{(0)}{e},\overset{(1)}{e},\overset{(2)}{e})$ is a set of orthogonal unit four-vectors with $\overset{(0)}{e}$ timelike and $\overset{(1)}{e},\overset{(2)}{e}$ spacelike. The set (e) is constructed from the momenta in an arbitrary manner. We also introduce the pseudovector

$$\overset{(3)}{e}_{\alpha} = \varepsilon_{\alpha\beta\gamma\delta}\overset{(0)}{e}{}^{\beta}\overset{(1)}{e}{}^{\gamma}\overset{(3)}{e}{}^{\delta} \tag{1}$$

which is spacelike. The spinor functions $W(e,m,\chi)$ fulfil the invariant equations

$$\overset{(3)}{I}(e)W(e,m,\chi) = m\,W(e,m,\chi) \tag{2}$$

$$P(e)W(e,m,\chi) = (-1)^{s}\chi W(e,m,\chi)\ . \tag{3}$$

---

*Abstract of Seminar given at X. Internationale Universitätswochen für Kernphysik, Schladming, March 1 - March 13, 1971.

Here the invariant operators $\overset{(3)}{I}$ and P are defined as:

$$\overset{(3)}{I}(e) \equiv iYM_{\alpha\beta}\overset{(o)}{e}{}^{\alpha}\overset{(3)}{e}{}^{\beta} \tag{4}$$

$$P(e) \equiv Z(|\overset{(3)}{\underset{\sim}{e}}| + \overset{(3)}{e}{}^{o})\, e^{2\gamma\underset{\sim}{\Sigma}\cdot\overset{(3)}{\underset{\sim}{e}}/|\overset{(3)}{\underset{\sim}{e}}|\; i\pi\underset{\sim}{\Sigma}\cdot\overset{(3)}{\underset{\sim}{e}}/|\overset{(3)}{\underset{\sim}{e}}|} \tag{5}$$

where $M_{\alpha\beta}$ are the generators of Lorentz rotations in our $2\times(2s+1)$- dimensional representation $D_{s,o} \oplus D_{o,s}$. The $M_{\alpha\beta}$ are expressed in terms of Σ^1, Σ^2, Σ^3 and Y as explained in [1], Chapter I. The operator Z is the space reflection operator.

Equations fixing relative phases and normalization of the W's are also investigated, as well as the discrete symmetries of the W's. Under Lorentz rotations they have the following transformation property:

$$W(e',M,\chi) = D(e',e)W(e,m,\chi), \tag{6}$$

where D(e',e) is the representation of that Lorentz rotation which changes the set (e) into (e'). It is possible to express D(e',e) as a finite sum depending only on the components of (e') and (e).

The M-function is now given as (the two sides of the reaction are separated by a semicolon):

$$M(p_1' \alpha_1' \ldots p_f' \alpha_f'; p_1 \alpha_1 \ldots p_i \alpha_i) =$$

$$= \sum_{\substack{r_1'\ldots r_f' \\ r_1\ldots r_i}} A(r_1'\ldots r'_f; r_1\ldots r_i) W_{\alpha_1'}(r_1')\ldots W_{\alpha_f'}(r_f')\bar{W}_{\alpha_1}(r_1) \times \ldots \bar{W}_{\alpha_i}(r_i), \tag{7}$$

where (r) is a shorthand notation for (m,χ) and $\bar{W}$ is defined in a similar way as in [1], equation (II. 11).

In equation (7) the spinor indices are given but momenta are suppressed both in the invariant amplitudes $A(r_1' \ldots r_f'; r_1 \ldots r_i)$ and in the W's.

We have now the tools to study all kinds of symmetries of the M-function, including crossing symmetry and the Pauli principle. Application of these two symmetries means that we have to change the set (e). We can do that with the aid of equation (6). Throughout the formalism is expressed in a manifestly covariant way in terms of the momenta.

REFERENCE

[1] Enflo, B. O. and Laurent, B. E., Fortschr. der Physik 16, 373 (1968).

CHIRAL INVARIANCE, DILATATION INVARIANCE AND THE BOOTSTRAP CONDITION IN QUANTUM FIELD THEORY*

BY

P.L.F.HABERLER

Max-Planck Institut für Physik und Astrophysik

München

Scale invariance and chiral invariance are studied in the framework of Quantum electrodynamics (QED) and pseudoscalar meson nucleon theory. Both field theories have the remarkable property that a simple relation between the bare and physical fermion mass exists when the boson to which the fermion is coupled has vanishing mass:

$$M_o = M\,Z_2 \tag{A}$$

Z_2 is the wave function renormalization constant of the fermion. This relation is crucial as to whether scale invariance and chiral invariance are good symmetries of the theory in question.

For QED we find: Due to relation (A) and the Källen result $Z_2=0$, which we interpret as compositeness criterion for electron and muon, the Gell-Mann Low function vanishes for a fixed value of the unrenormalized coupling constant.

---

*Abstract of Seminar given at X. Internationale Universitätswochen für Kernphysik, Schladming, March 1 - March 13, 1971.

As a result scale invariance is a good symmetry of QED. The mass scale of the theory is fixed by a cutoff determined by $Z_2=0$ and we further propose that the continuous mass representation of the dilatation group is important in QED. This means that all physical masses are multiples of one mass scale as given by the cutoff. Due to $Z_2=0$ an anomalous dimension for the fermion emerges. This means that renormalization is important and has an observable physical effect. On the contrary due to the Adler anomaly the axial current is not conserved in QED, even when all masses vanish. We suggest that the physical reason behind this anomaly lies in the fact that charged particles cannot be massless.

For ps-ps meson nucleon theory we find: Due to eq. (A) and the condition $Z_2=0$ again the Gell-Mann-Low condition is fulfilled. We extend the $Z_2=0$ condition to all wave function renormalization constants and find in that theory a spontaneous creation of chiral symmetry which at the same time is spontaneously broken as scale invariance. This is contrary to the caseof QED.
To summarize: the main difference between QED and ps-ps theory is the following:

In QED there exists a unique physical vacuum, scale invariance is not spontaneously broken but probably realized in a continous mass representation. The unrenormalized coupling constant is <u>not</u> independent of the physical coupling constant.

In ps-ps theory the vacuum state is degenerate, both scale invariance and chiral invariance are spontaneously broken. The unrenormalized coupling constant is independent of the physical coupling constant.

However in both theories the Gell-Mann Low function vanishes.

EXACTLY SOLUBLE MODELS FOR INTERACTING VECTOR AND SCALAR FIELDS*

BY

Z. HORVATH
Institute for Theoretical Physics
Roland Eötvös University, Budapest

We discuss Deo-type models 1, 2 for interaction of vector and scalar particles with derivative coupling. The Lagrangian of the general model is the following

$$L(A_{i\nu}, B_A) = -\frac{1}{2}(\partial_\mu A_{i\nu}\partial^\mu A_i^\nu - \partial_\mu A_{i\nu}\partial^\nu A_i^\mu - M_{ij}A_{i\mu}A_j^\mu) +$$

$$+\frac{1}{2}(\partial_\mu B_A \partial^\mu B_A - M'_{AB}B_A B_B) + R_{Ai}\partial_\mu B_A A_i^\mu$$

where i, j=1,...,n; A, B=1,...,m. M and M' are positive definite n×n and m×m symmetric matrices, respectively. R is a matrix of m×n elements. This Lagrangian is diagonalized if we express the bare fields $A_{i\nu}$ and B_A by means of the real free fields $U_{j\nu}$ and ϕ_B in the following way

$$A_{i\nu} = O_{ji}U_{\nu j} - \sum_{B=1}^{m} (M^{-1})_{ij}R_{Aj}F_{AB}\sqrt{\lambda_B}\,\partial_\mu\,\phi_B$$

---

*Abstract of Seminar given at X. Internationale Universitätswochen für Kernphysik, Schladming, March 1 - March 13, 1971.
A more detailed analysis of this subject will be published in Acta Physica Austriaca.

$$B_A = \sum_{B=1}^{m} F_{AB}\sqrt{\lambda_B}\ \phi_B \ .$$

Considering that $U_{i\nu}$ and ϕ_B are quantized as free fields, one can compute the commutators of the bare free fields $A_{i\nu}$ and B_A; they will be c-numbers. Thus, $A_{i\nu}$ and B_A give a new example of Greenberg's generalized free fields [3]. The asymptotic behaviour of the bare fields can be evaluated, and it shows that $A_{\nu i}$ and B_A simultaneously tend to several vector and scalar outgoing fields in weak limit. Within this frame the validity of the mass-squared sum rules [4] is shown. Furthermore, we present some particular examples, which insure the incorporation of PCAC and give an occasion to analyse the connection of the approximate and the exact symmetries, too.

REFERENCES

1. B. B. Deo, Nucl. Phys. 28, 135 (1961).
2. K. L. Nagy, T. Nagy and G. Pócsik, Acta Phys. Hung. 19, 91 (1965).
3. O. W. Greenberg, Ann. Phys. 16, 158 (1951).
4. G. Pócsik, Il Nuovo Cimento 43, 541 (1966).

PROGRESS IN RELATIVISTIC WAVE EQUATIONS OF HIGHER SPIN*

BY

Wm. HURLEY
Department of Physics,
Syracuse University
New York

While the free relativistic wave equations for spins 0, 1/2, and 1 are well known and understood, these equations have caused difficulties for spin s>1 for many years. The most serious of these difficulties are the lack of causality exhibited by their solutions in the presence of interaction, the lack of uniqueness due to the fact that two equations which are equivalent for free fields can yield entirely different results in the presence of interaction (e.g. in minimal coupling), and inconsistencies upon second quantization. Dr. Hurley, Syracuse University, recently derived Galilean invariant equations for s>1 which are free of such difficulties and exhibit various desirable properties. This work has now been generalized to the relativistic case with very promising results.

---

*Abstract of Seminar reported by F. Rohrlich at X. Internationale Universitätswochen für Kernphysik, Schladming, March 1 - March 13, 1971.

The wave equation which requires no subsidiary conditions refers to a unique mass $m>o$, fixed spin s (integer of half-odd integer), and is linear and first order. The minimum number of components possible are 4, 10 and $2(6s+1)$ for spins 1/2, 1 and $s>1$. The representation is the reducible representation $(s,o) \oplus (s-1/2,1/2)$ augmented by the minimum additional irreducible representations which are needed to make it parity invariant. Thus, one obtains the Dirac and Kemmer-Duffin results for $s=1/2$ and 1, respectively. The equation has a hermitizing matrix so that a conserved current exists, it is P, C, T invariant, and - most importantly - it is free of the causality problems for minimal electromagnetic coupling which beset so many previously proposed equations. Also, there seems to be no inconsistencies upon second quantization. In the Galilean limit it reduces to the equations previously proposed by Dr. Hurley. For $s>1$ the equation involves parity doublets; whether this feature causes difficulties is still under investigation.

MULTIPERIPHERAL APPROACH TO THE ANALYSIS OF HIGH-ENERGY HIGH-MULTIPLICITY PION-NUCLEON INTERACTIONS*

BY

A.JUREWICZ
Institute of Nuclear Research, Warsaw

In the theoretical studies there have been very few attempts to describe quantitatively the main bulk of high-energy, high-multiplicity pion-nucleon interactions. Among the models which aim at a rather complete description of experimental data the most popular is the Chan-Loskiewicz-Allison model which was shown to describe rather well all many-body channels. A recently proposed generalization of the Veneziano model seems to be a very promising attempt in the study of many-particle problems, but practical applications are still limited to the three body final states. Both above mentioned models in their present form do not include spins and isospins of interacting particles, do not allow for proper symmetrization in the final pion states etc.

In order to include proper spin and isospin structure in the matrix element, to take account of symmetrization in the final state, an attempt was made [1] to

---

*Abstract of Seminar given at X. Internationale Universitätswochen für Kernphysik, Schladming, March 1 - March 13, 1971.

make calculations with the aid of the Amati-Fubini-Stanghellini-Tonin (AFST) model.

The discussion was confined to the channel $\pi^+p\to p3\pi^+2\pi^-$ at one definite energy - 8 GeV/c incident pion momentum, for which there were at disposal experimental data of the Warsaw Bubble Chamber Group.

In this calculation only pion exchanges have been taken into account. To describe $\pi\pi$ and πp vertices presently available $\pi\pi$ and πp phase shifts were used. Off-shell corrections were taken into account according to the method proposed by Dürr and Pilkuhn.

Results are in fair agreement with the experimental data. Protons are emitted too much backwardwise due to the lack of the neglected contributions due to baryon exchange. Production of resonances ρ^o and Δ^{++} is well reproduced. Interference terms are large and positive, and contribute about 50% of the total cross section, which seems to be one of the most important resuts of the calculation.

REFERENCE

1. A. Jurewicz et al., AFST Model with $\pi\pi$ and πp Phase Shifts and Dürr-Pilkuhn Form Factors Applied to the Reaction $\pi^+p\to p3\pi^+2\pi^-$ at 8 GeV/c., Nucl. Phys. (in press).

TACHYON QUANTIZATION*

BY

JOHN R. KLAUDER
Bell Telephone Laboratories, Incorporated
Murray Hill, New Jersey

and

LUDWIG STREIT
Physics Department, Syracuse University
Syracuse, New York

As emphasized by Schroer [1] and by Tanaka [2] representations of

$$[A(x), A(y)] = i\Delta(x-y,m)$$

for imaginary mass m will give rise to canonical tachyon fields with causal propagation properties. To investigate these representations systematically we make the ansatz

$$\langle o|A(x)A(y)|o\rangle = \left(\frac{1}{2\pi}\right)^3\int d^3\underset{\sim}{k}[\rho_1(\underset{\sim}{k})e^{ik(x-y)}+\rho_2(\underset{\sim}{k})e^{-ik(x-y)}]$$

with $k^o=\omega(k)=\sqrt{\underset{\sim}{k}^2+m^2}$, Im $\omega\geq o$. Imposing first commutation relations and self-adjointness we find that <u>any</u> such (translation invariant) quantization must give rise to an

---

*Abstract of Seminar given at X. Internationale Universitätswochen für Kernphysik, Schladming, March 1 - March 13, 1971.

indefinite metric. The field can be realized in terms of creation and annihilation operators a_T ($\omega^2>o$, "tachyons") and $a_\pm$ ($\omega^2<o$, "pseudotachyons") acting on the direct product of Fock spaces $H_T \otimes H_+ \otimes H_-$ with, however,

$$[a_-(\underset{\sim}{k}), a_-^*(\underset{\sim}{k}')] = -\delta^3(\underset{\sim}{k}-\underset{\sim}{k}')$$

to accommodate the indefinite metric in H_-. Depending on the choice of $\rho_i(\underset{\sim}{k})$, the cyclic representation of the canonical algebra may be irreducible or reducible; in fact, reducibility of the algebra is <u>necessary</u> for rotation invariance of the ground state, $\rho_i(\underset{\sim}{k})=\rho_i(k)$. Interestingly, the algebra of space <u>and</u> time smeared fields is also reducible in this case. Relativistically invariant quantizations are obtained by further restricting the ρ_i. The corresponding two-point functions are linear combinations of the usual homogeneous Green's functions, analytically continued to imaginary masses.

REFERENCES

1. B. Schroer: The Quantization of $m^2<o$ Field Equations, Phys. Rev., in print.
2. S. Tanaka: Prog. Theor. Phys. 24, 171 (1960).

SOME PROPERTIES OF THE MESONS FROM A RELATIVISTIC SCALAR QUARK MODEL WITH STRONG BINDING*

BY

M. BÖHM, H. JOOS, M. KRAMMER
DESY, Hamburg

The quantum numbers of the established mesons and baryons fit well into the classification scheme of the nonrelativistic quark model which allows for $q\bar{q}$ and qqq bound states [1]. The explanation of the mass spectra, form factors etc., however, requests the solution of a dynamical problem. If quarks should exist, then the unsuccessful searches put a lower limit of 3 GeV on their mass. This has the consequence that hadrons are extremely strongly bound systems, completely different from atoms and nuclei where the binding energies are small compared to the mass of the constituents and where the Schrödinger equation is applicable. We believe that relativistic quantum field theory is the appropriate framework [2] for studying the dynamics of strong binding. The mesons, as qq systems, are then described by the two-field amplitudes

---

*Seminar given at X. Internationale Universitätswochen für Kernphysik, Schladming, March 1 - March 13, 1971; based on the report DESY 71/10 by M. Böhm, H. Joos and M. Krammer, reported by M. Krammer.

$$\tilde{\Phi}(x,P) = (2\pi)^{-3/2} \langle o|T(\phi(\tfrac{x}{2})\bar{\phi}(-\tfrac{x}{2}))|P,M\rangle$$

or their Fourier transforms $\Phi(q,P)$ which satisfy a Bethe-Salpeter equation [3]

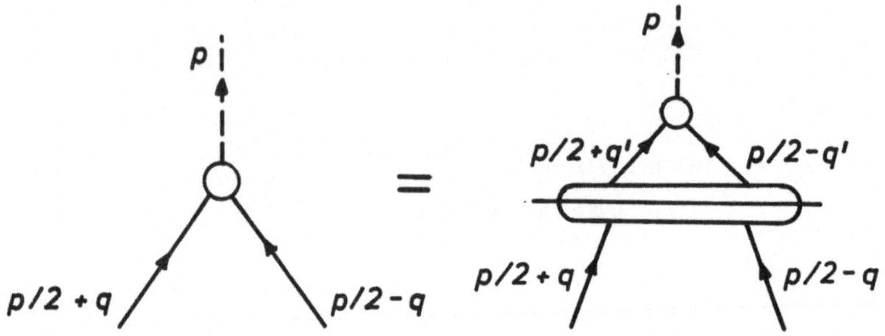

The kernel in the BS equation is assumed to depend on q-q' only. Simple one-particle exchange does not give an acceptable mass spectrum [4]. We use a kernel which simulates a $q\bar{q}$ potential which has the shape of a smooth well of finite range. Smooth potentials are also suggested by the nonrelativistic quark model [5]. The relationship between the BS equation and the nonrelativistic Schrödinger equation is unclear in the case of strong binding. We neglect the spin of the quarks in this attempt to understand the gross features of quark dynamics.

These physical ideas are represented by a BS equation for the vertex function $\Gamma(q,P)=i(2\pi)^4[(\frac{P}{2}+q)^2-m^2] \times$ $\times\ [(\frac{P}{2}-q)^2-m^2]\Phi(q,P)$

$$\Gamma(q,P) = \frac{i}{\pi^2}\int d^4q' \frac{1}{[(\frac{P}{2}+q')^2-m^2+i\varepsilon][(\frac{P}{2}-q')^2-m^2+i\varepsilon]} V(q-q')\Gamma(q',P) \tag{1}$$

with smooth kernels of the form

$$V_k(q) = \frac{\lambda}{(q^2-\mu^2)^k} = \frac{1}{(k-1)!}\left(\frac{\partial}{\partial\mu^2}\right)^{k-1} V_1 . \tag{2}$$

Kernels V_k with $k\geq 2$ can formally be understood as the limiting case of a superposition of attractive and repulsive one-particle exchanges. Starting with k=3 the potential in configuration space is finite at the origin. For $k\geq 4$ one obtains the shape of a smooth well.

In order to handle the integral equation (1) we perform the Wick "rotation" [6] which is allowed for the kernels eq. (2). The Lorentz vector q_μ thereby gets replaced by a Euclidean 4-vector: $q^2=-(q_4^2+\vec{q}^2)=-q^2_{Eucl}$. Furthermore it is convenient to choose the center of momentum frame $\vec{P}=0$.

It is known that the BS equation has O(4) symmetry in the case of vanishing mass of the boundstate. Therefore an expansion of the vertex function into hyperspherical functions leads to almost decoupled one-dimensional integral equations for $\frac{M}{2m}\ll 1$.

Introducing polar coordinates, q, β, θ, φ, in the 4-dimensional Euclidean space, we expand Γ into O(4) spherical functions [7] $S^n_{\ell m}(\beta,\theta,\phi)$

$$\Gamma(q;\beta,\theta,\phi) = \sum_{n\ell m} \gamma^n_{\ell m}(q^2) S^n_{\ell m}(\beta,\theta,\phi) \quad n \geq \ell \geq |m| .$$

From eq. (1) we get for the "hyperradial" vertex function $\gamma^n_\ell(q^2)$ a system of 1-dimensional integral equations with the non-diagonal terms being at most of order

$$\varepsilon(q') = \frac{q'^2+m^2 - \frac{M^2}{4} - \sqrt{(q'^2+m^2 - \frac{M^2}{4})^2+q'^2M^2}}{q'^2+m^2 - \frac{M^2}{4} + \sqrt{(q'^2+m^2 - \frac{M^2}{4})^2+q'^2M^2}} .$$

In the case of strong binding, $M/2m\ll 1$, we have

$$|\varepsilon| \approx \frac{q'^2 M^2}{4(q'^2+m^2)^2} \leq \frac{M^2}{16m^2} << 1 \quad \text{for all } q'^2$$

and obtain to order ε^2 the hyperradial equation*

$$\gamma_\ell^n(q^2) = 2\lambda \int_0^\infty dq' q'^3 \frac{H_k^n(q,q')}{q \cdot q'} \times$$

$$\times \frac{2}{(q'^2+m^2 - \frac{M^2}{4})(q'^2+m^2 - \frac{M^2}{4} + \sqrt{(q'^2+m^2 - \frac{M^2}{4})^2 + q'^2 M^2})} \times$$

$$\times \{1 + \frac{n^2+2n-2\ell^2-2\ell}{n(n+2)} \varepsilon(q')\} \gamma_\ell^n(q'^2) \qquad (3)$$

where the kernel H_k^n is defined as

$$H_k^n(q,q') = \frac{(-1)^{k-1}}{(n+1)\ (k-1)!} \left(\frac{\partial}{\partial \mu^2}\right)^{k-1} \times$$

$$\times \left[\frac{q^2+q'^2+\mu^2 - \sqrt{(q^2+q'^2+\mu^2)^2 - 4q^2 q'^2}}{2qq'}\right]^{n+1} .$$

It is seen that the spectrum is determined by the quantum number n. The term proportional to $\varepsilon(q')$ causes a small splitting of the levels with different ℓ. The solutions are either even or odd functions of q_o, i.e. they have positive or negative "time" parity. Solutions with $n>\ell$ correspond to "daughters". The vertex function of the odd daughters ($n-\ell$=odd) which have negative time parity vanishes if continued on the mass shell of the quarks. Since they do not occur as physical poles in the S-matrix, they may be considered as unphysical solutions.

---

*For $n=\ell=o$ the curly bracket has to be replaced by 1.

Solutions of eq. (3) which correspond to hyperradial excitations, characterized by a further quantum number N, were not considered because their physical meaning is still unclear.

For the numerical solution of eq. (3) we have used the variational principle. The trial functions were chosen in such a way that the small q and large q behaviour obtained by analysing eq. (3) is satisfied

$$\gamma_\ell^n(q) = \frac{q^n}{\prod_{i=1}^{k+n} (q^2+m_i^2)} \cdot \prod_{1,..r} \left[\frac{q^2+m_j^2}{q^2+m_{j'}^2}\right] .$$

The additional poles and zeroes approximate the cut structure. The parameters m_i, m_j, $m_{j'}$ are determined by the variational principle which minimizes the coupling constant λ for a given boundstate mass M.

Choosing a smooth potential eq. (2) with k=5 we solved eq. (3) by the above mentioned method on the computer for n=0,1,2 and 3. - The coupling constant was adjusted to a ground state mass of 150 MeV. - The results for the orbital excitations ($n=\ell$) are shown in a Chew-Frautschi plot in Figure 1. Within the accuracy of the eye the trajectories are linear. The dependence on the range parameter μ is qualitatively the same as in the Schrödinger case. The dependence on the quark mass m reflects the scaling property of the BS equation. Especially we may choose m=4 GeV, μ=130 MeV, so that we obtain

$$M_{\ell=0}^{n=0} = 150 \text{ MeV}, \; M_1^1 = 1000 \text{ MeV}, \; M_2^2 = 1430 \text{ MeV}, \; M_3^3 = 1750 \text{ MeV}$$

which correspond to a "pion" trajectory with "conventional" slope α'=1.0 GeV^{-2}. The first radial excitation ($n=2, \ell=o$) occurs at M_0^2=1480 MeV, only slightly above M_2^2.

As a consequence of the smoothness of the potential, the vertex functions are found to decrease rapidly with q^2. The width of the $\ell=0$ wave function in configuration space is approximately equal to $2\cdot(m\cdot\mu)^{-1/2}$. This means that the two quarks are sitting at the bottom of the smooth potential. There it can be approximated by a parabola, that is a 4-dimensional generalization of the harmonic forces. We think that this fact clarifies the meaning of models based on the relativistic harmonic oscillator [8,9] which, otherwise, violates fundamental principles of quantum field theory.

Finally, we wish to quote the result for the electromagnetic form factor of our groundstate particle. This is a sensitive test of the model insofar, as a fully relativistic description does not suffer from ambiguities, like choice of reference frames, as does the non-relativistic quark model. The form factor

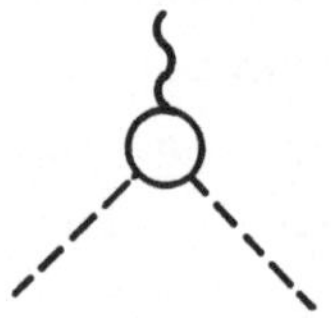

is decomposed into the quark form factor and the simplest two-particle irreducible quark-meson scattering diagram with the full off-shell meson-quark vertex functions inserted

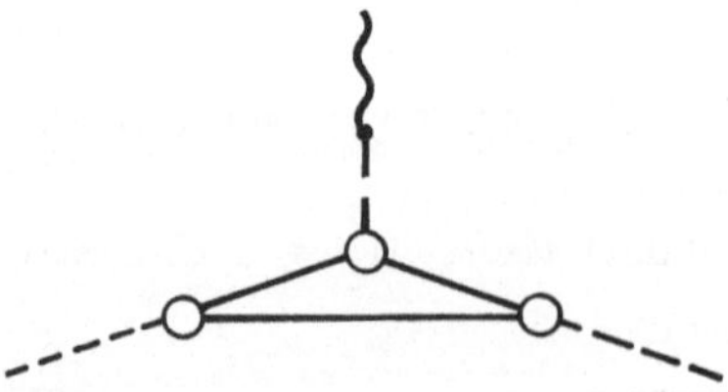

The resulting "pion" form factor is shown in Figure 2 for the timelike and spacelike region 1 $GeV^2 \geq t \geq -2$ GeV^2.

It is seen that the composite structure of the "pion" gives a sensible modification of the pure vector dominance form factor [10] which points into the same direction as the experimental results [11,12] do.

REFERENCES

1. M. Gell-Mann, Phys. Letters 8, 214 (1964);
G. Zweig, CERN preprints TH 401, 412 (1964), (unpublished);
J. J. J. Kokkedee, "The Quark Model" (W. A. Benjamin, Inc., New York (1969).
2. T. Gudehus, DESY-Bericht 68/11 (1968),(unpublished);
C. H. Llewellyn Smith, Ann. Phys. (N.Y.) 53, 521 (1969).
3. H. A. Bethe and E. E. Salpeter, Phys. Rev. 84, 1232 (1951);
K. Symanzik, J. Math. Phys. 1, 249 (1960).
An extensive review of the theory of the BS equation is given by:
N. Nakanishi, Progr. Theor. Phys., Suppl. 43, 1 (1969).
4. A. Pagnamenta, Nuovo Cim. 53A, 30 (1968).
5. R. H. Dalitz, "Symmetries and the Strong Interactions", Proceedings of the XIIIth International Conference on High Energy Physics, Berkeley, 215 (1966).
6. G. C. Wick, Phys. Rev. 96, 1124 (1954).
7. M. Gourdin, Nuovo Cim. 7, 338 (1958).
8. M. K. Sundaresan and P. J. S. Watson, Ann. Phys. (N.Y). 59, 375 (1970).
9. R. P. Feynman, M. Kislinger and F. Ravndal, CALT-68-283 (1970).

10. We have chosen the form proposed by:
G. J. Gounaris and J. J. Sakurai, Phys. Rev. Lett. 21, 244 (1968).
11. J. Perez-Y-Jorba, Proceedings of the 4th International Symposium on Electron and Photon Interactions at High Energies, Liverpool, 213 (1969).
12. W. K. H. Panofsky, Proceedings of the 14th International Conference on High-Energy Physics, Vienna, 23 (1968).

FIGURE CAPTIONS

Fig. 1: Regge trajectories as functions of the quark mass m and the range parameter μ of the potential.

Fig. 2: Electromagnetic form factor of the ground state particle ("pion") in the timelike and spacelike region. The dashed-dotted line denotes the binding form factor. The ρ-dominance form factor corresponds to the Gounaris-Sakurai form with M_ρ=765 MeV, Γ_ρ=125 MeV and is represented by the dashed line. The true form factor corresponds to the product of both and is given by the full line.

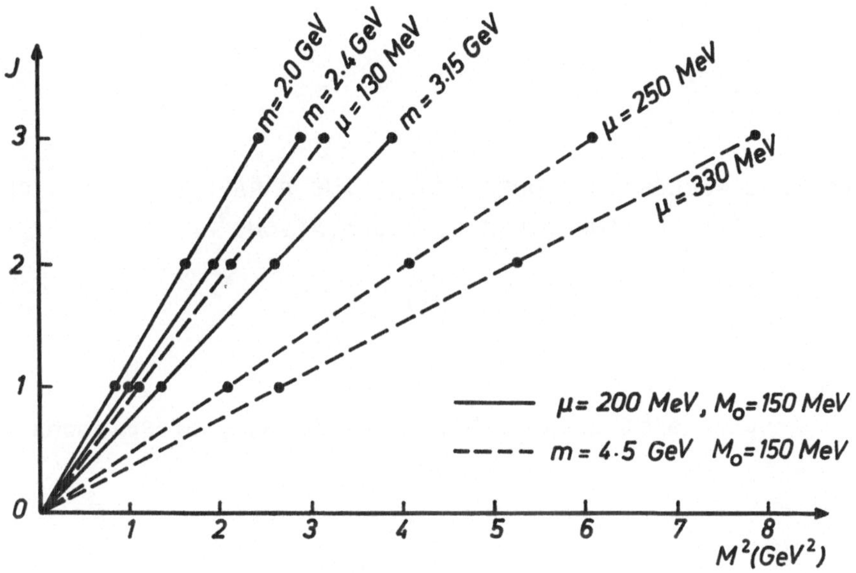
J
0
1
2
3
1
2
3
4
5
6
7
8
$M^2(GeV^2)$
m=2.0 GeV
m=2.4 GeV
μ=130 MeV
m=3.15 GeV
μ=250 MeV
μ=330 MeV
μ= 200 MeV, M_o=150 MeV
m = 4.5 GeV M_o=150 MeV

Fig. 1.

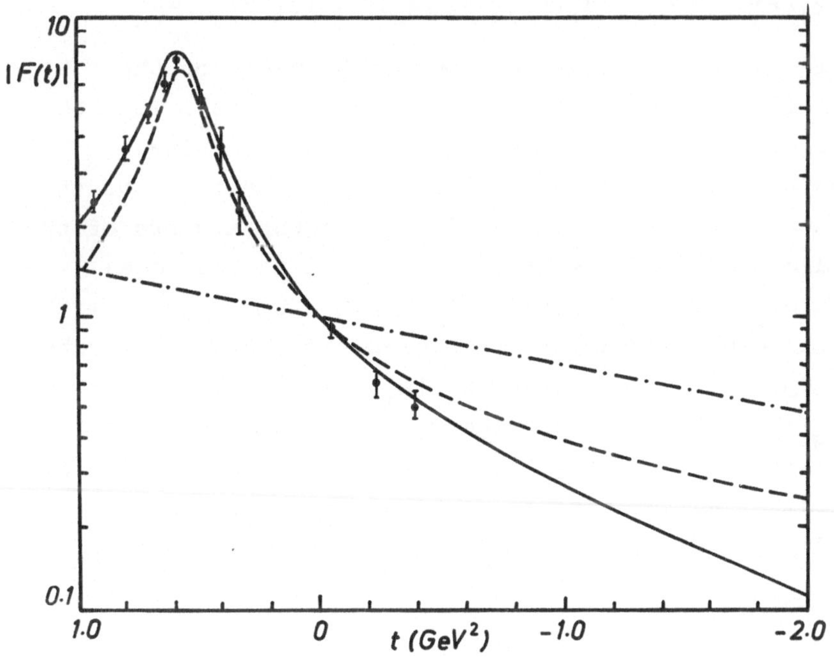
10
|F(t)|
1
0.1
1.0
0
t (GeV^2)
-1.0
-2.0

Fig. 2.

ON SOME ALGEBRAIC PROPERTIES OF THE PARA-FERMI OPERATORS AND THEIR REPRESENTATIONS*

BY

T.D. PALEV
Institute of Physics, Bulgarian Academy of Sciences
Sofia

The para-Fermi operators (PFO) are defined by the relations [1]

$$[[a_i,b_j],a_k] = 2\delta_{jk}\cdot a_i, [[a_i,a_j],a_k]=o, i,j,k=1,\ldots,n \qquad (1)$$

and all relations obtained from (1) by means of the Jacobi identity and a formal conjugation $a_i^{\dagger}=b_i$. The PFO can be embedded in a natural way into an infinite dimensional associative algebra V_n over the field of the real numbers. The algebra V_n is the factor algebra of the free algebra T_n of the PFO with respect to the ideal J generated by the relations (1). V_n is a Lie algebra with respect to the commutator $[a,b]=a\cdot b-b\cdot a, a,b\varepsilon V_n$. The subspace $U_n \subset V_n$ with a basis $a_i, b_i, [a_i,b_j]$, $[a_p,a_q]$, $[b_p,b_q] (p<q)$ is a Lie subalgebra of V_n isomorphic to $O(n,n+1)$. Therefore any irreducible representation of the PFO defines an irreducible representation of $O(n,n+1)$ (or of any other real form of the classical algebra B_n)

* Abstract of Seminar given at X. Internationale Universitätswochen für Kernphysik, Schladming, March 1 - March 13, 1971.

and vice versa. If $a_i^\dagger = b_i$ ($a_i^\dagger$ is the hermitian conjugate of a_i), then there is a one to one mapping between the hermitian representations of O(2n+1) and the representations of the PFO. As is known [2] all these representations are labeled by n integer or halfinteger numbers $(m_1, m_2, \ldots, m_n)$. It is always possible to choose the Cartan subalgebra S⊂O(2n+1) in such a way that $a_1 \ldots, a_n$ are positive root vectors. Therefore in every irreducible representation space V there exist a vector $|o\rangle$ such that $a_i|o\rangle = o$, $i=1,\ldots,n$.

Therefore all representations of the PFO are Fock type representations. They extend the class of the representations found by Greenberg and Messiah [3]. The subspace $V_o \subset V$ annihilated by $a_1, \ldots, a_n$ has a dimension more or equal to one. In the case $a_i^\dagger \neq b_i$ there exists a continuous class of infinite dimensional representations of the PFO.

REFERENCES

1. Green, H. S., Phys. Rev. 90, 270 (1953).
2. Gel-fand I. M., Zetlin M.L. Dan, 71, 1017 (1950), (in russian).
3. Greenberg O. W., Messiah, Phys. Rev. 138B, 1155 (1965).

A MATHEMATICAL NOTE ON THE CONFORMAL GROUP OF A PSEUDO-ORTHOGONAL VECTOR SPACE*

BY

H. TILGNER
Institut für Theoretische Physik (I)
Universität Marburg

For a (not necessarily linear) Lie transformation group of a (finite-dimensional) vector space, the concept of a Lie algebra of infinitesimal transformations is defined and applied to the conformal group of a pseudo-orthogonal vector space, using the multiplication law of the conformal group. For an arbitrary polynomial of the Weyl algebra (i.e. of the associative polynomial algebra generated by the q_i and p^k with the canonical commutation relations) a transformation is defined in the vector space spanned by the q_i and p^k which is non-linear, unless this polynomial is a symmetrized one of second degree or the identity. The mapping of polynomials into transformations is shown to be natural in the sense, that in the case of the conformal group, it gives an isomorphism of the Lie algebras of generators and infinitesimal transformations, if restricted to the

---

*Abstract of Seminar given at X. Internationale Universitätswochen für Kernphysik, Schladming, March 1 - March 13, 1971.

momentum vector space. It is proved that every conformal Lie algebra is isomorphic to a (non-compact) pseudo-orthogonal Lie algebra and vice versa. This cannot be generalized to the groups. Some remarks are added, concerning a class of algebraic structures related to the conformal Lie algebras.

COMPENSATION THEORY[*]

BY

M. WELLNER[**]
Department of Physics, Syracuse University
Syracuse, New York

This is a report of the progress to date of a most unconventional attempt in constructing an elementary particle Lagrangian for the purpose of field theory. This work is due to Professor M. Wellner of Syracuse University.

The basic assumptions are:

(a) lowest order perturbation theory is somehow meaningful even when strong interactions are involved,

(b) in each order of perturbation theory (and especially in lowest order) all divergences of observable transition probability amplitudes cancel provided all interactions of the relevant elementary particles are included.

The secondary assumptions are a nonet of baryons (including the singlet L which has the quantum numbers of a negative parity Λ) and a nonet of pseudoscalar mesons. The interaction Lagrangian includes SU(3) type V-A and V currents, and $SU(2)\times SU(2)$ type V+A currents, a total of 18 current which form a Lie Algebra.

---

[*] Abstract of Seminar given at X. Internationale Universitätswochen für Kernphysik, Schladming, March 1 - March 13, 1971.

[**] reported by F. Rohrlich.

The equations resulting from the compensation of divergences are an overdetermined set of equations for the coupling constants. The consistency requirement leads to the four equations for the baryon masses:

$$\frac{1}{N} + \frac{1}{\Xi} + \frac{1}{\Sigma} = \frac{6}{N+\Xi} \tag{1}$$

$$\frac{3}{\bar{\Lambda}} + \frac{1}{\Sigma} = \frac{2}{N} + \frac{2}{\Xi} \tag{2}$$

$$\bar{\Lambda}L = \Sigma^2 \tag{3}$$

$$\Lambda\cos^2\omega = \frac{2N\Xi}{N+\Xi} \tag{4}$$

where $\bar{\Lambda}$ stands for $\Lambda\cos^2\omega - L\sin^2\omega$. With N and Ξ as inputs, the masses for Σ and Λ come out in excellent agreement with experiment; one also predicts $L=1342$ and $\omega=8.5^\circ$. The strong coupling constants are found to be

$$g_{NN\pi} : g_{\Sigma\Sigma\pi} : g_{\Xi\Xi\pi} : g_{\Lambda\Sigma\pi} = \frac{1}{N} : \frac{1}{\Sigma} : \frac{1}{\Xi} : 0 .$$

The life times of $\pi^\pm$ and π^0 can now be computed; they are found to be within experimental error.

DIPOLE TYPE FORMULAS FOR ELECTROMAGNETIC FORM FACTORS*

BY

N. ZOVKO
Institut Ruder Bosković
Zagreb

The nucleon form factors can be parametrized over the measured range of the spacelike (momentum transfer)^2 by an extremely simple dipole formula [1]

$$G_D(t) = \frac{1}{(1-t/t_D)^2}, \quad t_D = 0.71 \text{ GeV}^2. \tag{1}$$

It was a great puzzle how to interpret this result; is the dipole a new physical phenomenon closely related to the rapidly decaying vector mesons, or does some dynamical mechanism generate it just in the particular case of the nucleon electromagnetic structure? Any answer to that question would also be of interest in possible implications of dipoles in weak interactions [2].

We shall here show that the usual analyticity properties, supplemented by the asymptotic behaviour $G(t) \sim t^{-3}$ [3], lead to the dipole formula within a certain saturation hypothesis for the dispersion integral.

---

*Abstract of Seminar given at X. Internationale Universitätswochen für Kernphysik, Schladming, March 1 - March 13, 1971.

We expect that the zero width vector meson dominated dispersion relation

$$G(t) = \frac{1}{\pi}\int \frac{\rho(t')}{t'-t}\,dt' = \sum_{i=1}^{n} \frac{a_i}{t_i-t} =$$

$$= \frac{N(t)}{(t_1-t)(t_2-t)\dots(t_n-t)}\,, \tag{2}$$

$$N(t) = (-t)^n \sum a_i + (-t)^{n-1}\left((\sum a_i)(\sum t_i) - \sum a_i t_i\right) +$$

$$+ \sum_{k=0}^{n} (-t)^{n-k-3}\, C_k(a_i, t_i)\,,$$

should describe most of the properties of the form factors and actually reproduce the dipole formula (1).

It is obvious that the coefficients of the two leading powers in N(t) will vanish if we impose the following constraints on the residues a_i

$$\sum a_i = \sum a_i t_i = 0\,. \tag{3}$$

These are just the zero and the first momentum superconvergent sum rules

$$\int \rho(t)\,dt = \int \rho(t)\,t\,dt = 0\,,$$

which follow from the assumed asymptotic behaviour. In particular, the two-pole approximation to the properly normalized form factor, according to (3), is just the dipole formula

$$G(t) = \left(\frac{t_1 t_2}{a_1 t_2 + a_2 t_1}\, \frac{a_1 t_2 + a_2 t_1 - t(a_1+a_2)}{(t_1-t)(t_2-t)}\right)_{\substack{a_1=-a_2 \\ t_1=t_2}} = \frac{1}{(1-t/t_1)^2}\,. \tag{4}$$

The correct order of taking the limit $a_1 \to -a_2$, and then $t_1 \to t_2$ is justified by the statement that the zero momentum sum rule is more accurately fulfilled within the same saturation hypothesis than the first momentum sum rule.

Although the mean mass of any pair of the three known vector mesons is rather close to the dipole mass, for a successful comparison with experiments additional mesons like ρ' are needed. One should keep in mind that the three-pole model still contains no free parameters if the experimental masses are used.

In the case of the pion form factor, which seems to behave in the same way at low values of t as the nucleon form factor, the question is: What is the second meson? If we use the ρ' (m_ρ^2,-1.1 GeV^2), an excellent agreement with the existing data is obtained.

REFERENCES

1. G. Buschhorn, this volume.
2. H. Pietschmann, this volume.
3. T. Massam and A. Zichichi, Lettere al Nuovo Cim. 1, 387 (1969).